Building Agricultural Extension Capacity in Post-Conflict Settings

FSC
www.fsc.org
MIX
Paper from
responsible sources
FSC® C013604

Building Agricultural Extension Capacity in Post-Conflict Settings

Edited by

Paul E. McNamara and Austen Moore

University of Illinois at Urbana-Champaign, Urbana, Illinois, USA

CABI

CABI is a trading name of CAB International

CABI
Nosworthy Way
Wallingford
Oxfordshire OX10 8DE
UK

Tel: +44 (0)1491 832111
Fax: +44 (0)1491 833508
E-mail: info@cabi.org
Website: www.cabi.org

CABI
745 Atlantic Avenue
8th Floor
Boston, MA 02111
USA

Tel: +1 (617) 682-9015
E-mail: cabi-nao@cabi.org

British Library Cataloguing-in-Publication Data

A catalogue record for this book is available from the British Library, London, UK.

Library of Congress Cataloging-in-Publication Data

Names: McNamara, Paul E., editor. | Moore, Austen, editor.
Title: Building agricultural extension capacity in post-conflict settings / [edited by] Paul McNamara and Austen Moore.
Description: Boston, MA : CABI, [2017] | Includes bibliographical references and index.
Identifiers: LCCN 2016057955 | ISBN 9781786390592 (hbk : alk. paper)
Subjects: LCSH: Agricultural extension work--Developing countries. | Postwar reconstruction--Developing countries.
Classification: LCC S544.6 .B85 2017 | DDC 630.71/5--dc23 LC record available at https://lccn.loc.gov/2016057955

ISBN: 978 1 78639 059 2 (hardback)

Commissioning editor: David Hemming
Editorial assistant: Emma McCann
Production editor: Lauren Povey/Alan Worth

Typeset by SPi, Pondicherry, India.
Printed and bound in the UK by CPI Group (UK) Ltd, Croydon, CR0 4YY

Contents

Contributors

Ian Christoplos, Danish Institute for International Studies, Østbanegade 117, 2100 Copenhagen, Denmark. E-mail: ian@glemdev.com

Kirk Edney, Department of Agricultural Leadership, Education, and Communications, Texas A&M University, 232 Agriculture and Life Sciences Building, 2116 TAMU, College Station, TX 77843, USA. E-mail: kc-edney@tamu.edu

Hélder R. Gêmo, National Directorate of Agrarian Extension, Ministry of Agriculture and Food Security—Mozambique, Rua da Resistência, N.1746, Cidade de Maputo, Moçambique. E-mail: gemohelder@gmail.com

Roger Hanagriff, Department of Agricultural Leadership, Education, and Communications, Texas A&M University, 242 Agriculture and Life Sciences Building, 2116 TAMU, College Station, TX 77843, USA. E-mail: rhanagriff@tamu.edu

M.W.A.P. Jayatilaka, Department of Agricultural Extension, University of Peradeniya, Res: 72/8, Aniwatte Circular Road, Pushpadana Mawatha, Aniwatte, Kandy, Sri Lanka. E-mail: wijayajayatilaka@yahoo.co.uk

Joseph King, Center for Conflict and Development, Texas A&M University, 408G Agriculture and Life Sciences Building, 2124 TAMU, College Station, TX 77843, USA. E-mail: kingjn@condevcenter.org

Linda R. Klein, LRK Consulting, 440 SE Hill Street, Pullman, WA 99163, USA. E-mail: linda.klein@wsu.edu

Nicholaus M. Madden, International Programs, University of California at Davis, 1103 Environmental Horticulture Building, 1 Shields Avenue, Davis, CA 95616, USA. E-mail: nmmadden@ucdavis.edu

Paul E. McNamara, AgReach Program, University of Illinois at Urbana-Champaign, 341 Mumford Hall, 1301 W. Gregory Drive, Urbana, IL 61801, USA. E-mail: mcnamar1@illinois.edu

Austen Moore, AgReach Program, University of Illinois at Urbana-Champaign, 212 Mumford Hall, 1301 W. Gregory Drive, Urbana, IL 61801, USA. E-mail: acmoore@illinois.edu

Christopher D. Pannkuk, International Research and Agricultural Development, Consultant, 801 E. 3rd Street, Moscow, ID 83843, USA. E-mail: chrispannkuk@gmail.com

Edwin C. Price, Center for Conflict and Development, Texas A&M University, 408G Agriculture and Life Sciences Building, 2124 TAMU, College Station, TX 77843, USA. E-mail: price.condev@gmail.com

Catherine Ragasa, International Food Policy Research Institute, 2033 K Street NW, Washington, DC 20006, USA. E-mail: c.ragasa@cgiar.org

Joshua Ringer, Department of Horticulture, Oklahoma State University, 358 Agricultural Hall, Stillwater, OK 74078, USA. E-mail: joshua.ringer@okstate.edu

Anastasiya Shtaltovna, Centre for International Studies (CÉRIUM), University of Montreal, 3750 Rue du Frère André, Apt. 6, Montréal, Québec, H3V1B3, Canada. E-mail: ashtaltovna@yahoo.com

Vickie A. Sigman, Freelance Consultant, 830 Red Fern Drive, Middleton, ID 83644, USA. E-mail: vsigman@sigran.com

Robert Strong Jr., Department of Agricultural Leadership, Education, and Communications, Texas A&M University, 231 Agriculture and Life Sciences Building, 2116 TAMU, College Station, TX 77843, USA. E-mail: r-strong@tamu.edu

John Ulimwengu, International Food Policy Research Institute, 2033 K Street NW, Washington, DC 20006, USA. E-mail: j.ulimwengu@cgiar.org

Don Van Atta, Center for Slavic, Eurasian, and East European Studies, University of North Carolina, Chapel Hill, NC 27599-5125, USA. E-mail: davanatta@gmail.com

Kate Whitney, Texas AgriLife Extension, Texas A&M University, 104 S. Fuller Street, Meridian, TX 76665, USA. E-mail: klwhitney@ag.tamu.edu

Acknowledgments

It is a pleasure to have the opportunity to recognize the efforts, talents and support of many important individuals who facilitated the production of this collective volume.

First, our sincere gratitude to the United States Agency for International Development (USAID) for developing the Modernizing Extension and Advisory Services (MEAS) project, the driver behind this book, as a vehicle for developing impactful learning products in the field of agricultural extension. Special thanks to our Agreement Officer Representatives—Gary Alex, Suzanne Poland, and John Peters—for their constant support and for providing the flexibility to reflect and analyze the experience of the MEAS project in pursuit of this crucial topic.

Many thanks also to our outstanding cohort of authors who gave their time, considerable efforts and intellectual energy to produce case studies that can benefit a varied audience of readers. We are very fortunate to have been able to engage individuals and experts with such extensive experience in extension, development, conflict and related fields. This book truly represents some of the best minds currently active in this sphere.

Our sincerest appreciation also to Andrea Bohn and Donna Sites at the University of Illinois for their administrative support through this extended process. Thanks also to Becky Mitchell and her team at Green Ink and David Hemming and his team at CABI for their exceptional work in editing and preparing this book for publication, and in developing a product that best represents the importance of this topic but also the depth of knowledge contained within these pages. We appreciate your diligence and attention to detail.

Lastly, Paul would like to express thanks and appreciation to Margaret and his family for their support in writing and editing this collective volume. Austen also thanks Terezinha and his family for their flexibility and understanding towards his fieldwork, editing and writing. This book would not be possible without the support of our families.

Our utmost gratitude and appreciation to all those mentioned above and any who might have been omitted. We are very proud of what we accomplished together.

Sincerest thanks.

Paul McNamara
Director
MEAS Project/AgReach Project
University of Illinois at Urbana-Champaign

Austen Moore
Research Specialist
MEAS Project/AgReach Project
University of Illinois at Urbana-Champaign

Introduction: The Intersection of Conflict, Agriculture and Extension

Paul E. McNamara* and Austen Moore*
University of Illinois at Urbana-Champaign, Urbana, Illinois, USA

Conflict, Agriculture, Poverty and Hunger

Armed conflicts have existed throughout human history, and are as varied as the contexts in which they occur. Conflicts can be short or long term, intense or sporadic, localized or widespread (Strand and Dahl, 2010). However, evidence suggests that the dynamics of conflict have changed in recent decades. Whereas many conflicts have historically occurred between governments (Lacina, 2004), internal conflicts are now the most common 'model of mass violence' since World War II (UNEP, 2002: 1).

The emergence of internal conflict has direct links to poverty, hunger and agricultural development. First, it is essential to understand that poverty, hunger and agricultural development are closely related and that gains in agricultural productivity and food security lead to improvements in poverty and hunger indicators. Global trends in poverty and hunger, disaggregated by region and country, clearly show this relationship (FAO, 2015). The International Fund for Agricultural Development (IFAD) found that a 10% increase in yields created a 7% reduction in poverty across the African continent and that a 1% increase in per capita agricultural GDP had more than five times the impact on poverty reduction than GDP increases in other sectors (IFAD, 2013). Furthermore, the *2008 World Development Report* showed that the impact of agricultural growth on poverty increased as incomes decreased, or that the poorest segments of the population benefit the most from agricultural development (Ligon and Sadoulet, 2008). Hunger indicators closely track those of poverty (World Bank, 2015).

Furthermore, progress in agricultural development and corresponding impacts on poverty and hunger are drivers of economic growth, contribute positively to a range of social factors and have been shown to reduce the probability of internal conflict (Collier, 2007). In contrast, failure to develop agricultural sectors contributes to worsening poverty and hunger indicators. Often these factors lead to perceptions of social injustice, lack of social mobility and few opportunities to address social inequality, which 'make society a fertile terrain for conflict' (UNEP, 2002: 3). In fact, Collier (2003) found that a 5% decrease in yearly economic growth and the corresponding poverty increases led to a 12% increase in the probability of internal

* Corresponding authors' e-mail: mcnamar1@illinois.edu and acmoore@illinois.edu

conflict. Accordingly, the world's poorest countries are typically those most prone to conflict (Waters *et al.*, 2007).

The majority of the poor are concentrated in agriculture, and correspondingly most armed conflicts occur in regions and nations with very high dependence on the agricultural sector (Zaur, 2006). Recent episodes in Rwanda and Côte d'Ivoire closely followed major drops in production and the agricultural GDP (Zaur, 2006). This demonstrates the importance of the productivity of the sector. Similarly, where food security is low, people become increasingly desperate and willing to turn to violence to feed themselves and their families. Food price spikes in 2008 and 2011 resulted in protests and sporadic violence in many developing countries (Lagi *et al.*, 2011). This correlation verifies that 'hunger anywhere threatens peace everywhere' (Swaminathan, 1994: 104).

Post-Conflict Periods and the Cycle of Conflict

Countries emerging from conflict face considerable challenges. The effect of conflict on nations' development is striking, erasing 'years of positive development in the blink of an eye' (Zaur, 2006: 3). This leads to a regression of development indicators in virtually all areas, including education, healthcare and infrastructure (UNEP, 2002). Poverty levels also often increase considerably as a result, with rural populations typically the most affected (UNEP, 2002).

Despite impacts in many sectors, the detrimental effects on agriculture have perhaps the direst consequences, especially where populations are primarily dependent on agriculture for their food supply and livelihoods. During conflicts, farmers are displaced from their crops and livestock, supply chains collapse and local/regional markets become inaccessible due to security concerns, and much of the agricultural labor force—particularly young men—are conscripted or recruited into conflicts that take them away from food production (UNEP, 2002). On average conflict leads to an annual 12.3% decrease in productivity (Zaur, 2006), with smallholder farmers generally the most severely affected (UNEP, 2002).

Decreases in the food supply then exacerbate hunger, especially among displaced and rural peoples (Wiggins and Leturque, 2010). Conflict is therefore considered 'the underlying cause of acute or lasting food insecurity situations in many countries' (UNEP, 2002: 3). When conflicts are longer lasting, more intense, widespread and destructive, countries can face food insecurity impacts that can last for years beyond the conflict itself and require considerable time to rebuild infrastructure and return to pre-conflict production levels.

Conflict creates a vicious cycle. The conditions that foster conflict (e.g. poverty, hunger) are often worsened by the violence itself (USAID, 2009), creating a dynamic that is difficult for nations to escape. Collier (2007: 177) writes 'around half of all civil wars are post-conflict relapses', illustrating the magnitude of this challenge. Post-conflict development that addresses the underlying factors causing conflict is essential to breaking this cycle and moving countries towards development, peace and stability.

Agriculture as a Post-Conflict Development Strategy

Development in the post-conflict period requires multifaceted strategies that address a range of areas and sectors affected by conflict. Post-conflict development needs to target the socio-economic conditions (e.g. poverty, hunger) that contributed to and were exacerbated by the conflict (USAID, 2009), but also be 'conflict sensitive' in its programming to promote reconciliation, rehabilitation and stability (Zaur, 2006). As a result, agricultural development plays a central role in post-conflict development strategies (UNEP, 2002).

Often the primary agricultural goal of post-conflict governments as well as donors—such as IFAD, United States Agency for International Development (USAID), World Bank—in the early post-conflict period is stimulating rapid agricultural productivity and food production gains. This approach reduces the issues of food insecurity and hunger that often impede progress towards development and peace, and alleviates the need for (and eventually replaces) emergency food assistance (UNEP, 2002; USAID, 2009). Interventions that promote large-scale production schemes for staple crop production, agricultural intensification, mechanization and the use of improved technologies are common (Longley *et al.*, 2007; USAID, 2009). Food production interventions may also include a transition from food aid to seed aid to maintain food security and promote increases in local food supplies.

A second approach targets poverty through the development of agricultural livelihoods and requires a range of complementary strategies to improve incomes (Scoones, 1998; Longley *et al.*, 2007). Commonly, efforts to improve agricultural livelihoods are coupled with interventions that strengthen value chains and improve farmers' access to markets. Value chain strengthening programmes are designed to allow domestic production to replace high food aid and food importation, which proliferate during the post-conflict period. Governments and donors seek to facilitate a transition to foodstuffs that are produced domestically by rebuilding local markets and marketplace processes, linking producers or farmer-based organizations to buyers, restricting dumping of cheap foreign foods and creating pro-producer policies (Maconachie and Binns, 2007).

Third, agricultural development is targeted at addressing issues of displacement, reintegration and unemployment, which are frequent problems for post-conflict countries. Agriculture has been shown to be the sector most capable of reincorporating displaced peoples and former combatants, thereby promoting stability and peacebuilding (Maconachie and Binns, 2007). Because rural peoples are most commonly displaced and combatants are often recruited from agrarian areas, reintegration requires creating livelihood options and employment opportunities for groups that have little experience outside of agriculture (Longley *et al.*, 2007; Blattman and Annan, 2011). Programmes that help displaced peoples return home and resume agricultural production can stimulate agriculture after conflicts (Maconachie and Binns, 2007). Other development initiatives seek to create agricultural opportunities for women—especially widows and female-headed households—in the agricultural sector (Maconachie and Binns, 2007).

Reincorporation of youth and especially former combatants is especially challenging. In many cases, former fighters relocate to urban areas where they find other livelihood options or face high unemployment instead of returning to rural and agricultural lifestyles (Maconachie and Binns, 2007). Depletion of youth in the agricultural workforce can affect short-term agricultural productivity, food security, poverty and peace during post-conflict rehabilitation (Collier, 2003). High concentrations of unemployed youth can also threaten long-term stability. Programs that encourage youth and former fighters to re-join the agricultural workforce can address these issues and are commonly implemented in post-conflict settings (USAID, 2009; Blattman and Annan, 2011).

Agricultural Extension in Post-Conflict Development

Agricultural extension plays a fundamental role in advancing all of the aforementioned objectives of post-conflict agricultural development (Waters *et al.*, 2007). Traditionally, extension focuses on four broad objectives: (i) disseminating new technologies and agronomic practices around the national staple food crops; (ii) increasing farm incomes through increased agricultural intensification and small-scale value addition activities; (iii) developing farmers' social capital, especially in the form of functioning groups, associations, cooperatives and farmer-based organizations; and (iv) training and capacity development in natural resource management (Swanson and Rajalahti, 2010). These objectives closely mirror the agricultural development goals of the post-conflict period.

At the end of a conflict in a poor country, there is usually a dire need to reach smallholder farmers with inputs, advisory services, credit and marketing support, and to re-establish agricultural infrastructure and services eroded through the conflict. Agricultural extension services are often pressed into service for this objective. In the absence of conflict, extension systems involve many different extension actors—including Ministries of Agriculture, nongovernmental organizations (NGOs), private sector organizations, research institutes and universities—delivering services, information, technologies, capacity development and adult education to farmers using a variety of methods and approaches (Swanson and Rajalahti, 2010). Post-conflict extension more commonly occurs either through direct delivery of services on donor-funded projects delivered by Ministries of Agriculture, or through NGOs that may in some cases utilize seconded Ministry of Agriculture extension staff to deliver projects. Thus, public sector extension services in a post-conflict setting are often on the frontline of reaching rural people with public services and providing initial post-war contact with government programs.

This provision of public extension represents an opportunity for new governments to establish legitimacy and reinforce their reputation by providing needed social services. In poor countries, extension has many of the characteristics of a public or 'merit' good, a good deemed important for people to receive despite their inability to pay for it. Thus, extension services not only become one of a post-conflict country's primary anti-poverty programs—especially in the context of high poverty where many of the poor people have a connection to agriculture—but also contribute to reconciliation and peacebuilding (Umali-Deininger and Schwartz, 1994; Collier, 2006; Swanson and Rajalahti, 2010).

In total, agricultural extension services can improve the productive capacity of food-insecure farmers, advance sustainable livelihoods at a crucial time, facilitate reintegration and reconciliation among rural peoples, and promote peace and stability (Scoones, 1998; Collier, 2006; Longley *et al.*, 2007). Establishing effective and stable extension institutions and processes are therefore common goals of many post-conflict agricultural development strategies (e.g. Blaikie *et al.*, 1977; UNEP, 2002; Zaur, 2006; USAID, 2009; USIP, 2013).

Background and Rationale for the Collective Volume

Extension plays a crucial role both in promoting agricultural development and in preventing conflict and supporting post-conflict recovery. Increasing development attention has therefore been paid to improving agricultural sectors and rural livelihoods. At the 2009 L'Aquila Summit, world leaders pledged US$20 billion to agricultural development (United States State Department, 2012). The US Feed the Future initiative was created largely in response to the US commitment to the L'Aquila accord and placed agricultural development at the center of the Obama administration's global development agenda. USAID is tasked with implementing Feed the Future, and agricultural development programs are central to its strategy in all 19 countries served (Ho and Hanrahan, 2011). Other major donors—including the UK's Department for International Development (DFID), the German Corporation for International Cooperation (GIZ) and the Japanese International Cooperation Agency (JICA)—have followed suit in prioritizing agricultural development (DFID, n.d.; GIZ, n.d.; JICA, n.d.). Agricultural extension is a key component of these efforts.

The Modernizing Extension and Advisory Services (MEAS) project was awarded in 2010 and ran until 2016, under the umbrella of USAID's Feed the Future initiative. The goal of the MEAS project was to help transform and modernize extension and advisory systems to increase farm incomes and enhance the livelihoods of the rural poor. MEAS worked closely with both public and private sector extension systems and services in more than 50 countries, promoting cost-effective as well as financially and institutionally sustainable extension services, supporting extension policy design and reform, building the capacities of extension professionals and service providers, conducting action-oriented research designed to improve extension service delivery and facilitating field-level interventions intended to demonstrate innovative extension approaches and practices.

Although MEAS did not have a mandate to work with specific countries, the project ultimately discovered that countries with recent histories of conflict—where extension systems were particularly weak, service provision mechanisms were disrupted and needed strengthening, and institutions required support and rehabilitation—tended to request MEAS assistance. This realization—along with the aforementioned connections between agricultural development and extension in preventing and addressing issues of conflict—provided the impetus for this book.

This collective volume is intended to: (i) explore the role and potential of agricultural extension in post-conflict countries; (ii) investigate the experience and issues involved with rebuilding extension systems in post-conflict settings; (iii) examine the impact of different extension policy approaches and practice in such settings; and (iv) identify the key elements needed to effectively rebuild agricultural extension systems and programs in post-conflict contexts. Exploration of these topics can help inform governments, practitioners and academics in extension policy making and programming, and contribute to post-conflict, political science, international development and agricultural extension literature.

The following chapters seek to capture learning from the MEAS project but also to engage experts in the field of global extension, in particular those with extensive experience in countries in various stages of emergence from conflict. Each chapter is written as a country-specific case study that provides a descriptive account but also analyzes strategies, successes and failures, and lessons learned. The book concludes with a global synthesis chapter that ties together the learning from the individual case studies into a set of cohesive global themes and commonalities.

References

Blaikie, P., Cameron, J. and Seddon, D. (1977) *The Effect of Roads in West Central Nepal: Part One.* Ministry of Overseas Development, London.

Blattman, C. and Annan, J. (2011) Reintegrating and employing high risk youth in Liberia: lessons from a randomized evaluation of a Landmine Action agricultural training program for ex-combatants. Report of Innovations for Poverty Action. Yale University, New Haven, Connecticut.

Collier, P. (2003) *Breaking the Conflict Trap: Civil War and Development Policy.* World Bank and Oxford University Press, Washington, DC.

Collier, P. (2006) Angola: options for prosperity. Oxford University Research Archive. Available at: https://ora.ox.ac.uk/objects/uuid:6c15dba8-728f-4ac4-9e68-9120b4e89b26

Collier, P. (2007) *The Bottom Billion: Why the Poorest Countries are Failing and What Can Be Done About It.* Oxford University Press, Oxford, UK.

DFID (Department for International Development) (n.d.) DFID: what we do. Available at: https://www.gov.uk/government/organisations/department-for-international-development/about (accessed 18 January 2017).

FAO (Food and Agriculture Organization) (2015) The state of food insecurity in the world 2015. Available at: http://www.fao.org/3/a-i4646e.pdf (accessed 18 January 2017).

GIZ (German Corporation for International Cooperation) (n.d.) Global agendas. Available at: https://www.giz.de/en/ourservices/1410.html (accessed 18 January 2017).

Ho, M.D. and Hanrahan, C.E. (2011) The Obama administration's Feed the Future Initiative. CRS Publication 7-5700. Available at: https://www.fas.org/sgp/crs/misc/R41612.pdf (accessed 18 January 2017).

IFAD (International Fund for Agricultural Development) (2013) Smallholders, food security, and the environment. Available at: https://www.ifad.org/documents/10180/666cac24-14b6-43c2-876d-9c2d1f01d5dd (accessed 10 April 2017).

JICA (Japanese International Cooperation Agency) (n.d.) Thematic issues. Available at: http://www.jica.go.jp/english/our_work/thematic_issues/index.html (accessed 18 January 2017).

Lacina, B. (2004) From side show to centre stage: civil conflict after the cold war. *Security Dialogue* 35(2), 191–205.

Lagi, M., Bertrand, K.Z. and Bar-Yam, Y. (2011) The food crises and political instability in North Africa and the Middle East. Available at: http://necsi.edu/research/social/food_crises.pdf (accessed 18 January 2017).

Ligon, E. and Sadoulet, E. (2008) Estimating the effects of aggregate agricultural growth on the distribution of expenditures (Background paper for the World Development Report 2008). Available at: http://siteresources.worldbank.org/INTWDR2008/Resources/2795087-1191427986785/LigonE&SadouletE_EstimatingEffectsOfAggAgGr.pdf (accessed 18 January 2017).

Longley, C., Christoplos, I., Slaymaker, T. and Meseka, S. (2007) Rural recovery in fragile states: agricultural support in countries emerging from conflict. Available at: http://www.odi.org.uk/sites/odi.org.uk/files/odi-assets/publications-opinion-files/10.pdf (accessed 18 January 2017).

Maconachie, R. and Binns, T. (2007) 'Farming miners' or 'mining farmers'? Diamond mining and rural development in post-conflict Sierra Leone. *Journal of Rural Studies* 23, 367–380.

Scoones, I. (1998) Sustainable rural livelihoods: a framework for analysis. Working Paper 72. Available at: http://www.ids.ac.uk/files/dmfile/Wp72.pdf (accessed 18 January 2017).

Strand, H. and Dahl, M. (2010) Defining conflict-affected countries. Available at: http://unesdoc.unesco.org/images/0019/001907/190711e.pdf (accessed 18 January 2017).

Swaminathan, M.S. (1994) *Uncommon Opportunities: An Agenda for Peace and Equitable Development: Report of the International Commission on Peace and Food.* Zed Books, London.

Swanson, B.E. and Rajalahti, R. (2010) Strengthening agricultural extension and advisory systems: procedures for assessing, transforming, and evaluating extension systems. Agriculture and Rural Development Discussion Paper 45. World Bank, Washington, DC. Available at: http://siteresources.worldbank.org/INTARD/Resources/Stren_combined_web.pdf (accessed 18 January 2017).

Umali-Deininger, D. and Schwartz, L.A. (1994) *Public and Private Extension: Beyond Traditional Frontiers.* World Bank, Washington, DC.

UNEP (United Nations Environment Programme) (2002) Agricultural development and the cost of conflict.

United States State Department (2012) L'Aquila Food Security Initiative final report. Available at: http://www.state.gov/s/globalfoodsecurity (accessed 10 April 2017).

USAID (United States Agency for International Development) (2009) Understanding the post-conflict country analytical template. Available at: http://pdf.usaid.gov/pdf_docs/pnadp089.pdf (accessed 18 January 2017).

USIP (United States Institute of Peace) (2013) A new opportunity: agricultural extension as a peacebuilding tool. Available at: https://www.usip.org/publications/2013/01/new-opportunity-agricultural-extension-peacebuilding-tool (accessed 10 April 2017).

Waters, H., Garrett, B. and Burnham, G. (2007) Rehabilitating health systems in post-conflict situations. Paper 2007/06. Available at: https://www.wider.unu.edu/sites/default/files/rp2007-06.pdf (accessed 18 January 2017).

Wiggins, S. and Leturque, H. (2010) Helping Africa to feed itself: promoting agriculture to address poverty and hunger. Available at: http://www.odi.org.uk/sites/odi.org.uk/files/odi-assets/publications-opinion-files/6265.pdf (accessed 18 January 2017).

World Bank (2015) Poverty and equity data. Available at: http://povertydata.worldbank.org/poverty/home/ (accessed 18 January 2017).

Zaur, I. (2006) Agriculture and conflict: a conceptual framework for development. Master's thesis. Available at http://dl.tufts.edu/catalog/tufts:UA015.012.DO.00148 (accessed 18 January 2017).

1 Agricultural Extension in Post-Conflict Liberia: Progress Made and Lessons Learned

Austen Moore*
University of Illinois at Urbana-Champaign, Urbana, Illinois, USA

Background to the Conflict

Conflict and agriculture have been intertwined in the Republic of Liberia since the country's formation by repatriated slaves from the USA and subsequent independence in 1847 (Pham, 2004; CIA, n.d.). Early in its history, Liberians of American descent dominated the political process and began to annex lands that traditionally belonged to indigenous groups (Humphreys and Richards, 2005; MoA, 2007). The Americo-Liberian minority eventually controlled many of the nation's most productive agricultural lands and natural resources (Unruh, 2009).

Many Americo-Liberian landowners practiced a plantation-style agricultural model, while most indigenous farmers either served as labor on large-scale plantations or practiced small-scale farming. The plantation model led to relative productivity in staple and cash crops into the late 1970s (MoA, 2007). Liberia became a net exporter of sugar cane, cocoa, palm oil and rubber.

This period also represented the peak of agricultural extension in Liberia. During the late 1970s, the World Bank heavily funded training-and-visit extension to spread Green Revolution technologies in Africa (Swanson and Rajalahti, 2010). Liberia received World Bank funding to expand its extension workforce and place officers in every county, district and *klan* (township), dramatically increasing the number of farmers reached by extension information (MoA, 2007). Funding was also used to enhance domestic agricultural research through the Central Agricultural Research Institute (CARI) in Bong County and Cuttington University, Liberia's leading agricultural institution (FARA, n.d.).

However, benefits were not felt by all Liberians. Highly productive plantations and lucrative export agreements allowed the Americo-Liberian minority group to accumulate further wealth and prosperity (Sawyer, 2005), while indigenous Liberians became disproportionately poor (Pham, 2004; Humphreys and Richards, 2005; GRC, 2007). Notably, the Liberian government authorized the lease of 100,000 acres of prime agricultural land to the US-owned Firestone Company in 1926, an arrangement that was subsequently criticized for benefitting 'elite' Americo-Liberians while displacing indigenous farmers (Saha, 1988).

Similarly, extension services did not reach all Liberians equally. The Green Revolution model and corresponding technology

* Corresponding author e-mail: acmoore@illinois.edu

transfer approach encouraged large-scale agriculture, and services were often directed towards plantation-style farming systems (Saha, 1988; Pham, 2004). Indigenous farmers' role as laborers or practitioners of small-scale subsistence agriculture meant that extension services were seldom tailored to their needs (MoA, 2007). This disparity contributed to mounting tensions, especially in rural areas where food security was lowest.

Political, economic and ethnic tensions ultimately led to full-scale conflict in 1980, when indigenous military leaders led a coup that toppled the Liberian government. The following 25-year period saw ongoing internal conflict, widespread human rights violations and devastating destruction (Humphreys and Richards, 2005; CIA, n.d.). Peace was re-established in 2003 under a transitional government, followed by elections in 2005 (MoA, 2007; World Bank, 2013).

The Liberian civil war devastated the country, erased years of development (UNDP, 2005) and crippled the Liberian economy (Humphreys and Richards, 2005). Post-conflict Liberia had few options to promote development, economic growth and poverty reduction or to address other effects of the war (Blattman and Annan, 2012).

Liberia's recent Ebola epidemic further complicated post-conflict challenges. A total of 10,666 cases and 4806 deaths were recorded during the outbreak, which occurred from 20 March 2014 to 9 May 2015 (WHO, 2015). The epidemic had considerable implications for Liberia's development, agricultural productivity and economic growth. In fact, Ebola is expected to cause development indicators (e.g. Human Development Index, poverty, life expectancy) to decline for the first time since 2005 (UNDP, 2015).

Post-Conflict Agricultural Context

As other post-conflict economic sectors struggled to rebuild, many Liberians reverted to small-scale agricultural production. In a 2007 report, the Ministry of Agriculture (MoA) described food crop production as Liberia's 'most important source of livelihood' (p. 13) and reported that agriculture was the main income-generating activity of 74% of Liberians. In 2013, the agricultural sector still made up 76.9% of the national gross domestic product (GDP) (CIA, n.d.).

Liberia also has significant potential for agricultural growth. The country contains huge amounts of arable lands and fertile swamps due to topsoil accumulation (FAO, n.d. b). Liberia receives 240 cm (94 inches) of precipitation annually (FAO, n.d. a), an extremely high amount relative to other West African countries. Improved water management and usage of arable land could increase production, food security and livelihoods for Liberian farmers (MoA, 2007).

Despite its importance and potential, the agricultural sector was also devastated by the Liberian civil war. Overall production levels of rice, cassava and maize were much lower than before the war (FAO, n.d., c), and food security dropped to crisis levels for many Liberians. During the early post-conflict period, food aid was provided to address emergency shortages, but little was done to improve domestic production. By 2007, food insecurity still affected 80% of rural households (MoA, 2007), with people displaced by conflict up to 20% more food insecure (Ghimire *et al.*, 2013). Female-headed households, the number of which increased due to conflict, also showed lower productivity per hectare and correspondingly lower food security and income levels (Ghimire *et al.*, 2013).

Staple crop production declined for several reasons. Large landholdings and plantation systems were targeted during the conflict as symbols of oppression and marginalization (Unruh, 2009). Plantations were often abandoned as owners and laborers were displaced, and post-war land-rights issues made restarting plantation agriculture problematic.

Efforts to restart large-scale production have centered on farmers' groups and producers' organizations. However, virtually all existing farmers' groups dissolved as farmers were displaced during the conflict, and a slow and difficult reintegration and reconciliation process made their revival and rehabilitation during the early post-conflict period difficult (Fearon *et al.*, 2009). Farmers' groups are more viable today, but are still affected by community dynamics and post-conflict tensions.

The conflict also changed the characteristics of rural labor. Rebel movements primarily originated in rural areas and also in the three counties (Bong, Lofa and Nimba) considered the nation's breadbasket (Humphreys and Richards, 2005; Stack and Brabazon, 2008). Since these were the epicenter of fighting, rural peoples in this region were displaced and fled towards neighboring Guinea or the coastal capital of Monrovia. Nearly half (45.5%) of Liberians displaced from these areas fled to Guinea, where many resettled or remain in refugee camps (Ghimire *et al.*, 2013). As a result, many rural areas were underpopulated when the fighting ceased.

Prior to the war, young men had comprised much of the agricultural workforce. However, many rural youth were displaced or mobilized into armed groups (Stack and Brabazon, 2008). Post-conflict youth are largely uninterested in agriculture, preferring economic activities in other sectors or in urban areas, and often lack the requisite skills to succeed in the sector, even if they do express a desire to join the workforce (Blattman and Annan, 2012).

As a result, the overall demographics of the rural agricultural workforce changed. The majority of farmers in post-conflict Liberia are older than 50 (USAID, 2008b), and most are women (MoA, 2008b). These groups are less suited to labor activities common to plantation systems (MoA, 2007). Furthermore, female-headed households are generally poorer (MoA, 2007) and less able to employ workers (Moore, 2014). Because the workforce made reinstituting plantation-style agriculture untenable, the majority of post-conflict agricultural production is small scale and largely for subsistence purposes (Ghimire *et al.*, 2013).

Agricultural production was also compromised by the widespread destruction of agricultural and non-agricultural infrastructure. Fighting damaged bridges and roads needed to bring agricultural products to markets, as well as limiting penetration of wholesale buyers and increasing the transport costs of seeds and inputs during the post-conflict period. Even in 2012, 94% of roads in Liberia remained unpaved, and many become impassable during the rainy season (Shor, 2012). Currently a major road project is underway connecting Liberia's 'breadbasket' counties to the capital (Moore, 2014). However, in the meantime farmers struggle to take produce to market, which reduces their returns and acts as a disincentive to large-scale production.

Other infrastructure was also affected. Irrigation infrastructure was widely destroyed, especially in swamp-based rice systems in the rural north. In 2013, only 5% of farmers had access to irrigation infrastructure (Ghimire *et al.*, 2013). Government buildings, including those of CARI and other agricultural extension facilities, were looted and burned. Liberia's main energy source, the Mount Coffee hydroelectric plant in Montserrado County, was damaged in the conflict and is still being rebuilt (IFC, 2013), which limits processing and storage capacities and accelerates post-harvest loss.

Agricultural tools and inputs were stolen, lost or destroyed during the war, as people fled rural areas. Farmers operating after the war lacked even the most basic tools to prepare fields, weed and harvest. They also lacked the seeds, rootstock and plant material needed to quickly resume production. Donor agencies distributed agricultural inputs and equipment to farmers in the early post-conflict period, but these conditions slowed the redevelopment of the agricultural sector (Ghimire *et al.*, 2013).

Significance of Agriculture and Agricultural Extension

Despite the challenges, agricultural development is fundamental to reconstruction, peace and stability, and to overall development in post-conflict Liberia. The transitional government identified the importance of rebuilding agriculture in early policy documents. Liberia signed the 2003 Comprehensive Africa Agriculture Development Programme (CAADP) agreement, which committed the country to allotting 10% or more of the national budget to agriculture

(MoA, 2007; IFPRI, 2013). These early policies were subsequently reinforced by President Johnson-Sirleaf following her 2005 election victory. The new administration placed agriculture 'at the center of reconstruction and development efforts' (MoA, 2007: 1) and identified rice as a commodity as important to national development as oil and concrete (Shor, 2012). Liberia's Poverty Reduction Strategy also cited agriculture as crucial for economic revitalization (Zinnah and Perry, 2011). Furthermore, agriculture was viewed as the sector most able to reincorporate displaced people and former fighters, contributing towards long-term peace and stability (Blattman and Annan, 2012).

Specific strategies represent the two phases of post-conflict development. Early agricultural policies created and implemented during the emergency (or relief) phase focused mainly on food security. In addition to food aid, programs to rebuild large-scale production systems, use modern technologies and develop value chains and markets for Liberian produce dominated this period (MoA, 2007). While the Government of Liberia lacked the capacity to adequately implement this strategy, international donors contributed funding and programming towards production agriculture (USAID, 2008a). Many international agencies distributed seeds, tools and other inputs designed to help farmers restart food and cash crop production. Rice production was prioritized through several large-scale irrigation and swamp development schemes. Increasing production of Liberia's second staple, cassava, has also become a deliberate focus of the Government of Liberia and international donors (MoA, 2007; MoA, 2009; DAI, 2012; Moore, 2014).

Agriculture in the development (or rehabilitation) phase expanded to better align with changing conditions. Stakeholders sought to transform the sector to foster both large- and small-scale production models (MoA, 2008b). Policies began to include 'pro-poor' strategies designed to enhance the production capacity and yields of smallholder farmers, thereby improving rural livelihoods and increasing incomes (MoA, 2007; MoA, 2008a, 2009). This livelihood-centric focus was intended to address poverty, build local food security and promote stability in rural areas (MoA, 2007; MoA, 2008a; MoA, 2009). The Ebola epidemic led to a downturn in agricultural yields and an increase in food insecurity, leading to increased efforts that reinforced the focus on agricultural development (World Bank, 2015b).

Post-Conflict Agricultural Extension

Swanson *et al.* (1997) described food security and rural livelihood development as the two primary objectives of agricultural extension. Similarly, during the post-conflict period, Liberia prioritized both objectives and relied heavily on agricultural extension to reach its development objectives (MoA, 2007).

Whereas pre-war extension services were provided largely by the MoA (Moore, 2014), in the current post-war phase the system is composed of a range of actors. At the center is the MoA's Department of Rural Development, Extension, and Research (DRDER). A total of 134 extensionists work in the DRDER (Moore, 2014), including 72 field-based extension officers serving Liberia's 15 counties (McNamara *et al.*, 2011). Each county contains County Agricultural Coordinators (CACs) and District Agricultural Extension Officers (DAOs), supported by regional subject matter specialists responsible for multiple counties (MoA, 2007; USAID, 2008a; USAID, 2009).

The demographics of public extension personnel were substantially impacted by the conflict. The majority of extension administrators, but also most field-level officers, are over 50 years old. Many had served since before the conflict and are expected to retire in the next few years (USAID, 2008b; McNamara *et al.*, 2011). During the conflict itself, younger extension officers fled, due to security concerns and the inability to draw a salary as services were suspended (MoA, 2007). As a result of this 'brain drain' there was a need to restock the extension system after peace was re-established (Eicher, 2006). However, the MoA struggles to attract and retain young extension officers, despite wide agreement on the need to employ younger staff (Moore, 2014).

In addition, 89.3% of MoA extension personnel are male (Moore, 2014). Representatives from the Farmers Union Network of Liberia (FUNL) described this imbalance as a major impediment to serving female farmers and an indication of cultural bias and inadequate political will rather than a lack of qualified candidates (Moore, 2014). The MoA is currently undertaking actions to narrow the gender gap (Zinnah and Perry, 2011).

While the MoA occupies a central role, a large number of donors, international nongovernmental organizations (INGOs), local civil society organizations (or domestic NGOs) and a small contingent of private sector actors also provide extension services to farmers. Liberia experienced a sizeable influx of donor-funded projects immediately following the 2003 peace agreement, when the institutional and operational capacity of its government to provide essential services to farmers was at its lowest (USAID, 2008a). This phenomenon is common in post-conflict or post-disaster situations, where injections of donor funding are often needed to provide services in a vacuum of public capacity (Schuller, 2012). Many of these efforts included short-term food aid, but also the dissemination of inputs and basic extension services to rapidly restart agricultural production.

As Liberia moved from the emergency phase into the development phase, many organizations ceased operations and were replaced by others with different mandates. Still others (e.g. the Adventist Development and Relief Agency) have shifted from providing aid to conducting more traditional extension activities (Moore, 2014). Approximately 60 donors and NGOs (both international and domestic) were operating in agricultural extension in 2011 (McNamara *et al.*, 2011). Among these are ACDI/VOCA, BRAC, Care International, the Food and Agriculture Organization of the United Nations (FAO) and Winrock International.

Domestic NGOs and civil society organizations have also proliferated in the early development phase, and especially since 2010. During the conflict and early post-conflict period, church-based organizations were the predominant actors within civil society. More recently, organizations such as the Community of Hope Agriculture Project (CHAP), the FUNL and the Sustainable Food and Seeds Project have emerged to address specific needs of the post-conflict period; serve unique segments of the population, regions and crops; and increasingly work in partnership with the MoA, donors and INGOs as they grow in number, capacity and relevance (Moore, 2014). Meanwhile, the private sector remains minimally involved in extension service delivery.

Donor- and NGO-employed extension officers are typically much younger and more likely to be female than their MoA counterparts. These organizations have been proactive about hiring and developing recent agriculture graduates of the University of Liberia and Cuttington University (Feed the Future, 2011). Donor projects in particular attract younger and more educated workers due to higher salaries and better working conditions (Swanson and Rajalahti, 2010). Often personnel are recruited away from the MoA or other agricultural employers because the supply of trained extensionists is low, having been diluted by conflict and the 'brain drain' (Eicher, 2006). Donors and NGOs have proven far more likely to employ female extensionists, although even these organizations have rarely exceeded a rate of 40% women (DAI, 2012; Moore, 2014).

Experiences, Impacts and Issues

Extension in post-conflict Liberia is complicated and faces considerable challenges in serving farmers. A range of contextual factors, strategies and approaches have been used—with varying levels of success—in the post-conflict period. The following sections assess some key issues affecting the extension system.

Policy and planning

Post-conflict Liberia has undergone several major extension policy shifts between the emergency and development phases. Initially

Liberia, like many post-conflict countries, did not have much formal agricultural policy in place. In the first few years following the conflict, the MoA largely promoted extension policies to rebuild food security through speeches, press releases and other informal channels. Liberia later signed the CAADP agreement, and President Johnson-Sirleaf emphasized agricultural production in her inaugural policies (MoA, 2007). This approach did not always provide a consistent message or strategy for agricultural development. Without established policy, donors and INGOs implemented their own operational policies and many actors used disconnected or contradicting development approaches in the early post-conflict period.

However, from 2007 to 2009 the MoA created several key policies intended to clarify development priorities and shift the extension system towards longer-term development efforts. These policies discussed a pluralistic, participatory, value chain approach designed to be more appropriate to changing conditions than the largely top-down, technology transfer model used in the pre-war period (MoA, 2007; MoA, 2008a; MoA, 2009). Policies also included an emphasis on livelihood development and farmer empowerment strategies.

This policy transition is significant for several reasons. It represents a semi-resolution of an ongoing revitalization versus modernization debate between extension providers. The majority of MoA administrators and CACs were trained during the training-and-visit era when Liberia's agricultural sector was stronger. In several interviews, senior administrators cited nostalgia for Liberia's pre-war agriculture and suggested that returning the nation to that model of extension should be the MoA's goal (Moore, 2014). This mindset was common among the 'old guard' within the MoA.

In contrast, younger members of the MoA were eager to modernize the extension system, with support from donors, the INGO sector and other domestic partners. The need to adapt to post-war conditions has slowly gained traction, despite ongoing resistance from senior officials. The Modernizing Extension and Advisory Services (MEAS) project was invited to conduct a scoping mission as a step towards this objective (MEAS, 2011). These transitions have slowly occurred, although considerable support from within the Liberian government and from outside stakeholders is required (MoA, 2007; MoA, 2008a; MoA, 2009; McNamara *et al.*, 2011).

However, implementation of transitions in extension policy has proven slow to match the rhetoric. One such policy transition is decentralization (USAID, 2009). While decentralization of extension services to the county level was an objective of the 2008 and 2009 policies, administrative procedures remain highly centralized and bureaucratic years later (Moore, 2014). Activities such as identifying program foci and priority-setting responsibilities are largely undertaken at the MoA office in Monrovia. Requests for materials, and especially funding, are also relayed by officers through their CACs to the national level, where funding and supplies are managed and dispersed (Moore, 2014). Donor projects and INGOs are more decentralized. Many large programs—such as ACDI/VOCA's LIFE III program and Development Alternatives International (DAI)'s Food and Enterprise Development (FED) program—have county-level offices with relative autonomy (MEAS, 2011; DAI, 2012). BRAC has community-level programs that operate independently, depending on the needs of its respective sites (BRAC, n.d.).

The MoA's struggles to decentralize are likely to be a legacy of the Liberian conflict. Mutual distrust between the central government and local authorities is certainly a factor. Whereas donor agencies and multinational partners (e.g. USAID, World Bank) stress localization and a decentralized democratic process as a means of promoting good governance, the post-conflict Liberian government may not be willing to expand the rights and power of peoples or regions that previously fought against the prior regime, destroyed government property and targeted public servants. This mindset, however obscure, is likely to contribute to slow decentralization shifts in Liberia.

Another area of policy discord relates to participatory, capacity-building extension that promotes farmers' empowerment. The heavy influx of donor dollars and programs providing free seeds, tools and inputs to farmers in the emergency phase is thought to have created a 'dependency syndrome' among Liberian farmers that compromises current capacity-building programming (Chronic Poverty Research Center, 2011). Farmers have avoided programs that do not include material or financial compensation for participating and are reluctant to engage in training that might produce longer-term capacity building and more sustainable development (Moore, 2014). Changing this mentality has been a challenge to organizations implementing participatory extension models and requires a unified effort among service providers. This shift is particularly complicated by a 'victim mentality' or sense of powerlessness to control one's own well-being, which developed among many Liberians as a result of conflict (Bar-Tal *et al.*, 2009).

In addition, while the MoA and many donors/INGOs have committed to capacity-building strategies, a number of other organizations still actively distribute free inputs to farmers. These programs undercut long-term capacity building and sustainable development of the sector. Progress was being made after several years of post-conflict development and better coordination between extension actors (Moore, 2014), but emergency aid stemming from the Ebola epidemic has reversed this trend (World Bank, 2015b).

Extension policies lack formal review processes to allow for revision and updating as circumstances change (Moore, 2014). The MoA struggles to evaluate the impact of its policies and programs, as do many domestic NGOs (McNamara *et al.*, 2011). Larger donor projects and INGOs have effective monitoring and evaluation processes to guide their operations and strategic planning (DAI, 2012), but no mechanism exists for these evaluations to affect overarching policy at the national level. As a result, the extension system lacks timely revisions to extension policies.

Evidence suggests that poor policy review within public extension is directly tied to post-conflict socio-political dynamics (Moore, 2014). First, much like the policies themselves, revision procedures must be developed anew and policy review must be conducted by individuals with the requisite capacity or experience. Furthermore, policy review, and especially review practices that include feedback from diverse stakeholders, may also be viewed as a threat to stability. Inviting opposing parties (often headed by former warlords), leaders of different ethnic groups, rural farmers or other influential stakeholders to critique policy of the ruling government could be more contentious than beneficial in post-conflict Liberia. Given these conditions, reticence to actively pursue policy review is not unexpected.

Funding

Funding is a central issue affecting the quality and effectiveness of extension. The overall budget of Liberia in the post-conflict period is compromised by an economy that is in the rebuilding phase coupled with a poor tax base, while threats of instability deter foreign investment and poor institutional capacity and governance concerns limit donors' willingness to provide direct lending (MoF, 2013). As a result, the Liberian government faces considerable challenges related to funding its public advisory services. Despite committing 10% of the national budget to agriculture through the CAADP agreement, evidence suggests that the actual allocation for agriculture is approximately 3%, or US$14 million dollars (IFPRI, 2013; MoF, 2013). An even smaller amount is dedicated to agricultural extension, with rural infrastructure development (e.g. roads, irrigation systems) accounting for much of agricultural spending (Moore, 2014).

As a result, shortfalls in public agricultural funding restrict services to farmers. The MoA is unable to hire sufficient officers to provide adequate coverage and to rebalance the gender and age of its officers. Retention of talented personnel is also a challenge due to low salaries and high job-related expenses. McNamara *et al.* (2011) reported that nine of 81 officers left the MoA between 2008 and 2011, because of issues related to salary. Moore (2014) also cited multiple complaints from MoA officers about low salaries and the desire to leave for more lucrative opportunities elsewhere. Officers often seek other income-generating opportunities to supplement their salaries, yet these are few in post-conflict Liberia compared with other countries with greater stability and a stronger private sector (Moore, 2014). Those officers who remain with the MoA lack resources to conduct training and have insufficient fuel allocations for travel to engage with farmers (McNamara *et al.*, 2011).

Insufficient funding also impacts the MoA's technical capacity. The ability to reconstruct CARI, and thereby to conduct research domestically, is compromised (USAID, 2008a), and lack of computers and other technologies limits officers' access to web-based information (Swanson, 2011). As a result, many officers struggle to provide up-to-date information to their farmers. Also, funding limits the capacity to produce print materials or to effectively incorporate information and communication technologies (ICTs) into extension service delivery, despite the potential of these channels to address poor coverage and inform a wider audience (McNamara *et al.*, 2011; Zinnah and Perry, 2011; Asenso-Okyere and Mekonnen, 2012).

In contrast, sizeable investments were made by donor agencies to provide extension services in the post-conflict period. Donor-led initiatives proliferated in Liberia, as in many post-conflict and post-disaster countries, due to the vacuum in public services as the government rebuilt and the dire need for basic agricultural programming. A similar influx of donor funding and donor-led programming has followed the recent Ebola epidemic (World Bank, 2015b).

Several large donor projects and INGOs now operate in the country, with a combined annual budget of more than US$100 million (Moore, 2014). USAID's funding to Liberia has been reported as the highest total within Africa, despite the country's small size and population (DAI, 2012). The largest single initiative is the FED project managed by DAI. This project represented the Obama administration's Feed the Future initiative for global hunger and food security (Ho and Hanrahan, 2011) and focuses on providing extension services and developing rice, cassava, vegetable and livestock value chains (DAI, 2012). The budget of FED alone (US$76 million over 5 years) exceeded that of the Liberian government (Moore, 2014). Other donor agencies and organizations, such as ACDI/VOCA, FAO, Winrock International and the World Bank, have also operated highly funded projects to develop staple and cash crop production (World Bank, 2015a).

These funding levels allow donor projects and INGOs to hire and retain the most highly skilled extensionists available, and even to attract those displaced by conflict back to Liberia. These organizations can also provide pre- and in-service training to their workers, supply and maintain vehicles to allow operations in remote rural areas, and produce print and ICT-based training materials. All of these factors provide a significant advantage that is reflected in the quality of extension services and the corresponding preference by farmers for working with donor-led projects and INGOs, as opposed to with the Liberian government (Moore, 2014). Funding is therefore the single largest factor influencing extension service delivery in post-conflict Liberia.

Pluralism and coordination

As discussed, the early post-conflict period saw a range of extension providers acting without coordinated policy and strategic plans. More recent policies have promoted pluralism, and the MoA has sought to foster an effective pluralistic extension system. The 2007 Comprehensive Assessment of the Agriculture Sector stated this goal: 'The extension system needs to transform from the transfer of technology model to a pluralistic extension system that involves multiple public and private sector service providers' (MoA, 2007: xvii).

The Liberian MoA actively promotes itself as a central facilitator and monitor. The government holds Agricultural Coordination Committee (ACC) meetings that gather service providers at the national and county levels to help coordinate extension activities, avoid duplication of programs, share challenges and lessons learned, and maximize resources (Zinnah and Perry, 2011; Moore, 2014). Stakeholders from the MoA, donor-led projects, INGOs, domestic NGOs and the civil sector all cited the value of these forums in working towards an effective pluralistic system (Moore, 2014).

However, the practical realities of implementing pluralistic extension in post-conflict Liberia remain challenging. Power dynamics developed after the conflict that significantly affect pluralism in extension provision. Specifically, the MoA and the donor/NGO sector are not equal partners. International agencies took the lead in providing emergency food aid and extension services in the early post-conflict period (MoA, 2007). Even after a certain level of institutional capacity and governmental stability had developed, the MoA still relies heavily on international actors to serve Liberian farmers (Moore, 2014).

Funding differences in particular create this need. Simply put, the extremely well-funded donor-led and INGO programs have the capacity to hire and train talented officers, develop effective training curricula and materials, provide comprehensive services and inputs to farmers, and travel to serve farmers in remote areas, while the MoA struggles in all these respects.

Although policy asserts the need to work together at the field level, collaboration and partnership dynamics are quite skewed and heavily reliant on the donor/INGO sector to buoy the MoA in conducting extension activities. Limited resources and capacities leave most MoA officers unable to offer much assistance to their counterparts working with donor-led and/or INGO projects. In many cases, these officers and MoA officers have conducted joint training, but more commonly the capacity gap has repositioned officers with donor/INGO projects as leaders and MoA officers as learners at the level of client farmers (Moore, 2014).

Both the MoA and donor/INGO sectors are acutely aware of these power dynamics, and consequently have different levels of motivation to pursue pluralism. Because many donor-led or INGO extension providers began operating during a period in which national policy was absent and the government lacked the capacity to operate, many international organizations became accustomed to operating independently. At the same time, donors/INGOs recognize that the MoA is reliant on their involvement, that they are driving the extension system and that correspondingly the MoA does not have the authority to significantly affect their operations. Convincing these organizations to foster pluralism in a more active manner, share resources and information, coordinate with the national government and adhere to overarching policy or strategic plans is a major challenge. In fact, some donor-led projects and INGOs (e.g. ACDI/VOCA, FED) have questioned whether pluralism really benefits their organizations or whether it instead represents an unnecessary use of their time and resources (Moore, 2014).

Nevertheless, most donor programs have increased their efforts to collaborate with the Liberian government in recent years. Donor programs and INGOs now provide considerable support to the MoA. ACDI/VOCA, the FED program, ZOA and other INGOs frequently invite MoA personnel to participate in in-service trainings, provide technical information and offer transportation to project sites (Moore, 2014). Unfortunately, despite intentions of equal collaboration by administrators, unequal capacities and roles at the field level often create a top-down relationship between MoA and donor project/INGO officers, which leads to tension and further complicates partnerships (Moore, 2014). Similar dynamics were found between donor projects/INGOs and their domestic NGO partners, where again disparities in skills and funding can lead to a top-down arrangement. As a result, pluralism is not functioning properly in post-conflict Liberia.

Coverage

An estimated 1 million smallholder farmers are active in post-conflict Liberia (CIA, n.d.), creating a huge audience for extension and advisory services. However, poor coverage by extension providers remains a major weakness of the system. As discussed, the MoA employed only 72 field-level officers in 2014. All 15 counties employed a CAC but only half of Liberia's 68 districts had a dedicated DAO (USAID, 2008a). This results in extremely high farmer-to-officer ratios (1000:1 to 5000:1).

Poor coverage, while certainly not unique to post-conflict countries, is tied to Liberia's conflict. Vacant posts are often distributed along ethnopolitical lines, specifically with regard to the role of different groups in the conflict. The 'breadbasket' counties of Bong, Lofa and Nimba, the site of several rebel movements and home to a number of post-conflict parliamentarians, are generally well stocked with extension officers and receive considerable development attention. In contrast, southeastern counties, home to the Krahn ethnic group and the indigenous rebel leader Samuel Doe, who initiated the 1980 coup, have many fewer officers and receive considerably less development attention (Moore, 2014). Safety and security concerns also affect officer placement (Moore, 2014), largely depending on different regions' feelings of 'insiderness' versus 'outsiderness' with respect to the Liberian government.

Even in areas where its personnel are present, other factors have left MoA officers simply unable to serve all potential clients. Vehicles issued to CACs are used heavily to attend central administrative meetings and sparingly for projects (McNamara *et al.*, 2011). A handful of DAOs were issued motorcycles, although the low fuel allocations left many officers unable to use them. DAOs have reported spending their own salaries to hire transport to reach farmers or relying on INGOs for occasional transport to their client communities (Moore, 2014), which was not considered a sustainable operating model. As a result, Moore (2014) found that virtually all farmers interviewed never interacted with government extension officers, and that rural, female and youth farmers were the least likely to receive services from the MoA.

Donor-led projects and international and domestic NGOs play an important role in addressing coverage challenges. As a component of partnership and pluralism policies, some projects, INGOs and local NGOs coordinate activities with the MoA to operate in areas where the government lacks the capacity to serve. For example, officers employed by the FED program help to serve in remote rural areas of Nimba and Lofa County that are otherwise understaffed by the MoA (DAI, 2012). However, these organizations generally work intensively with small groups of farmers, meaning their overall contribution to coverage is small (MEAS, 2011).

Many local NGOs and civil society organizations (e.g. CHAP, FUNL) also provide short-term services. This group of service providers is still growing and has limited capacity to contribute, but has been successful in working with otherwise marginalized audiences, such as female farmers, youth and farmers in remote areas of the country (including the traditionally underserved southeastern counties) (Moore, 2014). Still, despite these efforts, there are not enough extension personnel to serve all Liberian farmers, a factor that severely compromises the development of the agricultural sector. The ongoing capacity to expand coverage may also change when donor funding tied to post-conflict reconstruction—and later to Ebola response, which is a primary source of operational funding for domestic NGOs—is reduced in the future.

Extension Approaches/Models

Demand-driven extension (participatory extension models)

Making the transition to demand-driven and participatory models, based on farmer-led priority setting, hands-on and demonstration-based teaching and peer-to-peer learning, has proven difficult for the Liberian MoA (MEAS, 2011; Swanson, 2011). As previously described, an unwillingness to empower regions or ethnic groups with ties to the Liberian conflict to 'demand' services from the government is possibly a product of the civil war, even though the Liberian government verbally promotes such bottom-up extension approaches.

Placing emphasis on participatory and learner-centric extension has also been a challenge. Consequently, many MoA personnel, especially older officers who served in extension before and through the conflict and were trained during the technology transfer era when workshops and lectures were emphasized, have struggled to adapt to this philosophical shift (USAID, 2008b).

Liberia's agricultural institutions (e.g. Cuttington University, University of Liberia) similarly fail to prepare younger officers with participatory skills (McNamara *et al.*, 2011; Moore, 2014). Agricultural institutions in countries emerging from conflict often lack quality instructors due to displacement and attrition, use outdated curricula and struggle with issues of social cohesion—all factors that compromise the quality and relevance of the education received by graduates (Buckland, 2005). This phenomenon also affects the knowledge and skills of Liberian graduates, who learn production and technical skills but lack capacity in participatory extension (Moore, 2014).

Lack of professional development is also a factor. While recent policies and administrative decisions have repositioned field-level MoA officers as facilitators, community mobilizers and communicators (Sulaiman and Davis, 2012; Ganpat, 2013), no corresponding pre- or in-service training has been provided to retool these officers for their new responsibilities, due to funding and logistical limitations. MoA officers therefore lack capacity in participatory extension (McNamara *et al.*, 2011).

The legacy of conflict may also create reluctance on the part of field-level officers to empower farmers to demand services. Some officers indicated that this repositioning of power dynamics affected perceptions of safety and security but also social standing (Moore, 2014). Combined with lack of training, this mindset severely compromises attempts to move towards demand-driven extension.

In contrast, donor-led projects and INGOs have proven to be very successful at incorporating participatory extension methods into their work with farmers:

1. These organizations do not face the same post-conflict concerns. Empowering farmers to better dictate extension programming is seen as a positive endeavor to promote democracy, transparency and good governance, whereas the MoA has shown reluctance to fully engage in demand-driven models for the same reasons (Moore, 2014).

2. These organizations benefit from an operational strategy that is more conducive to participatory extension. Whereas the MoA is tasked with serving as many Liberian farmers as possible, donors and INGOs strategically focus on providing regular, comprehensive and longer-term services to fewer farmers. This level of interaction and responsiveness allows officers to develop trust with their farmers and to better understand their challenges and needs, which is essential for participatory extension to be effective (Swanson *et al.*, 1997).

3. Donor-led projects and INGOs also have the resources to better prepare their personnel in participatory extension methodologies. Field-level FED officers receive thorough pre- and in-service training that includes not only crop-specific technical information but also facilitation, agribusiness and cooperative management skills (DAI, 2012). They do not rely on the aforementioned agricultural universities and colleges to develop these skills and are therefore more capable and effective when working with farmers (Swanson, 2011; Moore, 2014).

In recognition of this gap, many donor projects and INGOs (e.g. ACDI/VOCA, FED) have made officer training available to members of the MoA. Unfortunately, without funding for transportation, lodging and daily expenses, the participation levels remain low and the potential benefits of free, high-quality professional development training are not maximized by the Liberian government (Moore, 2014).

This capacity gap also impacts the use of demonstrations and hands-on teaching. The use of demonstrations is heavily advocated by extension service providers (MoA, 2007). MoA administrators and field-level extensionists consistently acknowledge the benefits of 'learning by doing' in retaining information and adopting new practices. However, the MoA lacks the tools, seeds and other resources to perform effective demonstrations, making the model difficult to implement for most field-level officers (Moore, 2014).

Land for demonstration sites is limited. In 2011 the MoA intended to create demonstration farms to accompany each county office, for conducting farmer field days and trialing new crop varieties and farming methods (McNamara *et al.*, 2011; Swanson, 2011). However, by 2014 only one county had a functional demonstration farm (Moore, 2014), suggesting the approach was more easily verbalized than operationalized.

The donor/INGO sector took a different approach that proved more successful. Rather than creating and operating their own sites, donors and INGOs identified 'lead farmers' from within their client base and obtained permission to use a portion of those individuals' land to conduct demonstrations and trials (Moore, 2014). This model placed the demonstration site central to where farmers worked, helped promote local buy-in by allowing the lead farmer to benefit from the innovations tested without risk to his/her own production and reduced maintenance costs to implementing organizations.

The demonstration farm approach also tied in closely with the promotion of peer-to-peer learning and information sharing within farmers' groups. The formation of strong and functional farmers' groups has been complicated in post-conflict Liberia. Social cohesion issues arising from the displacement and reintegration of rural peoples present challenges (Fearon *et al.*, 2009), as do concerns that farmers' groups are more easily mobilized (Unruh, 2009). Early aid programs that provided seeds, equipment and other inputs frequently resulted in the ad hoc creation of groups of farmers who were not traditionally affiliated, only to separate after input distribution was complete (Moore, 2014). This tendency made it difficult to conduct longer-term group projects that required cohesion and continued participation.

Certain regions also have cultural tendencies to work together or work apart. For example, the counties of Bong, Lofa and Nimba practice a *kuu* system whereby farmers arrange themselves into groups and rotate between farms to complete otherwise time-consuming tasks. *Kuus* have leaders and share information among members more readily (Moore, 2014). On the other hand, there is a cultural reluctance in some regions, such as the Krahn-dominated Liberian southeast, to be 'grouped' by the government or other extension providers. The perception is that farmers in this region view extension information as a competitive advantage and do not readily share with neighbors (Moore, 2014). Inability to practice group-based approaches has deterred some providers from working there. However, post-conflict ethnopolitical factors may also be involved, especially since the heavily indigenous region produced Samuel Doe and was central to initiating the Liberian conflict. Negotiating group dynamics and tailoring extension programming to cultural differences has been a challenge and remains a point of focus in the post-conflict period.

Another prominent peer-to-peer learning strategy is the Farmer Field School model, which the MoA promotes heavily (MoA, 2007; USAID, 2008a). Farmer Field Schools involve central training sites that enroll and train lead farmers in a range of production skills to share with their respective communities, thereby spreading information to new audiences (Davis *et al.*, 2010).

Where social cohesion and group-based farming is effective, the Farmer Field School approach has been quite successful in Liberia. In an analysis of farmers' perspectives, Moore (2014) found that participant farmers described considerable knowledge and skill acquisition and routinely expressed their intention to train their neighbors. Efforts to tie Farmer Field Schools to existing *kuus* has particular promise, as recent efforts to identify *kuu* leaders as lead farmers and encourage their participation in Farmer Field Schools have been extremely effective in spreading new varieties and methods (Moore, 2014).

However, the MoA has limited capacity to implement the approach. Lack of demonstration sites, poor technical capacity of trainers and insufficient resources to operate the schools remain problematic (USAID, 2008a). Again, donors, INGOs and specifically the FED program are the leaders in the Farmer Field School model in Liberia (Moore, 2014).

Market-driven extension (value chain approach)

Efforts to modernize extension in post-conflict Liberia also include broadening the focus on production to include a full value chain approach (Zinnah and Perry, 2011). Following the absence of cohesive policy and development strategy in the early post-conflict period, this objective is now a near consensus among public, donor-led and NGO service providers. Market-driven extension was included as a priority of the MoA's 2008 policy (MoA, 2008b). Value chain development is also a key component of the Feed the Future initiative and therefore the FED program (Ho and Hanrahan, 2011; DAI, 2012). With the two most influential service providers promoting 'agriculture as a business' (Moore, 2014: 148), other donor projects, INGOs and civil society organizations have refocused their strategies to follow suit.

Agreement on this priority helps to avoid competing agendas in service delivery. Coordination at the national level, in policy decisions and through the ACC meetings has allowed extension providers to prioritize specific value chains to avoid duplication (Moore, 2014). The government is focusing on rice and cassava value chains, supported by the FED program, which focuses on the two main staples plus vegetables and goats (DAI, 2012; Moore, 2014). BRAC programs center on poultry, while ACDI/VOCA, Winrock International and the World Bank emphasize cocoa, coffee, palm oil, rubber and other cash crops for international markets.

However, verbal consensus on value chain development leads to different outcomes when the capacities of providers are considered. Despite administrative support, the MoA does not have a clear understanding of how to implement the value chain approach at the field level. Similar to capacities in participatory extension, MoA officers do not receive adequate training on value chain development and are therefore unable to incorporate key components (e.g. marketing, post-harvest handling, record-keeping) needed by farmers (Swanson, 2011). Most commonly, MoA officers promote higher quantity and quality of production with the assertion that this product will be attractive to buyers. However, lack of pre-established markets is a major obstacle. Liberia, like many post-conflict countries, struggles to access export markets due to a poorly developed trade infrastructure, minimal private sector involvement, and post-conflict and now Ebola-related stigma facing Liberian produce (Agwu *et al.*, 2012). Farmers who target domestic markets often overproduce and overestimate demand, which leads to unsold and spoiled produce and causes financial loss (Moore, 2014). Ebola further compromised the viability of local markets as buyers and producers alike lacked the mobility to trade (World Bank, 2015b).

One District Agricultural Extension Officer pre-arranged a buying agreement with the World Food Programme (WFP) to distribute beans grown by his farmers' groups to regional refugee camps. This relationship closely mirrored the WFP's Purchase for Progress (P4P) program, which exists in Liberia but on a small scale. Social cohesion concerns, inability of farmers' groups to meet production quotas and lack of rural infrastructure were cited as reasons (WFP, 2014). As a result, this case is unfortunately the exception and not the rule among MoA value chain programs. This individual had participated in a unique MoA exchange program that sent him to India to study agribusiness, and the experience provided the motivation to pursue this buying arrangement.

Donor projects and the INGO sector were far more effective at promoting the value chain approach. Officers receive pre- and in-service training in the range of skills needed to teach different components of the value chain (Moore, 2014). Again, the MoA is invited to send its officers to these trainings. ACDI/VOCA, FED and other projects/INGOs also train specialists to emphasize different areas of the value chain. FED's business extension officers teach record-keeping, marketing and other agribusiness skills on a rotational basis, while crop production specialists guide farmers' groups through growing specific crops (DAI, 2012).

Donors'/INGOs' model of long-term involvement with smaller groups of farmers also allows different trainings and interventions held at different stages of the agribusiness chain, from planning, predetermining markets and record-keeping to post-harvest processing, storage and sales (Moore, 2014). When possible, organizations collaborate to maximize different agencies' relative strengths within the value chain. For example, while FED is effective in teaching agribusiness skills to farmers, the overall value chain is backstopped by equipment and processing training supplied by the FUNL, German Agro Action and the World Bank. The result is that farmers are linked to markets (Moore, 2014).

Institutional Capacity Building

In any post-conflict situation, strengthening the capacity of local institutions allows more effective and sustained development while also promoting governmental stability (Arthur, 2011; Cunguara and Moder, 2011). When the MoA was incapable of providing extension services during the early emergency phase, donors and INGOs operated in parallel with, or sometimes counter to, the MoA in addressing pressing needs (MEAS, 2011). These actors prioritized emergency aid and service provision and did not emphasize or devote much time or resources to building institutional capacity. Without direct support, Liberia's public extension capacity increased slowly but still remains low. Following 2008 policies that called for collaboration and partnership, many large donor projects and INGOs (e.g. ACDI/VOCA, FED) have included a mandate to build MoA capacity in their programming (DAI, 2012; Moore, 2014). The impending end to many project and funding cycles may also have served as a call to action.

However, strategic planning towards this objective is largely informal, and concrete activities to build the administrative capacity of the MoA are lacking. In addition, disparities in skills and funding lead to a top-down arrangement. Many donor organizations and INGOs attempt to include the MoA in their operations as a teaching strategy. Other efforts include disseminating technical information to the MoA, inviting MoA personnel to participate in in-service trainings and joint planning sessions, and providing transport assistance when feasible (McNamara *et al.*, 2011; Moore, 2014). However, donors and INGOs still remain the drivers of planning and operations.

In many ways, these efforts echo earlier emergency-phase models of distributing inputs to farmers. Providing access to technical information does not increase the MoA's capacity to conduct its own research, and distributing donor/INGO extension materials does not enable the MoA to create its own print or online resources. These strategies may pay short-term dividends, but their long-term impact is questionable.

Stronger efforts to incorporate MoA personnel into professional development trainings could have far greater long-term benefits. This training could potentially build sustainable administrative and technical skills within the public sector. Indeed, administrative training in Monrovia is effective and frequently attended, since transportation and lodging are not concerns for MoA participants. Unfortunately, MoA personnel are far less able to participate at the field level due to the aforementioned financial and logistical barriers. Donors and INGOs have not shown adequate efforts to remove barriers and maximize capacity-building opportunities (Moore, 2014).

Donors and INGOs also have a loose mandate to develop the capacity of domestic NGOs and civil society organizations. The Feed the Future initiative and major multilateral donors (e.g. Gates Foundation, World Bank) have advocated partnership between international and domestic NGOs in serving farmers (Ho and Hanrahan, 2011). In Liberia, many donors and INGOs delegate field-level operations to local organizations but retain management over funds and materials. This model places domestic NGOs between clients and donor agencies, and sometimes makes aligning donor and client agendas difficult (Schuller, 2012; Moore, 2014). Furthermore, access to technical information and inclusion in professional development trainings was insufficient to build the capacity of domestic NGOs, which struggled greatly with administration, financial management, fundraising and advocacy skills that were not adequately addressed by donor partners (Blagescu and Young, 2006).

Implications and Recommendations

Liberia provides several important lessons and implications for other post-conflict countries. While many of the specific details are relative to the Liberian context, the overall themes are likely to be common to other nations practicing agricultural extension following violence and warfare. Similarly, key elements can be identified to effectively rebuild agricultural extension systems and programs in post-conflict settings. These are:

- Flexibility to adjust to changing policy environments and willingness to engage in policy creation and review.
- Sufficient, consistent and creatively utilized funding to support effective extension strategies and programs.
- Genuine commitment to and action towards pluralism in service delivery.
- Innovation in addressing coverage gaps and balancing personnel to meet farmers' needs.
- Conflict-sensitive extension approaches.
- Appropriate pre- and in-service training to support transitions in extension philosophies.
- Balanced coverage of all value chain elements to promote livelihood development.
- Commitment to and efforts towards institutional capacity building by international agencies.

Each component requires individual discussion, as follows:

1. Extension providers in post-conflict settings must be capable of working in an environment without set policies, willing to engage in collaborative policy creation and review processes, and flexible enough to adhere to new national policies and strategic plans when these are put in place. Post-conflict countries should expect a certain lack of synchronicity in extension during the early period as policies are developed, especially when donors and INGOs act autonomously and independently of governments to deliver emergency services. As government capacity allows, all stakeholders involved in extension service delivery should be prepared to engage in policy creation and planning efforts.

Inclusive policy making is critically important to post-conflict extension. It is essential to synchronize philosophies, strategies and objectives to avoid competing agendas that undercut one another. Governments that do not include other providers in policy making may face problems when policies are installed that do not align with existing donor/INGO strategies. For example, models of giving inputs versus building capacity affected outcomes in Liberia (Moore, 2014), and similar instances should be avoided. Also, while post-conflict governments should be central to policy making, donors and INGOs can help to create policy that is mutually beneficial, while simultaneously building institutional capacity in policy creation processes. Likewise, input from beneficiaries can allow policies to better represent the needs of farmers (Swanson, 2013).

In addition, donor organizations and INGOs must be receptive to policy shifts and support the implementation of policy, even if this means changing operational strategies. Failure to do so undercuts the efficacy of extension policy, leads to counterproductive power dynamics, undermines burgeoning governments and affects stability (Collier, 2006; Arthur, 2011).

Policy also needs to be flexible and allow for revisions as circumstances change.

Regular and systematic policy reviews must be planned and implemented in post-conflict extension systems. This requires governmental, non-governmental, private and civil society providers to conduct regular evaluation and to use adaptive management strategies, so that policies and year-to-year operating plans best represent the rapidly changing context found in post-conflict settings (Swanson *et al.*, 1997; Rossi *et al.*, 2004).

2. Funding levels and trends are central to the effectiveness and sustainability of extension systems. In Liberia, and likely in other post-conflict settings, funding levels determined coverage, capacity of personnel, access to technical information and other aspects of the system (USAID, 2008a; MEAS, 2011). Funding trends are also tied to the 'bubble' phenomenon of emergency aid, which is extremely high immediately following conflict or a major disaster but then declines (Schuller, 2012). When governments are underfunded, extension services available to farmers wax and wane with donor funding levels. For example, anticipated reductions in donor funding in Liberia called into question the sustainability of present service levels.

Large funding disparities between extension actors also create issues of power and accountability (Swanson and Rajalahti, 2010). These are particularly problematic when governments and domestic agencies are disempowered within their own countries. Post-conflict governments face additional challenges soliciting direct funding due to minimal track records of responsible borrowing, perceptions of poor governance and low institutional capacity. Multilateral donors are often hesitant to lend to post-conflict governments due to stability concerns and uncertain economic growth potential (Collier, 2006), even though these nations are likely the most in need of capital. To avoid directly funding government ministries, one option is to support the operations of field-level officers by addressing MoA funding shortfalls (e.g. transportation, materials), thereby leveraging funding to provide better public services while also building local capacity.

3. Commitments to pluralism must be extended beyond the boardroom and into field-level operations. Providers must do a better job of identifying and utilizing the relative strengths and weaknesses of actors within the system. For example, the Liberian MoA recognized donors'/INGOs' technical advantage but did not adequately leverage this resource to its own benefit (USAID, 2008a; MEAS, 2011). Partners must seek ways to address any detrimental weaknesses. This may involve donors/INGOs and the MoA working together to remove barriers to participation in professional development trainings.

Ministries of agriculture should also be wary of deferring too much of service delivery to donors, INGOs and the private sector. In Liberia this led to problems when the MoA attempted to implement policy and strategic planning in an environment where donors and INGOs had previously operated autonomously (Moore, 2014). Also, many international organizations operate on short-term funding cycles (Schuller, 2012), which affects the availability of services when agencies' funding decreases or contracts end.

Expecting that privatized extension will become available to fill this void is not a viable strategy. Private sector development is often slow in post-conflict countries, due to instability, inefficient governments, poor infrastructure and a suboptimal business climate. In Liberia, the current model is to quickly commercialize groups of farmers so that they can eventually pay for private services (MoA, 2007). Unfortunately, there is little evidence to suggest that the private sector will be ready to take over in the near future.

Instead, ministries of agriculture should be more involved in activities that foster a sustainable pluralistic system with multiple service providers. Facilitating effective coordination and removing barriers for outside actors to operate is a vital role for post-conflict governments faced with low capacities to provide direct services. Monitoring better-funded donor projects or INGOs can help to ensure that extension activities follow national policy and move towards development objectives (Swanson *et al.*, 1997). Profit-driven private extension services should be encouraged through the removal of barriers to their operation, but must also be monitored to prevent them from disproportionately benefitting financially successful farmers with the capacity to pay, while underserving poor, female or other marginalized farmer populations (Feder *et al.*, 1999). Without governmental oversight, this could potentially lead to the same levels of disparity that contributed to conflict in the first place.

4. Considerable coverage limitations and high farmer-to-officer ratios are common to extension in the developing world and are particularly unlikely to be addressed in post-conflict countries (Birner *et al.*, 2011). Even with donor and INGO funding and personnel numbers at peak levels, most Liberian farmers do not receive direct services from providers. Alternative means are needed to minimize coverage gaps in Liberia and other post-conflict settings.

Although direct funding that allows ministries to employ more officers is unlikely, one potential avenue is to use community-based extension volunteers. The FED program employs youth in this role (Moore, 2014), and a number of other countries advocate the coverage benefits of farmer-to-farmer extension approaches (Franzel, 2015). Of course, issues of social cohesion, safety and security, funding sustainability and individual capacity must be considered before mobilizing community extensionists in post-conflict settings.

Similarly, informal youth officers, especially those who are lightly subsidized, can be a major boon. They are already present in the community and know the context, people and local language. They can also receive basic information from central extension personnel/institutions by phone, which they can then disseminate to farmers for a small fee. Also, employing rural youth as extension officers may attract these young people to remain in or return to agriculture. Providing them with nominal employment can also reduce their susceptibility to mobilization and willingness to engage in further conflict.

The use of ICT to spread information messages remains underutilized across the developing world but especially in post-conflict countries (Asenso-Okyere and Mekonnen, 2012). While mobile phones represent a way to reach huge numbers of farmers in Africa and across the developing world with important extension information, post-conflict countries still face challenges. Often this results from the destruction of infrastructure during conflict. In Liberia, destruction of the national hydropower plant has made the availability of electricity a limiting factor. A potential solution in contexts of extreme infrastructure limitations is radio. Rural radio has been expanded in recent years and now provides extension messaging to multiple counties (Swanson, 2011). Distribution of low-cost solar radios may be a feasible alternative and has been successful in other post-conflict settings (e.g. Timor Leste) where physical infrastructure was destroyed.

While not addressing overall coverage, post-conflict countries that rebalance personnel can better serve target demographics. Extension officers who work with youth, women and other marginalized farming populations are essential to promoting development, peace and stability (Blattman and Annan, 2012). In Liberia, many MoA extension personnel are older and nearing retirement age. At the same time, a large pool of talented young extension officers may become available as donor programs withdraw, including many highly trained female officers who are effective in working with women farmers (Meinzen-Dick *et al.*, 2011). A short-term influx of money from the Liberian government could fund retirement packages and signing bonuses, to entice younger officers to join the MoA. Younger officers can better relate to youth farmers.

5. Public extension systems in the developing world often strive to be more modern, decentralized and participatory, and to provide demand-driven services to farmers (Swanson and Rajalahti, 2010). While these approaches may be beneficial in stable countries, decentralization and demand-driven extension in particular may not be appropriate in the short term for post-conflict countries. Empowering local leaders and farmers to form groups and have a greater capacity to make ‘demands’ of newly established governments may be seen not as a path towards more effective extension but instead as a threat to peace and stability. In Liberia, this may contribute to differences in the actions taken by the MoA and those taken by the international community. Therefore, hesitation by extension administrators in post-conflict Liberia to embrace decentralization and demand-driven extension is not only understandable but expected. Policy makers and agenda setters must take these dynamics into account when selecting extension models and pursuing extension reform.

6. Philosophical shifts in extension approaches cannot be effective at the field level without corresponding retraining of extension personnel. Training of extensionists often ends during conflicts, meaning that officers in post-conflict periods may possess skills and information that are years out of date (Birner *et al.*, 2011). Additionally, extension officers undergo a process of skill deterioration termed 'forgetting by not doing' when extension services are suspended. This further exacerbates capacity decline (Collier and Duponchel, 2013: 67).

In Liberia, government extension officers need to be retrained in both participatory and market-driven extension to address these objectives, yet efforts towards this remain insufficient (MEAS, 2011; Swanson, 2011). Some MoA administrators assume that trained officers would migrate to the MoA after donor projects in Liberia end, thereby closing the capacity gap without requiring the MoA to provide its own training (Moore, 2014). However, this belief ignores the maintenance factors (e.g. low salary, lack of materials, poor upward mobility within the organization) that deter many of these individuals from working for the MoA in the first place. Instead, Liberia needs to take better advantage of donor or INGO training that is available, especially since this opportunity will not last forever. Other post-conflict countries should also capitalize on available avenues to contemporize the skills of their officers.

7. Market-driven extension can only be a step towards improved rural livelihoods if no gaps remain in the value chain approach. In Liberia, extension programs emphasized production of a few crops (e.g. rice, cassava, cash crops) without considering demand and market conditions. Export markets may be a more feasible long-term objective, with a focus on local markets a more realistic approach. Greater efforts to engage producers in P4P programs that supply refugee camps and school feeding programs would be especially beneficial in post-conflict settings where the demand for food aid is high (WFP, 2014). Coordinated local marketing can also avoid local markets being flooded with commodities, thereby driving down prices received by farmers while increasing the prices of other foods that those rural communities consume (FAO, 2013).

Market-driven extension must also provide adequate training on processing and storage while simultaneously implementing programs or lending schemes that allow farmers to purchase necessary equipment or materials. Transportation issues related to post-conflict infrastructure conditions, security concerns, low post-conflict private sector involvement and distance to markets must also be considered when advocating production schemes for farmers. These gaps in the value chain approach lead to high post-harvest loss in Liberia (Moore, 2014) and should be better addressed to truly enhance livelihoods.

8. Capacity-building agendas must focus on developing institutions and civil society organizations, rather than focusing exclusively on developing the capacities of beneficiaries. Donors must recognize that low governmental capacity is an inevitability of post-conflict settings, and that enhancing rather than circumventing local institutions can build institutional capacity to promote sustainable growth. Strong domestic actors can also provide a better and more sustainable level of service to farmers over the long term, rather than the peaks and valleys that accompany donor initiatives and funding cycles (Abi-Ghanem *et al.*, 2013). Also, developing the ability of ministries of agriculture and local NGOs to serve farmers effectively can foster positive farmer–government interactions that promote peace and stability (Collier, 2006; Arthur, 2011). Strengthening the public and civil sectors could have this effect in Liberia as well as in other post-conflict countries.

Clearly many challenges exist in post-conflict extension systems. However, successes and opportunities remain abundant even in this difficult period. Liberia has made considerable progress and has taken many important steps towards modern, pluralistic and farmer-driven extension that should provide dividends well beyond the agricultural sector. It is the responsibility and challenge of individual post-conflict countries to consider and adapt the Liberian case to local conditions. This will require innovation, flexibility, collaboration and political will, yet the benefits can be fundamental to promoting peace, stability and long-term development.

References

Abi-Ghanem, R., Carpenter-Boggs, L., Koenig, R.T., Ullman, J.L., Murphy, K.M. and Pannkuk, C. (2013) Access to agricultural inputs, technology and information, communicating with farmers, and the role of women in agriculture: perceptions of Iraqi extension agents. *Journal of International Agricultural Extension Education* 20(1), 6–18. DOI:10.5191/jiaee.2013.20101

Agwu, A.E., Egbule, C.L., Amadu, F.O., Morlai, T.A., Wollor, E.T. and Cegbe, L.W. (2012) Linkages among key actors in the climate change and food security innovation system in Nigeria, Sierra Leone and Liberia. *Journal of Agricultural Extension* 16(2), 34–51.

Arthur, P. (2011) Capacity development and reconstruction in post-conflict African environments. *World Journal of Entrepreneurship, Management, and Sustainable Development* 7(2–4), 267–306.

Asenso-Okyere, K. and Mekonnen, D.A. (2012) *The importance of ICTs in the provision of information for improving agricultural productivity and rural incomes in Africa.* United Nations Development Programme Regional Bureau for Africa, New York. Available at: http://www.zef.de/module/register/media/ece1_ICT%20Productivity%20(1).pdf (accessed 18 January 2017).

Bar-Tal, D., Chernyak-Hai, L., Schori, N. and Gundar, A. (2009) A sense of self-perceived collective victimhood in intractable conflicts. *International Review of the Red Cross* 91(874), 229–258.

Birner, R., Cohen, M.J. and Ilukor, J. (2011) Rebuilding agricultural livelihoods in post-conflict situations: What are the governance challenges? The case of Northern Uganda. USSP Working Paper 07. International Food Policy Research Institute, Kampala. Available at: https://pdfs.semanticscholar.org/8f93/7cc7284da9ba53c2233746921307cbe15168.pdf (accessed 18 January 2017).

Blagescu, M. and Young, J. (2006) Capacity development for policy advocacy: current thinking and approaches among agencies supporting civil society organisations. Working Paper 260. Overseas Development Institute, London. Available at: www.odi.org/sites/odi.org.uk/files/odi-assets/publications-opinion-files/156.pdf (accessed 18 January 2017).

Blattman C. and Annan, J. (2012) Reintegrating and employing high risk youth in Liberia: lessons from a randomized evaluation of a Landmine Action agricultural training program for ex-combatants. Report of Innovations for Poverty Action. Yale University, New Haven, Connecticut.

BRAC (n.d.) Liberia: agriculture and food security. Available at: http://brac.net/brac-liberia/item/790-agriculture-and-food-security (accessed 18 January 2017).

Buckland, P. (2005) *Reshaping the Future: Education and Post-Conflict Reconstruction.* World Bank, Rome.

Chronic Poverty Research Center (2011) Social assistance and the 'dependency syndrome'. Policy Brief 22. Available at: www.chronicpoverty.org/uploads/publication_files/PB%2022.pdf (accessed 18 January 2017).

CIA (n.d.) Liberia country profile. Central Intelligence Agency, Washington, DC. Available at: https://www.cia.gov/library/publications/the-world-factbook/geos/li.html (accessed 18 January 2017).

Collier, P. (2006) Angola: options for prosperity. Department of Economics, University of Oxford, Oxford. Available at: http://users.ox.ac.uk/~econpco/research/pdfs/Angola-OptionsforProsperity.pdf.

Collier, P. and Duponchel, M. (2013) The economic legacy of civil war: firm-level evidence from Sierra Leone. *Journal of Conflict Resolution* 57(1), 65–88.

Cunguara, B. and Moder, K. (2011) Is agricultural extension helping the poor? Evidence from rural Mozambique. *Journal of African Economies* 20(4), 562–595.

DAI (2012) Food and enterprise development program (FED). DAI, Bethesda, Maryland. Available at: http://dai.com/our-work/projects/liberia%E2%80%94food-and-enterprise-development-program-fed (accessed 18 January 2017).

Davis, K.E., Nkonya, E., Kato, E., Mekonnen, D.A., Odendo, M., Miiro, R. and Nkuba, J. (2010) Impact of farmer field schools on agricultural productivity and poverty in East Africa. International Food Policy Research Institute, Washington, DC. Available at: www.ifpri.org/sites/default/files/publications/ifpridp00992.pdf (accessed 18 January 2017).

Eicher, C. (2006) The evolution of agricultural education and training: global insights of relevance for Africa. World Bank, Washington, DC. Available at: http://www.umb.no/statisk/noragric/World%20Bank%20workshop%20October%202011/global_aet_insights_eicher.pdf (accessed 18 January 2017).

FAO (n.d. a) FAO Aquastat. Database. Food and Agriculture Organization, Rome. Available at: www.fao.org/nr/water/aquastat/data/query/index.html?lang=en (accessed 18 January 2017).

FAO (n.d. b) FAO soils portal. Food and Agriculture Organization, Rome. Available at: www.fao.org/soils-portal/en/ (accessed 18 January 2017).

FAO (n.d. c) FAOStat. Database. Food and Agriculture Organization, Rome. Available at: http://faostat.fao.org/site/339/default.aspx (accessed 18 January 2017).

FAO (2013) The state of food insecurity in the world 2013: the multiple dimensions of food security. Food and Agriculture Organization, Rome. Available at: www.fao.org/docrep/018/i3434e/i3434e.pdf (accessed 18 January 2017).

FARA (n.d.) CARI Liberia. Forum for Agricultural Research in Africa, Accra, Ghana. Available at: www.erails.net/LR/cari-liberia/ (accessed 18 January 2017).

Fearon, J.D., Humphreys, M. and Weinstein, J.M. (2009) Can development aid contribute to social cohesion after civil war? Evidence from a field experiment in post-conflict Liberia. *American Economic Review: Papers and Proceedings* 99(2), 287–291.

Feder, G., Willett, A. and Zijp, W. (1999) Agricultural extension: generic challenges and some ingredients for solutions. Available at: http://comunitat.crpa.it/media/documents/crpa_www/progetti/comunitat/documentazione%20at/feder.pdf (accessed 18 January 2017).

Feed the Future (2011) Liberia: FY 2011–2015 multi-year strategy. Feed the Future/US Agency for International Development, Washington, DC. Available at: http://feedthefuture.gov/sites/default/files/resource/files/LiberiaFeedtheFutureMultiYearStrategy.pdf (accessed 18 January 2017).

Franzel, S. (2015) Farmer-to-farmer extension: the quiet transformation. Presentation at the MEAS Symposium, Washington, DC.

Ganpat, W. (2013) Competencies needed for future extensionists. Presentation at the University of Florida, Gainesville, Florida.

Ghimire, N., Kibriya, S. and Price, E. (2013) *War-induced Displacement Spell and Food Security Nexus in Liberia.* Conflict and Development Center, Texas A&M University, Washington, DC.

GRC (2007) *The Way Forward: Land and Property Right Issues in the Republic of Liberia.* Governance Reform Commission, Government of Liberia, Monrovia, Liberia.

Ho, M.D. and Hanrahan, C.E. (2011) The Obama administration's feed the future initiative. CRS Publication 7-5700. Congressional Research Service, Washington, DC. Available at: https://www.fas.org/sgp/crs/misc/R41612.pdf (accessed 18 January 2017).

Humphreys, M. and Richards, P. (2005) Prospects and opportunities for achieving the MDGs in post-conflict countries: a case study of Sierra Leone and Liberia. Columbia University, New York. Available at: www.columbia.edu/~mh2245/papers1/HR.pdf (accessed 18 January 2017).

IFC (2013) Liberia: amended management contract for LEC. International Finance Corporation, Washington, DC. Available at: www.ifc.org/wps/wcm/connect/373438004f748b9d97f0ff0098cb14b9/PPPStories_Liberia_AmendedLEC.pdf?MOD=AJPERES (accessed 18 January 2017).

IFPRI (2013) CAADP 10 years out: how have countries fared in agricultural development? Press Release. International Food Policy Research Institute, Washington, DC. Available at: www.ifpri.org/pressrelease/caadp-10-years-out-how-have-countries-fared-agricultural-development (accessed 18 January 2017).

McNamara, P., Swanson, B. and Simpson, B. (2011) Rebuilding and strengthening the pluralistic extension system in Liberia. Report of the Modernizing Extension and Advisory Services [MEAS] Project, University of Illinois, Urbana-Champaign, Illinois.

MEAS (2011) Extension and advisory services in Liberia. Modernizing Extension and Advisory Services Project, University of Illinois, Urbana-Champaign, Illinois. Available at: http://www.meas-extension.org/meas-offers/country_studies/country-overview/liberia (accessed 18 January 2017).

Meinzen-Dick, R., Quisumbing, A., Behrman, J., Biermayr-Jenzano, P., Wilde, V., Noordeloos, M., Ragasa, C. and Beintema, N. (2011) Engendering agricultural research, development, and extension. Available at: www.ifpri.org/publication/engendering-agricultural-research-development-and-extension (accessed 18 January 2017).

MoA (2007) Comprehensive assessment of the agricultural sector in Liberia. Ministry of Agriculture of Liberia, Monrovia. Available at: ftp://ftp.fao.org/docrep/fao/010/ai562e/ai562e00.pdf (accessed 18 January 2017).

MoA (2008a) National food security and nutrition strategy. Ministry of Agriculture of Liberia, Monrovia. Available at: https://extranet.who.int/nutrition/gina/es/node/7965 (accessed 18 January 2017).

MoA (2008b) Smallholder agricultural productivity enhancement and commercialization. Ministry of Agriculture of Liberia, Monrovia. Available at: www.gafspfund.org/sites/gafspfund.org/files/Documents/Liberia_3_of_7_GAFSP_Proposal_Final_Sept_2010_0.pdf (accessed 18 January 2017).

MoA (2009) *New Policy for Agricultural Advisory Services in Liberia.* Ministry of Agriculture of Liberia, Monrovia.

MoF (2013) Fiscal year 2013/14 National Budget. Ministry of Finance, Liberia. Available at: https://docs.google.com/a/mopea.gov.lr/viewer?a=v&pid=sites&srcid=bW9wZWEuZ292LmxyfG10ZWYtYnVkZ2V0fGd4OjJlYTc5MmYxZTZiZWRiNGY (accessed 18 January 2017).

Moore, A. (2014) Challenges and opportunities for improving the delivery of agricultural extension services to small-scale farmers in post-conflict Liberia. PhD thesis. University of Florida, Gainesville, Florida.

Pham, J. (2004) *Liberia: Portrait of a Failed State.* Reed Press, New York.

Rossi, P., Lipsey, M. and Freeman, H. (2004) *Evaluation: A Systematic Approach.* Sage, Newberry Park, California.

Saha, S.C. (1988) Agriculture in Liberia during the nineteenth century: Americo-Liberians' contribution. *Canadian Journal of African Studies* 22(2), 224–239.

Sawyer, A. (2005) *Beyond Plunder: Toward Democratic Governance in Liberia.* Lynne Rienner Publishers, London.

Schuller, M. (2012) *Killing with Kindness: Haiti, International Aid, and NGOs.* Rutgers University Press, New Brunswick, New Jersey.

Shor, J. (2012) Liberia's rough road to rice production. *Yale Globalist.* Available at: http://tyglobalist.org/in-the-magazine/theme/liberia%E2%80%99s-rough-road-to-rice-production/ (accessed 18 January 2017).

Stack, J. and Brabazon, J. (2008) *Liberia: An uncivil war* [documentary]. New Video Group, New York.

Sulaiman, R. and Davis K. (2012) The 'new extensionist': roles, strategies, and capacities to strengthen extension and advisory services. Global Forum for Rural Advisory Services, Lindau, Switzerland. Available at: www.g-fras.org/en/gfras/157-the-new-extensionist.html (accessed 18 January 2017).

Swanson, B.E. (2011) Recommendations for strengthening the extension system in Liberia. Modernizing Extension and Advisory Services Project, University of Illinois, Urbana-Champaign, Illinois.

Swanson, B.E. (2013) Assessing pluralistic extension systems: Bangladesh, Liberia, Rwanda, and Tajikistan. Presentation at the Association of International Agricultural and Extension Education conference. Fort Worth, Texas, USA. Available at: https://www.aiaee.org/attachments/article/1528/paper_abstracts_jiaee20(2final%20revised.pdf [p.95]

Swanson, B.E. and Rajalahti, R. (2010) Strengthening agricultural extension and advisory systems: procedures for assessing, transforming, and evaluating extension systems. Agriculture and Rural Development Discussion Paper 45. World Bank, Washington, DC. Available at: http://siteresources.worldbank.org/INTARD/Resources/Stren_combined_web.pdf (accessed 18 January 2017).

Swanson, B.E., Bentz, R.P. and Sofranko, A.J. (1997) Improving agricultural extension: a reference manual. Food and Agriculture Organization, Rome. Available at: www.fao.org/docrep/W5830E/w5830e00.htm#Contents (accessed 18 January 2017).

UNDP (2005) *Human Development Report 2005: International Cooperation at a Crossroads.* United Nations Development Programme, New York. Available at: http://hdr.undp.org/sites/default/files/reports/266/hdr05_complete.pdf (accessed 18 January 2017).

UNDP (2015) Work for human development: Briefing note for countries on the 2015 Human Development Report. United Nations Development Programme, New York. Available at: http://hdr.undp.org/sites/all/themes/hdr_theme/country-notes/LBR.pdf (accessed 18 January 2017).

Unruh, J.D. (2009) Land rights in postwar Liberia: the volatile part of the peace process. *Journal of Land Use Policy* 26, 425–433.

USAID (2008a) Technical assistance to the Ministry of Agriculture (TA-MOA): assessment of the Department of Regional Development Research and Extension. USAID, Washington, DC. Available at: http://pdf.usaid.gov/pdf_docs/pdacw803.pdf (accessed 18 January 2017).

USAID (2008b) Technical assistance to the Ministry of Agriculture (TA-MOA): assessment of the Department of Technical Services. USAID, Washington, DC. Available at: http://pdf.usaid.gov/pdf_docs/pdacw807.pdf (accessed 18 January 2017).

USAID (2009) Technical assistance services in support of the Ministry of Agriculture (TASMOA): decentralization issues and strategy process for decentralizing the Ministry of Agriculture. USAID, Washington, DC. Available at: http://pdf.usaid.gov/pdf_docs/pdacw809.pdf (accessed 18 January 2017).

WFP (2014) P4P Liberia: building relationships and growing businesses. Available at: https://www.wfp.org/purchase-progress/news/blog/p4p-liberia-building-relationships-and-growing-businesses (accessed 18 January 2017).

World Bank (2013) Harmonized list of fragile situations FY13. World Bank, Washington, DC. Available at: http://siteresources.worldbank.org/EXTLICUS/Resources/511777-1269623894864/FCSHarmonizedListFY13.pdf (accessed 18 January 2017).

World Bank (2015a) Liberia. World Bank, Washington, DC. Available at: http://data.worldbank.org/country/liberia?display=graph (accessed 18 January 2017).

World Bank (2015b) The socio-economic impacts of Ebola in Liberia. World Bank, Washington, DC. Available at: www.worldbank.org/en/topic/poverty/publication/socio-economic-impacts-ebola-liberia (accessed 18 January 2017).

WHO (2015) Ebola situation reports. World Health Organization, Geneva, Switzerland. Available at: http://apps.who.int/ebola/ebola-situation-reports (accessed 18 January 2017).

Zinnah, M. and Perry, E. (2011) *Summary Report of Liberian Delegation's Participation in the 3rd Symposium and General Assembly of the African Forum for Agricultural Advisory Services [AFAAS]*. Ministry of Agriculture of Liberia, Monrovia, Liberia.

2 Strengthening Agriculture Extension in Post-War Sierra Leone: Progress and Prospects for Increased Extension Impact

Paul E. McNamara*
University of Illinois at Urbana-Champaign, Urbana, Illinois, USA

Introduction

More than a decade after the 1991–2001 civil war in Sierra Leone and with the country emerging from the Ebola crisis, agricultural extension in the country finds itself in a difficult condition. Despite the numerous agricultural development projects implemented since the war's ending, extension capacity remains low. Frontline extension workers receive low salaries and routinely lack the funds necessary to obtain transport to get to the field with farmers or to run programs. Along with extension workers from the Ministry of Agriculture, Forestry and Food Security (MAFFS), most districts have a number of extension and agricultural development programs run by non-governmental organizations (NGOs).

The nature of the coordination among the programs varies dramatically across geographic areas. Other operational challenges facing extension include weak linkages with agricultural research and a pipeline of improved agricultural technologies and practices; gender gaps in reaching women farmers and helping them to gain access to land, quality inputs and finance; enabling environment issues such as land access and tenure for smallholders; the lack of well-developed input markets; very poor road infrastructure in some rural areas; and a need to train frontline extension workers on agricultural technologies as well as on extension process skills and information and communication technologies (ICTs).

Since the ending of the war, Sierra Leone has worked to rebuild infrastructure, such as rural roads and agricultural processing capacity in rice and other value chains. Furthermore, the Government of Sierra Leone (GoSL) has pursued efforts to commercialize agriculture and help farmers move from the subsistence agriculture pursued during the war years to more market-oriented agriculture. In spite of the extensive efforts to rebuild agricultural infrastructure and commercial agriculture for smallholder farmers, much more remains to be done.

Areas where stronger extension services could improve things readily appear. For example, there are enormous numbers of inland valley swamps with the potential for producing multiple crops of rice per year if basic water management infrastructure were available. A better functioning extension system could help with simple village-level planning and the development of rice swamp

* Corresponding author e-mail: mcnamar1@illinois.edu

improvements, as well as providing technical backstopping on appropriate seed varieties and inputs to make the most from the investment. Another example would be regarding inputs: most smallholder farmers do not apply fertilizer because they lack finance, experience, understanding and trust in dealers and input marketers.

Effective extension services would include a basic soil fertility management program for farmers' groups and on-farm learning plots or community-based demonstration plots to show best practices and allow for comparisons of alternative applications. Many other examples exist where better extension services could increase agricultural productivity and reduce rural poverty. In sum, the economy realizes a substantial opportunity cost in terms of rural poverty and missed agricultural productivity growth because of the weak state of agricultural extension services.

This chapter presents the case of agricultural extension in post-war Sierra Leone. The writing is based on a review of literature related to the Sierra Leonean civil war and the agricultural sector in the country. The chapter draws on field research consisting of meetings with farmers, frontline agricultural extension agents, staff from the Sierra Leone Agricultural Research Institute (SLARI), farmers' organization leaders and MAFFS staff, which occurred in September 2013. The field research was a part of an agricultural extension policy review conducted for the GoSL's West Africa Agricultural Productivity Project (funded by the World Bank) and was organized by Enterprise Development Consultants of Freetown, Sierra Leone.

Additional interviews and conversations concerning agricultural extension occurred during visits to Sierra Leone, starting in January 2009. With colleagues at Njala University, the author spent many days leading and participating in training courses on extension and extension skills, speaking with MAFFS staff, travelling around the country to observe research and extension activities and working with NGOs involved in agricultural extension efforts. Additionally, this chapter benefits from fieldwork conducted by Festus Amadu, a PhD student in natural resources and environmental sciences at the University of Illinois, and the author, who together conducted research on the post-Ebola extension situation in Sierra Leone from December 2015 until March 2016 (Amadu and McNamara, 2016).

Agriculture and Extension Prior to the 1990s

In 1990, just prior to the beginning of the civil war, Sierra Leone's economy was imploding following years of disastrous economic practices during the presidencies of Siaka Stevens (1968–1985) and Joseph Momoh (1985–1992). In the late 1980s, the income per capita in Sierra Leone was assessed by some analysts to be lower than the figure for the 1960s, even as incomes increased globally over the same period (Fashole Luke and Riley, 1989). This long period of decline and economic stagnation resulted from a number of factors:

1. Like other oil-importing developing countries, Sierra Leone had experienced a strong decline in its terms of trade over the post-independence period. While prices for primary commodities produced in Sierra Leone —such as iron ore, cocoa, palm oil and rice—were soft, oil prices increased sharply from the price levels in the 1960s.

2. Macro-economic performance in terms of the budget deficit and money supply, stability of the exchange rate, availability of credit within the economy and inflation declined precipitously during the 1980s so that by the end of the decade the country's economy was shrinking in size.

3. The rule of Siaka Stevens and Joseph Momoh weakened the state through their approaches to governing, which informalized and personalized the state. While the diamond sector and the mining industry are perhaps the poster children for this issue, the fact remains that the apparatus of governing throughout the GoSL was used to provide benefits to senior figures and to political clients of the leaders during this period. These factors of declining terms of trade, macro-economic instability and weakness, and fractured governance and public performance all led to a severely weakened public sector extension system by 1990.

At the level of terms of trade and international economic links, the pre-war period saw exports from Sierra Leone decline for the agricultural commodities. Asenso-Okyere and Workneh (2008) analyze statistics from the Food and Agriculture Organization of the United Nations (FAO) and show how exports for cocoa beans, coffee and oil palm all dropped in the period from 1985 to 1992. Oil palm exports went from 600 metric tons in 1985 to zero tons in 1992. Cocoa bean exports were 10,220 metric tons in 1985 and 4000 metric tons in 1992. Coffee exports dropped from 10,240 metric tons in 1985 to 4320 metric tons in 1992.

The exchange rate for the national currency (the Leone) also weakened over that period and the GoSL used a system of exchange rate controls, import licenses and foreign exchange restrictions to allocate the limited foreign exchange. This led to further disincentives for agricultural production, particularly for export crops, since the exchange rate controls essentially amounted to a tax on sales of these crops through formal and legitimate channels. A black market for foreign exchange was established. The overall impact of these economic policies was to make commercial agriculture production less attractive and remunerative, and thus the policies worked to weaken the investment in agriculture and the production of export crops. These policies also served to push people towards subsistence agriculture simply as a way to feed themselves as the economy began to contract in size and the industrial and service sectors of the economy declined.

An example of how weak the state had become by 1990 can be seen in the total government revenue figures. In 1980 total government revenue amounted to 17.0% of gross domestic product (GDP), while in 1990 total government revenue amounted to only 3.9% of GDP. The experience of other countries in the region, such as the Gambia during its structural adjustment period in the mid- to late 1980s, shows that variations in revenue are often a function of corruption at the port and the performance of customs and revenue agents (McNamara and McPherson, 1995).

Kandeh (1999) emphasizes how the breakdown of the state directly affected citizens across the country, and particularly young people. He writes: 'Non-delivery of public services and collective goods had a devastating impact on the youth segment of the population' (p. 357). Kandeh goes on to highlight how under Momoh and Stevens the government actually moved away from the understanding that government had an obligation to provide citizens with services, including education. Momoh stated that education was a privilege 'rather than a right of the citizens' (Kandeh, 1999: 357). This in turn led to an educational system that distributed educational opportunities on the basis of political connections and not on scores or merit. Young people became disenfranchised and discouraged, and even those who graduated from university and received a government job later experienced disappointment when salaries were low or withheld. This further devalued education as an instrument for development when young people saw teachers going unpaid and without classroom materials or facilities in working order.

By the late 1980s, Sierra Leone was running a large budget deficit and addressing it by printing money. In 1989, because of the macro-economic and fiscal disarray, the GoSL turned to an International Monetary Fund (IMF) structural adjustment program to obtain funds in return for meeting a number of conditions. The conditions included reducing subsidies on basic commodities, rationalizing the exchange rate through devaluation, reducing the budget deficit and privatizing state-owned enterprises (Kandeh, 1999). This further deepened the poverty as government employees were let go and parastatal enterprises were closed and sold.

Extension was in no way immune to these fiscal and organizational pressures, which it experienced directly. Although the extension system was designed as a national system with control centralized in the MAFFS national headquarters, funding scarcity reduced the extent of coordination and control. Extension suffered from the limited availability of government funding; many field staff were not paid on time and failed to receive the funds necessary to travel to the field or to conduct field demonstrations and other activities with farmers. Salaries declined in real terms with each devaluation of the Leone. Over time, extension came to be more about distributing fertilizer and inputs, which could sometimes be distributed in a manner that would benefit the government politically, and less about a coherent program of community-based informal agricultural education and services.

With resources scarce for operation and mobilization, extension workers became available for informal employment by NGOs and for other tasks if they became available. By 1990, setbacks in the funding for agricultural extension programs led to the lack of a coherent and delivered extension program on the ground and declining skills because of a lack of in-service training of field-level extension staff. Increasingly, NGOs provided agricultural development programming as an alternative to government programs. Zack-Williams (1999) describes the increase in NGO operations as evidence of the fragility of the state. He notes that from the late 1970s until the civil war, NGOs increased their operations in areas such as education, health, food and agriculture, and democracy and governance at the same time that government's capacity and strength declined.

The Roots of the Sierra Leone Civil War

Scholars and observers debate the causes of the civil war in Sierra Leone but several explanations appear to have strong support. The war in Sierra Leone, while not entirely unique in the history of sub-Saharan Africa, had unique features that raise questions about the causes of the war. These features included:

- The use of terror through mutilation and the hacking of limbs.
- The widespread use of child soldiers (on both sides) in the fight.
- The fact that GoSL soldiers turned into robbers and rebels at night and became 'sobels' with the result that people could not trust either side in the conflict.
- The central role of diamonds both for the Revolutionary United Front (RUF) as a source of guns and funds, and for the GoSL to fund contracts with security contractors or mercenaries like Executive Outcomes.

Additional features of the war included:

- The way that combatants moved through territory for the purposes of pillaging and looting, not necessarily to establish firm control over the territory with active presence.
- The massive displacement of people and the shifts of people from rural villages to larger cities such as Bo and especially Freetown.
- The rise of alternative institutions to fill the gap left by the broken and failed state, such as the Karamajors (a traditional Mende hunting society turned militia).
- The sheer loss of life (estimates of more than 50,000 people killed).
- The fact that the conflict affected nearly everyone in the entire country through deaths, rapes, looting and robberies, and displacements.
- The ambiguous role of the international community, the Nigerian and Guinean troops in the Economic Community of West African States Monitoring Group (ECOMOG), and the UN troops and British Army commandos, who effectively shut down the war for good by breaking up the remaining criminal gangs of rebels operating around Freetown, Lunsar and Port Loko.

The breakdown in the state, starting from the time of Siaka Stevens and continuing through the time of Joseph Momoh, contributed significantly to the war and its unique nature. Without strong authority and governance in the GoSL army, fighting the insurgency became especially difficult as soldiers could be bribed and influenced by the rebels. Furthermore, the troops on the frontlines felt abandoned at times and did not trust their commanders to have their best interests at heart, particularly in terms of providing them with adequate supplies of arms, munitions and transport, as well as good quality emergency medical services and pay. Furthermore, the breakdown in the state and the endemic levels of corruption promoted distrust of the government by rural residents as well as international donors. This no doubt lengthened the war as governments abroad felt unsure about the wisdom of getting too close to Sierra Leone's incumbent government. Kandeh (1999) points out that since the state under Siaka Stevens and Joseph Momoh increasingly used gangs of thugs and young people to enforce aspects of its rule, it was a short jump from that violence to a broader violence that grabbed and looted in the name of a revolution instead of in the name of a contract with the government or the approval of a license.

Another explanation for the particular nature of the war in Sierra Leone has been advanced by Paul Richards (1996), who has written extensively about the war and post-war period in Sierra Leone. Richards (1996) emphasized that many young people joined and supported the RUF because of frustration and disappointment with traditional institutions, particularly local courts of justice involving the Paramount Chiefs and other Chiefdom officials. He also notes that in a battle fought without heavy weapons, terror becomes an important part of the 'performance' of war (Richards, 1996). Richards' (1996) first point poses a central question to our understanding of the war. Was the war primarily about gaining control of the diamond trade—as has been posited—or was it rooted in broader injustices that either mobilized young people or created a class or large group of young people who were disenfranchised and felt abandoned by or ostracized from their home communities? This question becomes important not only for understanding the war's causes, but it also may have policy implications for actions in the post-war period. Do traditional authorities and means of implementing local rule need to be changed? Is the current policy emphasis on district-level councils and local government working, at least at the level of provision of services? How effective are traditional authorities in advocating positively for their youth and other marginalized groups?

Extension in Post-Conflict Sierra Leone

When the war ended on 18 January 2002, agricultural extension found itself in a very different place than it had been in 1980. Hundreds of thousands of people were displaced within the country and abroad in camps. Agricultural production for key commercial crops had declined to a fraction of its pre-war levels. As an example, rice production in Sierra Leone was at 503,000 tonnes (t) in 1990; in 2001, it was at 360,000 t (Asenso-Okyere and Workneh, 2008). The war, which drove people into subsistence living patterns, served to increase the production of some crops, such as cassava, because it was found to be more conflict-resistant.

Along with declines in production and changed cropping patterns, enormous amounts of physical and human capital were lost to the war. Agricultural processing centers such as rice mills and oil palm mills were destroyed. Rural roads went for years without proper upkeep and repair. Key contributors to agricultural knowledge systems at Njala University and in the MAFFS were lost or had left the country. Major institutions such as SLARI and Njala University suffered extensive damage to their facilities and programs. More broadly for the agricultural economy of Sierra Leone, some important trading and marketing institutions and patterns were lost, such as the breakdown of agricultural marketing boards and the decline of cash crop businesses (e.g. palm oil mills) that had previously bought from local producers. Even rice swamps that had been in production before the war were lost owing to years of neglect and lack of maintenance. Each of these agricultural assets would take years and tremendous efforts on the part of government officials, donors, NGOs and community leaders to repair and rebuild.

Beyond the loss of obvious physical assets were the intangible assets that the war damaged severely: trust and human capacity. Trust is an essential part of community development, is required for agricultural development and is a key component in extension programs. Yet trust was destroyed in many communities. The war had driven communities apart, and many people had betrayed or done terrible things to their neighbors. If you cannot trust your neighbors not to steal your crops, why would you be willing to invest much in that farm? If you cannot trust your community group to manage your money fairly and honestly, why would you continue to contribute to the rotating savings group or village savings and lending association? Maconachie (2011: 1064) states: 'in societies that have been seriously affected by violent conflict, such as Sierra Leone, there are often complex challenges in rebuilding social capital, which remains vital in operationalizing collective action and mobilizing community-driven development'.

In the immediate period after the war ended, NGOs played a critical role in assisting people who had been displaced and were in the process of resettlement, as well as working to help people rebuild their livelihoods. Many different international NGOs from all over the world participated in this effort. Additionally, a number of capable local NGOs and firms also contributed to this work, often in partnership with the international NGOs. Further, MAFFS staff lent their support to NGO projects and operations in addition to their efforts with the MAFFS.

In the first few years following the war, the priority was to help people rebuild their lives and livelihoods through basic agricultural development programs. Important work was implemented by World Vision of Sierra Leone in the areas around Kenema, Kailahun and Kono, which were key centers for much of the fighting and whose communities were tremendously impoverished by the war. Under the Promoting Agriculture, Governance and Environment (PAGE) project, led by Agricultural Cooperative Development International and Volunteers in Overseas Cooperative Assistance (ACDI-VOCA) and funded by the United States Agency for International Development (USAID), which ran from 2008 to 2012, World Vision worked with farmer groups to develop their swamps and build their rice farms. Many of the groups consisted of vulnerable people, including the previously landless, who the project had provided with long-term access to land by brokering agreements with customary authorities. The groups then worked with World Vision staff to rehabilitate and develop the swamps and to learn and test agronomic practices, including simple yield increasing techniques from the System of Rice Intensification approach, an evolving set of principles and practices for increasing the productivity of irrigated rice by changing the management of plants, soil, water and nutrients.

Village savings and lending associations (VSLAs) were established in some communities, allowing members to contribute a small amount of funds each week. The associations make loans and earn interest as well as offering a kind of social fund to provide insurance to members in case of a medical crisis or other emergency. Building trust through a group process facilitated by World Vision helped the farmers (many of them women) to rebuild their agricultural activities and increase their incomes. By 2011, some of the groups had moved into a commercial phase during which they were selling rice by the truckload to be milled and sold to the World Food Program's local purchase program. Others were using their newfound incomes to develop new swamps, expand their farms or to invest in a marketing business.

The program did not eliminate poverty or the very real risks these farmers faced. However, the groups were in a more secure situation with regard to their livelihoods than when they began. This sort of work took place all over Sierra Leone to varying degrees and was implemented by NGO projects as well as by donor-funded government-implemented projects. The NGOs and development implementers (e.g. ACDI/VOCA) were appreciated by donors for their ability to adhere closely to donor guidelines for program implementation and financial management.

With regard to capacity and the skills aspects of human capital, the economy suffered a breakdown in the education system lasting over 10 years. Indeed, the breakdown in education began long before the war with the hollowing out of government capacity to deliver quality education. At one time, Sierra Leone was recognized as having one of the strongest education systems in West Africa. However, all of this was lost during the pre-war and war period.

The economists Collier and Duponchel (2012) studied the impact of the war on the economy of Sierra Leone by examining company-level data from before and after the war. They analyzed how variations in the intensity of the conflict affected productivity and examined the factors associated with losses in productivity. The authors found that the size of a firm was negatively impacted by the intensity of the war in their operating area. Firms in areas where the war was more intense tended to contract to a larger degree than firms in areas where the conflict was not as intense. Collier and Duponchel (2012) also found a long-term skills gap, supporting their theory 'that conflict results in a significant loss in human capital stock as a result of a "forgetting by not doing" phenomenon, broadly analogous to learning by doing' (p. 83).

While Collier and Duponchel (2012) did not examine data from across the economy, this effect of 'forgetting by not doing' may well affect agriculture and activities like extension or other components of the agricultural knowledge system. There is no doubt that the war in Sierra Leone had a strong negative impact on the stock of human capital, and its impact on the agricultural extension system can be seen in the current lack of in-service training opportunities and the levels of process and technical skills held by extension field workers and management.

Extension Services in Sierra Leone Today

Extension in Sierra Leone at the present time consists of a pluralistic system of extension providers ranging from the MAFFS in the public sector, to research demonstrations from SLARI, to major donor-funded projects (e.g. the IFAD Smallholder Commercialization Project), to NGOs and bilateral donors (Action Aid, Catholic Relief Services [CRS], GIZ, USAID, World Vision and many others), to private input suppliers and international research programs and farmers' associations. All of these actors play a role in the extension system and are involved to some degree in delivering extension services to farmers. The many actors pose a coordination challenge at the local and national levels. In many districts, the District Agriculture Officer holds a coordinating meeting on a monthly basis with projects in his or her district. Furthermore, many NGO projects coordinate with frontline extension workers on events and trainings and other field-level work.

At the national level, the Agricultural Extension Division within the MAFFS consists of the Director of Extension Services, Deputy Director of Extension Services and three Assistant Directors (Research and Extension Liaison, Field Operations and Cross-Cutting Issues). At the district level, the District Agriculture Officer directs the district program, which includes six Subject Matter Specialists (Crops, Livestock, Agricultural Engineering, Forestry, Extension, and Program Evaluation and Monitoring). There are Extension Supervisors for each block—or administrative sub-portions of a district—and up to 40 or so frontline extension workers per district, although in many districts some positions are unfilled. Along with these staff at the district level, the extension personnel also work with NGOs, farmer-based organizations, Agricultural Business Centers and development projects (IFAD, etc.).

MAFFS extension staff and programs face a number of serious constraints despite the strengthening activities that have occurred since the war. Good staff are hampered in their ability to deliver programs due to low salaries. In 2013, frontline extension workers were being paid less than 200,000 Leones, or roughly US$40 per month. Block Supervisors were being paid around 300,000 Leones a month. No ongoing in-service training program exists and no training occurs unless it comes through a project (NGO or donor-funded). While donor- and project-funded training can be useful, it does not often strengthen basic and core extension process skills and technical competencies. Transportation for extension staff is severely lacking, making it difficult for frontline extension agents to get to the field. Staff prefer to get seconded to donor-funded projects for the 'top-up pay' they receive or other perks.

Extension also faces challenges related to the technologies and approaches available to teach and distribute to farmers. Most smallholder farmers have a low level of literacy and numeracy, as well as a low level of basic scientific knowledge. This makes promoting technical innovations challenging and requires a time-intensive demonstrate-and-coach approach. Simple training methods like fact sheets need to be replaced by group demonstrations of new seeds and techniques. Radio has proved useful for the distribution of agricultural messages, and videos are being tested and deployed in the country. Hopefully these will allow some economizing on staff time and the costs of extension work.

To boost productivity, more innovative agricultural tools and equipment—such as walk-behind tractors for rice cultivation, small irrigation pumps, etc.—are needed that are not prohibitively expensive. The development of input markets and the means to finance inputs are needed to allow the adoption of better quality seeds, fertilizer and pest management chemicals. Extension is needed in the areas of post-harvest processing, storage and marketing. Further emphasis on marketing and farming as a business is needed to help farmers find profitable activities that boost incomes.

While extension in Sierra Leone faces the challenges noted above, some of the difficulties facing extension previously appear to have been reduced or eliminated. The exchange rate, a key factor in the economy that affects the terms of trade for agriculture, is no longer distorted as it was during the 1970s and 1980s. International price signals for commodities are able to influence investment activities and the allocation of efforts. Furthermore, while some in Sierra Leone would like to see the re-establishment of the Sierra Leone Produce Marketing Board for agricultural crop marketing, the marketing board is not currently operating, even as markets for agricultural inputs and output operate freely.

One market of interest in African agriculture is the fertilizer market, where government activity or actions can crowd out private sector input supply businesses. In Sierra Leone, the MAFFS distributes fertilizer and seeds (when available) to farmers and farmer groups, although farmers have complained about the lack of transparency and fairness in the distribution of fertilizer in their area.

Overall, the enabling environment appears to have substantially improved for agriculture, although major concerns and weaknesses exist in terms of the security of access to land by smallholder farmers, as well as weaknesses in public finances and the lack of support for transport and other program expenses.

The most common method of providing extension services in Sierra Leone appears to be the Farmer Field School approach, which was introduced and promoted by the FAO. MAFFS frontline extension workers and senior staff are familiar with this approach, and most have experience in implementing it in the field. Often a modified approach to the traditional Farmer Field School method is applied. Other approaches to extension emphasize group capacity building and participatory approaches, which are particularly seen in the programs of NGOs working in agriculture. Extension approaches such as on-farm demonstrations, on-farm farmer-led research, demonstrations, agricultural newsletters and newspapers, extension embodied in outgrower schemes, agriculture days and agricultural fairs, radio shows (especially conducted by District Agriculture Officers) and videos are all employed in Sierra Leone. Group approaches work well in a social setting where agriculture is often conducted within family groups and other strong networks at the village where the ages and genders of cohorts are similar.

Some extension services in Sierra Leone are farmer-led at this time. Farmers report being consulted in the design and planning of projects to be implemented in or around their villages. They also report going to the MAFFS and other extension providers with specific requests. In some NGO agricultural development projects, the farmers have a strong say in the direction of the group process and in the topics and focus of the agricultural services they are to receive. However, in most cases, farmer input into the topics covered by extension services at the local level is limited or non-existent. An important step will be for Sierra Leone to develop functioning block- and district-level advisory committees that can help guide program priorities. New means are being developed to incorporate farmer feedback on the quality of extension services and these may offer an additional channel for farmer input.

Despite their importance in agricultural development and their supporting role in providing and facilitating agricultural extension services, farmer-based organizations in Sierra Leone are in their early stages. The experience of cooperatives prior to the war was largely negative from the farmers' point of view as many cooperatives were manipulated for political gain, overwhelming the benefits received by members. Furthermore, in the context of Sierra Leonean political life during the Stevens and Momoh presidencies, an independent cooperative with a large number of members would have been seen as a threat and would have invited political intervention. Rebuilding trust and capacity in farmer-based organizations (FBOs) is happening at this time, and a number of FBOs have been established in the post-war period. Some notable FBOs are located in the Kono, Kailahun and Kenema areas, and have been established as a part of the USAID-funded PAGE project among others. These FBOs focus on cocoa marketing, rice production and marketing, and other commercial crops.

Extension Strengthening Efforts

While Sierra Leone has had many donor-funded agricultural development projects in the post-war period, only a few have focused on building the capacity of the extension system. The Research into Use effort aimed to strengthen research capacity and the translational capacity to deploy the results in programs for farmers. Other projects have had components that indirectly strengthened extension through the provision of some training opportunities and the use of novel methods, or by providing access to district-level agricultural office improvements and transportation. Nevertheless, much more remains to be done to create an effective extension system that benefits farmers.

To gauge the priorities of key extension system stakeholders on what reforms or measures should be taken to strengthen extension in Sierra Leone, researchers fielded a brief questionnaire on the topic to agricultural extension workers, farmer association leaders and other MAFFS staff including SLARI researchers. This took place in September 2013 as part of the background work for a report to the MAFFS and the World Bank's West African Agricultural Productivity Program. A total of 70 responses was received, representing a convenience sample of extension stakeholder opinions.

In terms of the largest obstacles to strengthening the extension system, the respondents reported that funding levels and political will to strengthen extension were the major impediments. Approximately 77% of the respondents indicated that funding levels were a significant barrier and another 44% (not mutually exclusive) indicated that political will was lacking for improved extension. On an open-ended question concerning suggestions for strengthening extension and extension policy in Sierra Leone, responses ranged widely from advocating complete devolution of extension to the District Councils to greater farmer participation in the country's extension and research program. One response noted that there ought to be a unified national extension system that would involve all extension providers (including NGOs) in a single unified system with the same conditions of service for employees. Other suggestions proposed the timely provision of extension funding for transport and program expenses; providing training manuals and program materials to field extension staff; supporting the ability of extension staff to get to the field; the necessity of the availability of good seed, fertilizer and other inputs; building the capacity of FBOs; increasing the educational requirements for frontline extension workers; allowing field staff to have greater program autonomy; improving salaries and terms of employment for extension workers; and moving to performance contracts for extension services at the local level. One respondent included a caveat that 'having worked with the (District) Council for over 5 years, the levels of corruption and power concentrated in the hands of a few, such as the finance officer and chairman, make it difficult for the Council to achieve results'. These statements show that the stakeholders have no illusions with regard to the challenges and barriers to improving extension in Sierra Leone.

In terms of extension approaches, the work carried out by World Vision of Sierra Leone under the PAGE project shows how extension efforts can help address root causes and contributing factors to the conflict. Richards (1996) argues that an important impetus for the war was the disenfranchisement of youth and other groups in rural Sierra Leonean society. The difficulties and challenges that young people in rural communities faced in accessing resources contributed to that sense of disenfranchisement. The World Vision work that brokered access to productive land for poor people provides an example of the type of extension program that directly addresses a root cause of the conflict. Many extension efforts that focus on agricultural technology dissemination or a pre-identified topic may fail to address a root cause of conflict in the community, such as access to productive resources. In fact, if not designed carefully, some extension efforts could actually exacerbate income inequalities in rural communities and lead to future conflicts.

Maconachie (2011) emphasizes how the civil war in Sierra Leone served to break down trust in rural communities. The World Vision work described above and the work of CRS (and CRS savings and internal lending communities [SILCs]) address this issue of trust through the development of village-level internally financed savings and lending institutions. Working together with facilitation from NGO staff, the savings and lending associations build trust across members in the group, and create pathways for job creation, income growth and access to capital. Additionally, social insurance funds create a risk management mechanism that helps families deal with health shocks, unexpected bills and needs for finance. By addressing the social capital building aspect of VSLAs and SILCs, the projects are rebuilding trust within rural communities in Sierra Leone.

Another project—conducted by IFAD in the Kenema area—has shown good impacts in working with rice farmers (women, men and youth) in inland valley swamp ecologies (Amadu and McNamara, 2016). The project mobilizes communities for swamp development and improvement through village-level meetings. The training and workshops place a significant emphasis on gender, and women are encouraged to participate in the project. Women are trained to participate in extension roles and to take leadership roles in community projects. This represents a marked increase in targeting women farmers and focusing on inclusivity in agricultural extension efforts.

Recommendations and Opportunities for Improvement

An observation of 'best-fit' approaches around the world yields suggestions for improving extension in Sierra Leone:

- Focus on program quality and impact rather than on breadth, coverage and scope. The quality of extension services in Sierra Leone is presently very low, and if good examples of well-functioning extension engagements were more common and observable, it would help the discussion around capacity strengthening and funding levels.

- Continue to work with NGO extension programs, but consider how they might be able to contribute to strengthening the public sector system to be more sustainable. Over time, the public sector system is most likely to have the scale and presence to have the long-term impacts that good extension programs can deliver. A weakness of the heavy reliance on NGOs to date has been the relatively short time frame of most of their agricultural development projects.
- Continue efforts to build up the staff quality of extension within the MAFFS and to boost the terms of service so that good staff do not leave so readily.
- Work to increase farmer input and control of extension programs through committees at the block, district and national levels.
- Increase the rigor of extension monitoring and evaluation so that impacts are readily communicated and measured, and justifications for public resources are provided.
- Work with ICTs to extend the extension impact of the MAFFS and NGOs.
- Define a small number of critical national agricultural extension programs, and provide training and program materials—including manuals and field support—for their deployment. Support these programs so that impact can be perceived by local government officials and other community and political leaders.
- Figure out how to remove staff who do not perform once they are provided transport and programming resources. Increase the performance accountability of the programs. Extension in Sierra Leone has the possibility to be a driver of poverty reduction and agricultural productivity improvement, leading to strong food security in one of the world's poorest countries. Increasing the prominence of extension in the hierarchy of development priorities is necessary to realize this potential contribution.

Conclusions and Lessons Learned

Agricultural extension in Sierra Leone continues to be largely influenced by the enabling environment in the country, including the weak capacity of state ministries and organizations, the dynamic of NGOs and a set of international donor-funded projects leading to a highly pluralistic setting for extension services. Several key lessons can be derived from the country's post-civil war and post-Ebola experiences:

1. The length of time and the breadth of engagement in agricultural extension services needed after a significant civil war are much longer and deeper than most immediate recovery programs plan on or are designed to handle.

2. As seen in the World Vision programs, which included access to productive resources and building social capital, as well as the IFAD inland valley swamp rice program, which focused on social inclusion for women and youth, it is possible to directly address the root causes of conflict in the design and implementation of an extension program. These programs may require more staff resources in terms of time and training, but they also may hold the potential for decreasing the chances of further conflict.

3. Knowledge of effective extension approaches in post-conflict settings is not based on rigorous empirical research but on standards and norms of professional development and disaster recovery practice. Many of these standards deserve more rigorous testing and scrutiny where possible.

References

Amadu, F.O. and McNamara, P.E. (2016) A rapid assessment report on the current situation for extension services in Sierra Leone. Draft MEAS Project Report, 1 March. University of Illinois at Urbana-Champaign, Illinois.

Asenso-Okyere, K. and Workneh, S. (2008) *Trends and Conditions of Agriculture in Sierra Leone.* Prepared for the Sierra Leone Agricultural Research Institute. International Food Policy Research Institute, Washington, DC.

Collier, P. and Duponchel, M. (2012) The economic legacy of civil war: firm-level evidence from Sierra Leone. *Journal of Conflict Resolution* 57(1), 65–88.

Fashole Luke, D. and Riley, S.P. (1989) The politics of economic decline in Sierra Leone. *Journal of Modern African Studies* 27(1), 133–141.

Kandeh, J.D. (1999) Ransoming the state: elite origins of subaltern terror in Sierra Leone. *Review of African Political Economy* 26(81), 349–366.

Maconachie, R. (2011) Re-agrarianising livelihoods in post-conflict Sierra Leone? Mineral wealth and rural change in artisanal and small-scale mining communities. *Journal of International Development* 23, 1054–1067.

McNamara, P.E. and McPherson, M.F. (1995) Customs reform. In: McPherson, M.F. and Radelet, S.C. (eds) *Economic Recovery in the Gambia.* Harvard Institute for International Development, Cambridge, Massachusetts, pp. 251–263.

Richards, P. (1996) *Fighting for the Rain Forest.* James Currey/Heinemann, Oxford, UK, and Portsmouth, New Hampshire.

Zack-Williams, A.B. (1999) Sierra Leone: the political economy of civil war, 1991–98. *Third World Quarterly* 20(1), 143–162.

3 Challenges in Rebuilding the Agricultural Extension System in the Democratic Republic of Congo

Catherine Ragasa* and John Ulimwengu
International Food Policy Research Institute, Washington, DC, USA

Introduction

This chapter considers the implications of conflict for the provision of agricultural extension services, and challenges in designing and implementing strategies to rebuild agricultural extension systems (AES) in the Democratic Republic of Congo (DRC). The DRC case offers a unique perspective and contribution to this theme on several fronts. First, the DRC has suffered the 'deadliest' war since World War II, creating an infrastructural, technical and institutional vacuum, which makes rebuilding lost capacity through training, extension services and rural education extremely important. At the same time, the intensity of impact makes the rebuilding process extremely challenging. Second, the DRC represents a case where there is not much observed commitment in terms of the rebuilding processes as far as the extension service is concerned. While new policies and plans have been formulated and new approaches have been adopted (e.g. from more top-down to more participatory approaches to policy making processes), there is still a lack of commitment by the government to fund and implement elements of the newly adopted policies and plans. While numerous donor projects have been, and are being, implemented, they die off and there is no follow-up and institutionalization of these strategies because of the lack of government commitment.

Information and data cited in this chapter are from literature reviews, interviews with key informants and surveys of 107 extension organizations and 162 field staff in western DRC (Bandundu, Bas-Cong and Kinshasa provinces), analyzed in Ragasa *et al.* (2013b).

Background of the Conflict

Beginning in 1996, and six years after the introduction of the multi-party system, the DRC experienced a succession of wars. The first Congolese war (1996–1997) started when Laurent Désiré Kabila, heading the *Alliance des Forces Démocratiques pour la Libération du Congo* (AFDL), contested Mobutu's leadership. The second Congolese war (1998–2003) had a much more international dimension because rival countries and factions saw in the conflict-hit DRC a convenient ground for waging proxy wars (Maystadt *et al.*, 2014). These wars, according to a survey by the International Rescue

* Corresponding author e-mail: c.ragasa@cgiar.org

Committee, caused more than 5 million deaths from 1998 to 2008 (IRC, 2007) and an estimated 1.7 million internally displaced people (IDMC, 2011), which makes it the world's deadliest war since World War II with approximately 45,000 deaths every month since 1998. During the mid-2000s, the DRC was often referred to as the world's most neglected humanitarian crisis.

The causes of the DRC wars were multiple, complex and intermingled (Vlassenroot and Raeymaekers, 2004; Prunier, 2009). The weakness and inefficiency of Mobutu's regime over 32 years, ethnic polarization, spill-over effects from the Rwandan genocide, regional control by foreign powers and natural wealth have all been listed among the key factors. Without doubt, the DRC's natural wealth in mineral resources has consistently been blamed as the main driver of the violence, either as a way to finance warring parties or as a warfare objective in itself (Congdon Fors and Olsson, 2004; Turner, 2007; International Alert, 2010; Gambino, 2011; Stearns, 2011). Nest *et al.* (2006) argue that the conflict events that have taken place in the DRC are mainly due to the weakness of the state; military, financial and logistical interventions from neighboring countries; ethnic political violence; regime survival; and contests for control over natural and mineral resources.

Repeated conflicts have profoundly disrupted local livelihoods (Peace Direct, 2014). This violence, and the turmoil it creates, has severely affected the agricultural livelihoods of the poor, reducing their capacity to produce and trade. Yet claims of a direct causal line from conflict to poverty should be treated with caution, as causality may be reversed or indirect (Ali *et al.*, 2015). The outbreak of civil wars is commonly attributed to poverty (Bannon and Collier, 2003; Justino, 2006), and the correlation between low per capita incomes and higher propensities for internal war is one of the most robust empirical relationships in the literature.

In his review of the literature, Pellillo (2012) indicates some studies on the microeconomic consequences of conflict which address the impact of conflict and political instability on human capital accumulation (Akresh and de Walque, 2010; Chamarbagwala and Morán, 2011; Shemyakina, 2011), local institutions (Bellows and Miguel, 2006; Bellows and Miguel, 2009), political participation (Blattman, 2009; Kyle, 2010), labor market outcomes (Kondylis, 2010) and health outcomes (Bundervoet *et al.*, 2009; Akresh *et al.*, 2011; Akresh *et al.*, 2012; Minoiu and Shemyakina, 2012). In the case of the DRC, it is obvious that all the areas mentioned above must have been affected by the repeated conflicts. The evidence presented by Pellillo (2012) suggests that the conflict events that have taken place across the DRC have had a significant negative impact on household-level durable ownership and living conditions.

In a recent study, Ali *et al.* (2015) found that conflict is highly detrimental to the wealth of households: a 10% increase in conflict fatalities in the past 5 years decreased wealth by 3.6%. Their findings confirm the high socio-economic cost of war in the DRC. Indeed, they find that conflict—measured as the number of fatalities from violent conflict around markets—has a statistically significant positive effect on multi-dimensional poverty: a 10% increase in the number of fatalities around markets increases the probability of being multi-dimensionally poor by about 6%. Agriculturally involved households, female-headed households and households in rural areas are more likely to be multi-dimensionally poor as result of prolonged conflicts. Larger households and those with more children in the age range 0–5 years are more likely to be multi-dimensionally poor; households with older heads are less likely to be multi-dimensionally poor.

Eastern Congo has been particularly affected by warfare and continuous violence, even after the signing of the Global and Inclusive Agreement in 2002 (Inter-Congolese Dialogue, 2002). Eastern Congo continues to suffer from direct violence resulting from fighting and insecurity, unlike the northern and western provinces, which have been able to return to a state of relative peace. Several factors contribute to the continuous unrest and perpetuation of conflicts in eastern Congo. Eastern regions hold by far the largest endowment of the DRC's abundant natural resources and also offer direct access to neighboring countries (Burundi, Rwanda and Uganda), which have been implicated in the quest for the various economic and political gains resulting from the civil war. The northern and western regions of the country are experiencing relative peace and stability while dealing with the challenges of poverty, but eastern DRC continues to host a diversity of armed groups and militias that constantly perpetuate the cycle of violence and challenge peacekeeping and reconstruction efforts in the region. One example is the insurgency of the M23 militias, defeated in 2012 due partly to efforts from a tougher United Nations (UN) peacekeeping mandate.

The regional difference relating to conflicts and their impacts on agriculture highlights the need for special considerations in rebuilding the agricultural sector and extension system. In the eastern regions, additional challenges include: (i) the levels of insecurity and violence towards farmers attempting to work in their fields; (ii) the inability of extension agents to reach farmers, as both trainees and trainers face constant situations of unrest; (iii) risks of gender-based violence as women face the risk of rape while collecting firewood, fetching drinking water and working in their fields; and (iv) exacerbated land and water access challenges due to constant displacements in the Kivu provinces.

Indeed, different armed conflicts affecting the country, in particular in the eastern regions, continue to take a heavy toll on the region. Conflict and the resulting displacement of the population has caused an important reduction in agricultural production and has weakened the livelihoods of households essentially based on agriculture and animal husbandry (WFP, 2011; UNOCHA, 2012). Paving the way for economic development and political stability is necessary not only for the DRC but also for the whole Great Lakes region, which in the 1990s experienced one of the bloodiest decades of its history.

There is an urgent need to improve food security, which has been at alarmingly low levels, particularly after the two Congolese wars. More than 70% of the Congolese population are undernourished, while around two-thirds of the population live on less than US$1 per day (Ulimwengu *et al.*, 2009). Such food insecurity contrasts starkly with the DRC's huge agricultural potential. Despite 80 million ha of arable land, climatic diversity and relative abundance of water, only 10% of the land is cultivated. The DRC is also known for its abundant mineral resources including gold, copper, uranium, diamonds and coltan. This natural wealth has the potential to serve as a catalyst for economic development, but instead has been argued to constitute a factor for fuelling conflict and further exacerbating food insecurity (Maystadt *et al.*, 2014).

Food security represents an important goal in post-conflict DRC and should be the main target of any post-conflict rebuilding program as there is an interaction between food security and resilience to conflict in the DRC. The impact of conflict on food security depends on the magnitude of damage to physical, human, financial, social and political capital, all of which affect households' access to the resources (including food) needed to ensure sustainable livelihoods. We argue that extension services should play a major role in the process of restoring households' livelihoods, especially in rural areas where 70% of the Congolese population live.

Characteristics of the Post-Conflict Agricultural Extension System (AES) in the DRC

The current AES in the DRC is characterized by drastic underfunding of the system without institutional reforms. The results are severe human resource management problems; unclear mandates and a lack of mission orientation; an elite capture problem (public extension agents serving primarily large-scale farmers); and largely underserved and illiterate rural producers. It is a largely defunct public extension system, with an extensive field staff still on the payroll, and numerous non-government organizations (NGOs), church-based organizations (CBOs) and rural producer organizations (RPOs) trying to fill the gaps mostly from ad hoc and project-based funding. Government extension agents and supervisors are often hired directly by these NGOs, CBOs and RPOs for their extension work, given the limited capacity of these organizations to staff projects. Numerous NGOs and other service providers operating alone, without coordination, have created a disorganized system with much inefficiency, redundancy and confusion due to conflicting messages given to farmers on specific issues or technologies. As described below, this has been a result of more than 15 years of repeated conflicts with a persistent lack of commitment or sustained funding by the government for the agriculture sector in general and the extension service in particular.

Structure

The Congolese extension system is supported by outdated laws adopted in 1988 and 1993. These decrees define the extension system in the DRC to comprise the National Extension Service (*Service National de Vulgarisation* [SNV]), with national headquarters and a coordinating team in each province that support structures, including researchers and specialists for training and technical backstopping. These structures include the National Agronomic Research Institute (INERA), the National Mechanization Agency (SENAMA), the National Seed Agency (SENASEM) and the National Fertilizer Agency (SENAFIC). The extension system also includes the national agricultural inspection system within the Ministry of Agriculture, Livestock and Fisheries (MINAGRI), with a total 18,500 staff, of whom 11,245 are said to be qualified extension agents (FAO, 2005), complemented by agricultural workers from NGOs for the actual delivery of extension services to rural communities (Fig. 3.1).

With these public sector field staff serving an estimated 6 million farmers, the ratio of farmers to public extension agents in the DRC is a comparatively low at 540:1 (Table 3.1). Anderson and Feder (2004) show that the farmer-to-agent ratio in most developing countries is over 1000:1. Thus, the number of agents may not be the major problem in the DRC's defunct extension system. The challenge seems rather to be in effectively managing the large number of field staff to perform their extension work effectively and to help address the constraints faced by rural farmers.

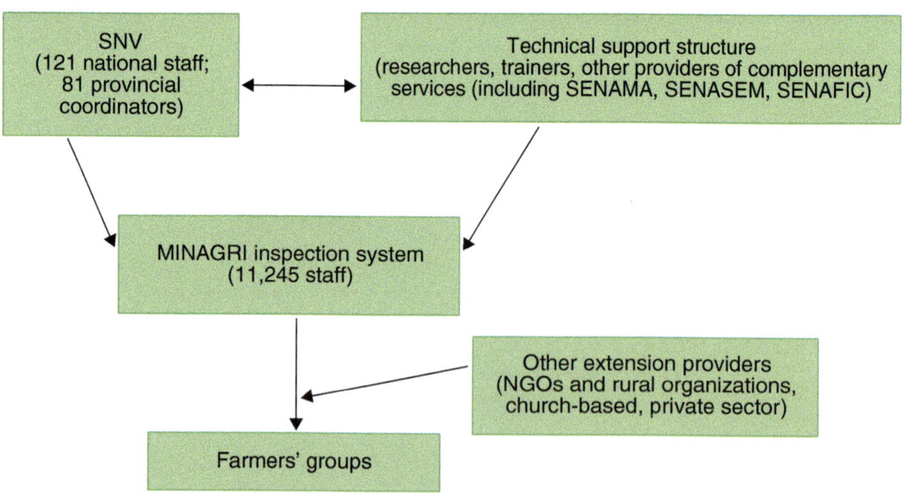

Fig. 3.1. The agricultural extension system in the DRC. MINAGRI, Ministry of Agriculture, Livestock and Fisheries; SENAMA, National Mechanization Agency; SENAFIC, National Fertilizer Agency; SENASEM, National Seed Agency; SNV, Service National de Vulgarisation (National Extension Service). From *Journal Officiel* (1988, 1993).

Table 3.1. Farmer:extension agent ratio in the DRC and selected countries. From Davis *et al*., 2010; Kaunda, 2011; Kamau, 2012; GoM, 2015; Ragasa *et al*., 2015a.

Country	Number of farmers per agent	Total number of agents
Ethiopia	480	60,000
DRC	540	11,000
China	620	800,000
Kenya	950	6,000
Indonesia	1,670	30,000
Malawi	2,100	2,000
Tanzania	2,500	7,000
Nigeria	3,330	5,000
India	5,000	60,000

To fill this huge gap in public extension service provision, international NGOs and donors have been actively involving local NGOs and producer organizations in extension service delivery. NGOs can offer greater incentives: the salary and benefits of an NGO extension agent are around US$500–1000 a month, while that of a government extension agent is about US$50 a month. The salaries of public sector extension agents are extremely low, even within the agricultural sector. To put this into context, the daily farm labor wage is $0.5–3.5, which on average is higher than the salary of an extension agent in the public sector.

However, the involvement of NGOs in extension service delivery has major limitations: (i) they lack technical capacity and often hire MINAGRI staff as extension agents and supervisors; (ii) similarly to government, they lack follow-up of targeted farmers, defined and feasible performance targets, mission orientation and accountability measures—there are also complaints of elite capture and leakages; (iii) the quality of services is rather questionable and many farmers are skeptical about whether extension agents have sufficient knowledge about new technologies that are of interest to them; and (iv) NGOs cover only a limited number of farmers, as confirmed by the available dataset on Bandundu, Bas-Congo and Kinshasa provinces (Ragasa *et al*., 2013b).

Defunct National Extension Service (SNV)

The SNV was created in 1988 as an independent unit within the National Rural Development Department, with its own budget, to institutionalize the earlier National Extension Program. Its aim was to coordinate, harmonize and support rural extension activities. In 1993 a decree was passed to grant administrative and financial autonomy to the SNV under the umbrella authority of the MINAGRI.

The headquarters of SNV is in the capital city of Kinshasa, but its area of activity is the entire country. During its early implementation (late 1980s), there were 121 staff at the national level and 82 coordinators and officers assigned to the different provinces. Support was provided by researchers, subject matter specialists and trainers from the government agencies, NGOs, the private sector and other projects. SNV and its supporting structures worked with the MINAGRI extension chief, sector chiefs and technicians to train supervisors, agronomists, veterinarians and agricultural monitors, who are the frontline staff in direct contact with rural producers. SNV was tasked to develop extension materials and production guides and to update them regularly; to provide training and backstopping based on these extension materials; and to coordinate and monitor extension service provision.

From 1989 to the onset of the conflict, SNV had been supported financially by donor and international organizations including the Food and Agriculture Organization of the United Nations (FAO), French Cooperation Agency (AFD), the United Nations Development Programme (UNDP) and the World Bank, as well as by the DRC government. By the mid-1990s donor support had ceased and government funds were substantially reduced. From 1997 to 2001, funding came solely from the government, and mostly to pay for salaries. Starting in 2002, there was no operational funding available from any source, and even staff salaries were not paid regularly. This makes it difficult for SNV to fulfil its function, as the staff of the agricultural inspection system are not under the supervision of

SNV, and without offering a budget SNV cannot train the staff of this system or provide other services to them. Nonetheless, the SNV continues to prepare and submit a budget every year. For example, in 2010 the SNV prepared a budget of US$100 million for extension activities, but did not receive any funds at all (SNV Director, pers. comm. to the head of IFPRI office in DRC, 2011). The budgetary resources needed for the salaries of SNV staff alone are approximately US$2 million.

Although SNV staff were unpaid for years, many remained at their posts, looking for any means of economic survival. Most remaining staff members are untrained in newer approaches to transmitting improved technologies to farmers and have passed retirement age. A large proportion of staff members in the provinces are uninformed about recent and pending policy changes, such as the Agricultural and Rural Management Councils (*Conseil Agricole et Rural de Gestions*, CARGs), the Agricultural Code (see below) and other measures related to decentralization and privatization. Those who are still able to work require new local leadership, catch-up training and resources to begin to work again.

A result of the defunct SNV is a serious issue of lack of coordination among the pluralistic extension service providers. Interviews conducted also suggest that many agriculture-related projects and activities do not work in coordination with local MINAGRI offices, SNV, INERA and SENASEM (Ragasa *et al.*, 2013b). This lack of coordination leads to serious inconsistencies, conflicting extension messages and duplication of efforts. There have been many cases of different service providers providing different and at times conflicting extension messages to producers. Not involving agents and key staff of relevant government offices is also an unsustainable approach, as capacity is not built and there is no continuity of knowledge and activities when these international NGOs leave at the end of their projects. Similarly, most training activities are conducted in an ad hoc, small, uncoordinated and unsustainable way by NGOs and CBOs with links to their international counterparts. The cadre of government field staff and other organizations are often not included and rarely receive regular skill upgrading. This has implications for a more regular, coordinated and sustainable provision of training and learning opportunities and for strengthening the whole agricultural education system in the DRC.

Large field staff with no clear mandate

The lowest level of field staff in the MINAGRI are referred to as agricultural monitors, which is the same title they had in the colonial period. They are the frontline professionals who are tasked to work directly with rural communities. At the sector level (the administrative level between territory and village), they are supervised by sector agronomists and sector veterinarians, who report to agricultural and veterinary inspectors at the territory level, who in turn report to the provincial directors and inspectors. In provinces where the decentralization policy is not fully implemented, there are also agricultural and veterinary inspectors at the district level.

Interviews conducted by Ragasa *et al.* (2013b) suggest that the field staff (agricultural monitors) in this system are not aware of their mandate to promote new technologies or to engage in other activities that would help farmers to increase their income. Instead, they view their role as collecting data and—as their title suggests—monitoring farmers' activities. One local expert with long-term field experience suggested that their role remains influenced by the colonial past and pointed out that 'they are still largely seen as policemen' by the farmers (Ragasa *et al.*, 2013b). Very few consider the current system as a truly functioning extension service. However, there are other field staff who believe that extension is part of their mission, and even without operating funds from government they have been working with a number of international organizations to provide extension services to farmers. This suggests a serious lack of a clear mission and mandate as perceived by those involved in the Congolese extension system.

To understand the current system of public agricultural extension service provision, it is useful to trace how the extension system worked in the past and to take account of its roots in the DRC's colonial history. The Belgian colonial rulers did not practice a system of 'indirect rule', but rather engaged directly in the management of local affairs, including the management of agricultural production. As Mills (n.d.: 2) points out:

> Agricultural officers (with their African assistants) kept a close eye on the local African population; with the development of large-scale mining and plantation activities, there was a large need for food for the laborers. There was a definite policy of using African peasant production to supply this local domestic market. The agricultural officers supervised and directed African farmers in what crops to produce.

It is also well known that the atrocities committed by the Belgian colonial rulers to exploit agricultural labor were far beyond those observed elsewhere in Africa (Mills, n.d.). To implement the policy of compulsory food crop production, the colonial rulers created an intensive, centrally guided and administered agricultural production system. The system that is in place today still features elements of the system set up by the colonial administration.

The lack of a clear vision and mandate for agricultural extension poses serious issues that impede the whole system. A related issue is the lack of operational targets and monitoring. There are no targets set for agents' and organizations' extension work, as reported by a large majority of agricultural workers interviewed and more than half of extension organizations' heads surveyed by Ragasa *et al.* (2013b). As a result, there is no effective monitoring and evaluation to assess the quality and performance of extension services. While there are more than 11,000 field staff at the MINAGRI, it is not clear to them whether they should be doing extension work in addition to their data collection and census activities.

Capacity of agricultural workers

The DRC's AES is characterized by ageing personnel in extension and in the whole agricultural ministry in general. Of the 162 field staff surveyed in western DRC 60% are over 50 years old. Low education levels and rare training opportunities are also limiting factors to effective agricultural extension. A quarter of field staff interviewed in western DRC have only a primary or secondary school agricultural degree. While 61% of agents had received other professional training in addition to formal education, the majority of agents with lower levels of formal education (primary or secondary school) had not received professional training. Without extensive training and retraining of agents and technicians, there could be a seriously limited technical capacity to provide extension and advisory services to farmers.

A major concern with skill development programs and capacity of field staff is the serious weakness in the country's agricultural training and education system. Within the Ministry of Education, the Agro or Agro-Veterinarian Studies Institute/College (ISEA/ISEAV) and Rural Development Institute (ISDR) are the government institutions that offer agricultural training and education and are responsible for training agricultural extension agents and rural development workers. ISEA/ISEAV trains agricultural technicians (agronomists and veterinarians), while ISDR trains rural development workers. Almost all territories have at least one ISEA or ISDR. In some cases, farmers' organizations also go to them directly to request training, often for a fee. Students carrying out their practicum (work placement) in the villages are also useful in disseminating technology packages and production techniques to farmers. In these cases, ISEA and ISDR are potentially useful pathways of technology dissemination and extension services delivery that can be explored further. However, visits to some of these institutes highlight several challenges in agricultural education and training in the DRC, including: a lack of sustained funding; no vision, strategic planning or forward-looking mentality (for ISDR/ISEA in particular and the education system in the DRC more generally); an outdated curriculum; problems of quality of education, starting at the primary school level; a lack of up-to-date training and skills development for staff; and a lack of links with the rest of the agricultural support system including extension, research and universities.

Extension approaches and delivery methods

The main approach is still heavily based on the training-and-visit (T&V) system, a remnant of the World Bank-supported programs of the 1980s. A survey of 162 field staff in western DRC (Ragasa *et al.*, 2013b) reveals that the most common method of delivering extension services is visits by field staff to farms or homes, followed by visits and information sharing at farmer-based organizations (FBOs), training sessions and demonstration farms. NGOs and RPOs more frequently use visits to and meetings with FBOs, training sessions and demonstration farms, while government, CBOs and the private sector more frequently use visits to farmers' fields or homes.

The survey results indicate that 74% of field staff hold special meetings, such as planning meetings, in which farmers can express which types of activities they expect to carry out. These meetings are usually held at the community, *groupement* (a group of villages belonging to the same clan) or sector levels. Field staff activities also include the formation, mobilization and support of RPOs.

Regarding the distribution of inputs together with the provision of extension services, a quarter of surveyed field staff from government extension in western DRC reported that they distribute inputs, while more than half of field staff from NGOs and all field staff from CBOs reported distributing inputs. This indicates that NGO field staff often have a more integrated approach to service delivery and tend to focus more on complementary inputs, technical knowledge and services for the farm management.

In eastern DRC, where numerous projects have been implemented, the use of radio and innovation platforms (IPs) are also common methods of disseminating information on agricultural management practices. The IPs assemble stakeholders to share information, identify opportunities, discuss problems and agree on joint activities related to a shared interest, often with a specific commodity focus.

Local participatory approaches have also been implemented, although at a small scale to date. The most common approach is Farmer Field Schools (FFSs; *champ-écoles de paysans* [CEP]). A FFS is a group of 20–25 farmers with associated land (usually 0.5–1 ha) who are trained in the field by following the various stages of development of a given crop, based on field observations and analyses. The objectives are to strengthen the capacities of farmers themselves to identify and solve problems encountered during crop production and to reinforce their organizational capacities to further manage their own development.

These approaches rely on higher involvement of farmers and producer organizations to identify problems at a local level, define adaptive research and on-farm trial needs, implement trials and disseminate appropriate technologies. In western DRC, half of the government field staff interviewed reported teaching or facilitating in FFSs, whereas 64% from NGOs and only 17% from CBOs use FFSs.

Efforts to rebuild the AES

To understand the evolution of the AES and lessons learned from the past, this section summarizes the policies, programs and status of AES in the DRC before the conflict, and then compares these with the efforts to rebuild the system afterward.

Before conflict

There are three distinct periods of policy interventions in the DRC agricultural sector: the colonial period (1885–1960), the post-colonial period (1960–1990) and the democratization period (1990–present). As pointed out by Tollens (2004), the actions by different governments to revive the agricultural sector and achieve rural development since independence have produced positive but insignificant results. Most of the policies failed to be comprehensive and lacked internal coherence and adequate articulation on the various interventions.

The colonial period was characterized by compulsory labor to produce cash crops for export and food crops for local consumption. This policy had a provision that allowed the colonial administration to collect tax against the sale of compulsory crops, which in turn was used for building rural infrastructure, debt repayment and paying for a system to monitor farmers. During the colonial times, agricultural workers were commonly referred to as agricultural monitors. They monitored farmers' activities, whether they followed the prescribed practices and whether they reported their harvests and revenues correctly. While there was adequate production, compulsory labor led to human and social injustices, as those opposing the prescribed practices were sanctioned through either imprisonment or sequestration of their land. To address these injustices and to continue to promote agricultural development, King Leopold III, during his address to the Belgian Senate in 1933, advocated for the inclusion of social considerations by providing the native population with ownership of their own land. This was the starting point of the peasantry period (1946–1960), which was characterized by: (i) the abolition of compulsory cultivation; (ii) the creation of the National Institute for Agronomic Study in Congo (INEAC), later renamed INERA, which undertook research on crucial agricultural issues including soil science and the selection and improvement of planting materials; and (iii) a strong focus on disseminating knowledge on agricultural management practices and mechanization for smallholder farmers. Groups and associations of farmers were also set up for easier organization of services and dissemination of technology messages.

In the 1960s, immediately following the colonial period, the agricultural sector went through a difficult transition. Agricultural research and extension systems collapsed, feeder roads deteriorated and waterways and railroads were disrupted. As a result, the share of agriculture in total exports dropped from 41% in 1958 to 16% in 1966. This period was characterized by extreme poverty, food insecurity and total discouragement of farmers.

To revive the agricultural sector and make it the driving force of economic development, in the early 1970s government authorities decided to start regulating the sector through more realistic policies. In 1973, the first post-colonial agricultural policy was implemented as part of the *Zairianization* movement (measures taken by Congolese authorities to transfer productive assets from private foreigners to private Congolese). It was followed later by many other policies, most of which were short-lived and ineffective. From 1973 to 1990, several agricultural policies were implemented but none achieved its objectives (Table 3.2).

The most notable was the Five-Year Plan (1986–1990), an ambitious long-term program implemented to induce economic and social development and to achieve food self-sufficiency. It targeted agricultural development through the increase in size and productivity of peasant farms, promotion of entrepreneurship in the sector, intensification of extension and applied agronomic research, maintenance of feeder roads and job creation, and increased incomes in rural areas. It was characterized by the move towards liberalism and the promotion of private initiatives; promotion of partnerships including the private sector, NGOs, peasants, small and medium-sized agricultural firms and donors; and the presence of a detailed 5-year investment plan supporting

Table 3.2. Major policies affecting the agricultural extension system following the colonial period. From Le Plae, 1939; Béguin, 1965; Jewsiewicki, 1979, 1980; Mokili Danga Kassa, 1998; Northrup, 2003; AfDB, 2004; Kakule, 2006; World Bank, 2010; Van Hoof, 2011; AfDB, 2012, 2013; World Bank, 2013, 2016.

Major policies and programs	Year	Objectives and major components	Role of extension	Outcomes
Zairianization	1973–76	Agricultural development, improvement of social conditions of farmers, food self-sufficiency	Not prominent	Difficult transition; enormous abandonment of fields; plummet in agricultural production; selection of new acquirers of agro-industrial units based on political criteria rather than management skills of new owners
Radicalization	1974–76	Transfer of strategic agro-industrial units and large fields mismanaged by unskilled acquirers either to the state or to more skilled new acquirers; new government agencies created to provide technical support to the sector and to ensure marketing of agricultural products	Not prominent	Failed to revive agricultural sector as expected; agricultural research and extension almost non-existent; several agencies created to support the sector non-operational; most agro-industrial units bankrupted; fall in production of all commercial crops except coffee
Retrocession	1976–78	Rehabilitation of bankrupt firms by providing financial and technical support		Limited success, many owners of nationalized agricultural firms declined the invitation; productivity and production stagnation continued
Economic Revival Plan or Plan Mobutu	1978–82	Revival of the agricultural sector and others (e.g. mining, transportation, energy) to improve food security, increase production of exportable crops and crops used as inputs by local industrial firms, and fund projects contributing to rural development, with support from the IMF	Prominent	Limited success even with 26 projects during the first phase (1979–81) and 22 during the second (1981–83), mainly due to funding shortage from the government
Development Conventions	1979–80	Inducing large industrial and commercial firms to contribute to agricultural development through an increase in agricultural productivity and supply of basic inputs to local agro-industry, via several instruments including a special tax	Prominent	Limited success: selective policy benefited a few rural areas where selected firms chose to operate; taxes led to increased prices of agricultural products, a disincentive for many firms to reinvest in agriculture
Agricultural Revival Plan	1982	Improving partnership between the government and the private sector for more rational management of the agricultural sector	Not prominent	Some success, but small farmers and other stakeholders in the agricultural sector ignored

Continued

Table 3.2. Continued.

Major policies and programs	Year	Objectives and major components	Role of extension	Outcomes
Interim Economic Rehabilitation Program	1983–86	Preparatory stage to the Five-Year Plan (see below); first time donors were seen as stakeholders; supported by the World Bank and IMF; involvement of donors in management of the sector included diagnosis of previous agricultural policies and identification of causes of failure		Some success, public investment plans funded under this program included 68 agricultural projects (forestry, livestock, fisheries); government defaulted on its financial commitments
Five-Year Plan (Plan Quinquennal)	1986–90	Agricultural and rural development through increase in size and productivity of peasant farms, promotion of entrepreneurship, job re-creation in rural areas, intensification of extension and applied agronomic research, maintenance of feeder roads	Prominent	Limited funds, government defaulted on its commitments
Master Plan	1990–2002	Achieving agricultural and rural development through participatory approaches; rehabilitation of rural infrastructure; improvement of supply of inputs and credit; funding of agronomic research and extension; improvement of agricultural markets and agro-processing	Very prominent	Political instability prevented implementation
Agricultural Policy Code (*Note de Politique Agricole*)	2009–	Reducing poverty and hunger by 50% by 2015 (Millennium Development Goal); specific objectives: 10% of government budget going to agriculture to meet Maputo commitment; MINAGRI restructuring; decentralized agricultural service provision; improvements for financial services; promoting agricultural entrepreneurship	Very prominent	Targets not achieved, funding for agriculture remains low, government commitment continues to be lacking

the general policy. Agricultural extension services were one of the major themes in this program. However, its implementation failed, as with the previous initiatives, due to a shortage of funds (the government defaulted on its financing commitments).

To support the government of the DRC, many externally funded programs operating in the 1970s and 1980s supported agricultural technology transfer and advice to small producers, among others. Most notable among these programs were the World Bank-funded projects based on the T&V method. In 1988 the National Extension Service (SNV) was created to harmonize approaches and methods previously developed for various extension services and to integrate them in the MINAGRI's day-to-day activities. However, starting with the unrest of 1991, which brought a nationwide crumbling of institutions and withdrawal of most donor support, SNV's institutional capacities were weakened and ceased to function almost immediately, giving way to an emergency model of extension services with NGOs stepping in to fill the gap, sometimes without the adequate experience and capacity previously brought by the SNV.

The political instability that characterized the country since 1990 has not allowed full implementation of any agricultural policy. In 1990 an agricultural strategy known as the 'Master Plan' was developed but has never been implemented. The Master Plan was a comprehensive and coherent agricultural and rural development policy involving participatory approaches with all stakeholders involved in the agricultural sector, and was the first time that the DRC had aimed to decentralize services and restructure the functions and staffing in the MINAGRI. Improvement in agricultural extension services provision, through the SNV, was a major component of this policy. Despite its coherence, the program was never implemented, as the country was entering a period of political and social unrest.

After conflict

In 2003, after the war, the DRC President announced his intention to revive and implement the Master Plan. However, the context was totally different at that point, politically and economically. New policies were implemented but at a rather slow pace. Moreover, as part of billions of dollars of international assistance for security, humanitarian, stabilization, peace-building and economic recovery interventions, many donor-supported projects were already being implemented targeting agriculture and the rural sector. Table 3.3 lists the major programs with prominent extension services provision.

Drafting of the Agricultural Code

In 2009 a new policy document known as the 'Agricultural Code' was drafted, picking up on some past planned reforms and paying attention to new challenges and the environment that the agricultural sector now faced. It aimed at reducing poverty and hunger by half by 2015 (a Millennium Development Goal) and allocating 10% of the government budget to agriculture to meet the Maputo commitment. Among other things, the three major components of this code are: (i) decentralization of agricultural services; (ii) restructuring of the MINAGRI; and (iii) implementation of the CARGs, which are platforms for consultation, dialogues, problem-solving, articulation of demand of rural services, monitoring of programs at the territory and local levels, and providing an avenue for engagement in policy making processes at the provincial and national levels. These three planned reforms have significant implications for the DRC's AES. FAO supported the design and plan for restructuring the MINAGRI, while the Belgian Technical Cooperation and various donors and NGOs have supported the establishment of more than 140 CARGs across the country. However, this policy document remains a draft and has not been implemented.

Agricultural and Rural Management Councils (CARGs)

There are high expectations that CARGs can be the solution to the ineffective AES in the DRC. The evaluation conducted by Badibanga *et al.* (2013) reveals several problems and challenges in the implementation of the CARGs and overall weakness in their operations, confirming earlier studies emphasizing the limitations of local management councils and multi-stakeholder platforms in effecting positive change in societies. Half

Table 3.3. Projects supporting agriculture and extension systems in the post-conflict Democratic Republic of Congo. From Pamuk *et al.*, 2012; Vanlauwe *et al.*, 2012; Kilelu *et al.*, 2013; Pamuk 2014; van Paassen *et al.*, 2014.

Project	Donor	Period	Target provinces	Objectives and major components	Extension approach	Lessons learned for extension services provision
Agricultural and Rural Sector Rehabilitation Support Project (PARSAR)	African Development Bank	2004–12	Bandundu and Bas-Congo	Strengthening food security and reducing rural poverty through stimulation of agricultural production	Work with village organizations; farm demonstrations; focused on seed multiplication and distribution	Not documented
Agricultural and Rural Sector Rehabilitation Project (PRESAR)	African Development Bank	2005–10	Katanga, eastern Kasai and western Kasai	Contributing to food security and poverty reduction in rural areas	Work with village organizations; farm demonstrations	Not documented
Western Growth Poles	World Bank	2013–19	Bas-Congo	Increasing productivity and employment in selected value chains among small-scale farmers	Value chain approach; institutional development of SNV	Ongoing project; no relevant status report and lessons yet
Food Production, Processing and Marketing (FPPM)	USAID	2010–15	Bas-Congo; Kinshasa and Bandundu	Increasing productivity, market efficiency and capacity of rural communities to respond to market forces	Strong focus on seed multiplication and distribution, not much on extension services	Project discontinued; major management issues; weak outcomes based on midline assessment
Eastern Recovery Project	World Bank	2016–20	Eastern provinces	Improving access to livelihoods and socio-economic infrastructure in vulnerable communities	Innovation platforms; value chain approach; focus on women in agricultural value chain	Design stage
Agriculture Rehabilitation and Recovery Support	World Bank	2010–17	Equateur	Increasing productivity and employment in selected value chains	Pluralistic service provision; use of groups and farmers' organizations; rehabilitation of INERA stations and the SNV	Ongoing project; no relevant status report and lessons yet

Continued

Table 3.3. Continued.

Project	Donor	Period	Target provinces	Objectives and major components	Extension approach	Lessons learned for extension services provision
Consortium for Improving Agriculture-based Livelihoods in Central Africa (CIALCA)	CIAT, IITA and Biodiversity International	2005–13	Eastern provinces	Improving livelihoods of rural households in Central Africa through increased productivity of banana-, maize-, cassava- and legume-based systems	Radio, FFSs, village meetings, demonstration trials, on-farm trials	Multipronged, targeted communication approach; holistic approach in looking at and addressing farmers' constraints
Sub-Saharan Africa Challenge Program	Implemented by FARA	2005–13	Eastern provinces	Introducing a new approach (IAR4D) to promote innovation and its diffusion in African agriculture	Innovation platforms	Mixed outcomes in terms of technology adoption, food security and poverty reduction
Improvement of Food Security in the Cross-Border Districts of Burundi, DR Congo, Rwanda and Uganda in Support of the Agricultural Modernization within the NEPAD/CAADP framework	FAO	2007–13	Eastern provinces	Improving access to income and standard of living of rural households in cross-border districts of the four countries through strategic support to profitable production systems, market-oriented and value-added activities	Value chain approach	Holistic approach in looking at and addressing farmers' constraints
Centers for Integrated Development	African Development Bank	2016–20	Nationwide (145 territories)	Promoting youth entrepreneurship while addressing poverty, youth unemployment and food insecurity by making land (which is state property) available to small and medium enterprises	Value chain approach; local hubs of socio-economic development; focus on youth entrepreneurship	Holistic approach in looking at and addressing farmers' constraints

CAADP, Comprehensive Africa Agricultural Development Program; CIAT, International Center for Tropical Agriculture; FARA, Forum for Agricultural Research in Africa; FFS, Farmer Field School; IAR4D, Integrated Agricultural Research for Development; IITA, International Institute of Tropical Agriculture; INERA, National Agronomic Research Institute; SNV, *Service National de Vulgarisation* (National Extension Service).

of the surveyed CARGs achieved results consistent with at least one of their main goals, which means the other half have been conducting activities that are not consistent with their objectives. While 60% of stakeholders interviewed were aware of CARGs, only 33% had attended a CARG-related meeting and perceived CARGs to be useful, and only 11% reported having benefited or knowing someone who benefited from CARGs. These responses suggest limited scope, coverage and relevance of CARGs based on stakeholders' experience.

The expectation that CARGs can be the major actor in agricultural service provision has been the subject of much controversy, which is partly explained by the lack of a clear definition of the CARGs' mission. While some stakeholders in the agricultural sector believe CARGs can handle any issue relating to agricultural strategies and policies, others do not see how a multi-stakeholder platform such as the CARG can deal efficiently with an issue that requires resource commitment and regular management operations, such as the provision of agricultural extension services.

The survey of 55 CARGs in western DRC (Badibanga *et al.*, 2013) shows that 33% (18 CARGs) indicated agricultural extension provision or monitoring as one of their main objectives. Of these CARGs 60% have activities consistent with these objectives. However, 33% (six) of the remaining CARGs did not achieve any extension-related goal. Further, 20% (11) of the CARGs surveyed did not have extension among their goals but achieved the extension outputs incidentally. Thus, the total of the surveyed CARGs achieving extension outputs with or without an extension objective was 42% (23).

The evaluation of the role of the CARGs undertaken by Badibanga *et al.* (2013) demonstrates that this platform can play a role as a facilitator of access to agricultural extension services. CARGs can also act as a platform for demand articulation and aggregation. It appears that CARGs are showing potential as an effective demand-side strategy and a bridging institution for demand and supply. However, they cannot play the role of service provider and therefore cannot function as a supply-side strategy. To operate effectively as a service provider, an extension organization needs: (i) to design and implement a coherent agricultural extension program with clearly defined targets; (ii) to use different means and methods for providing agricultural extension services; (iii) to use well-trained and experienced extension personnel; and (iv) to have substantial financial and material resources to complete its goals. The CARGs do not meet these conditions. In fact, a CARG is a multi-stakeholder platform for consultation and thus differs from any permanent structure or organization that can design a program and commit the resources needed to implement it. Supply-side strategies, including capacity building and training of pluralistic extension service providers, will be crucial to complement investments and support to demand-side strategies such as CARGs.

National Agricultural Investment Plan (NAIP)

In 2013 the DRC government and its partners adopted the NAIP as part of the Comprehensive Africa Agricultural Development Program (CAADP) process. Research, extension and agricultural training is the NAIP's main component. Overall, this program aims to boost research activities to generate and disseminate information on improved technologies for sustainable agricultural development and to improve agricultural productivity in the country. Specifically, it aims to strengthen public institutions and civil society organizations for advisory support through:

- Identification and labelling of NGOs and various actors working in rural development.
- Establishment of provincial and national directories of government agencies and private extension organizations offering extension services.
- Development of manuals on technical standards, procedures and management tools to inform extension officers at the provincial level.
- Training of extension services trainers.

- Construction and rehabilitation of extension services buildings and infrastructures at national and provincial levels.
- Equipping of extension services structures with means for transportation and logistical support to ensure an improved level of service delivery to farmers.
- Establishment of a monitoring and coordination hub to harmonize activities among NGOs and partners.
- Installation of an action-tracking database and advisory support at provincial level.

Dissemination and popularization of technological innovations include actions such as:

- Development of a national strategy for agricultural extension services and dissemination of research results.
- Dissemination of simplified and user-friendly versions of research results.
- Organizing open days for technological innovations.
- Training program leaders in strategic communication.
- Capacity strengthening of research and development journalists.
- A framework for consultation and information sharing linking researchers, extension workers and agricultural organizations at territorial, provincial and national levels.
- Publication and dissemination of research results through appropriate channels (radio, print, television, professional organizations).

These activities will substantially revitalize the AES in the DRC.

Of the US$857 million available for the NAIP, 93% (US$797.3 million) represents the contribution of development partners and 7% (US$60 million) is from the Congolese government. The breakdown of this amount, according to the projects already planned, indicates that US$504 million (59%) was mobilized during the 2013–2015 period and US$351 million (41%) will be released in 2016–2020. The distribution of planned funds by sub-programs is given in Fig. 3.2.

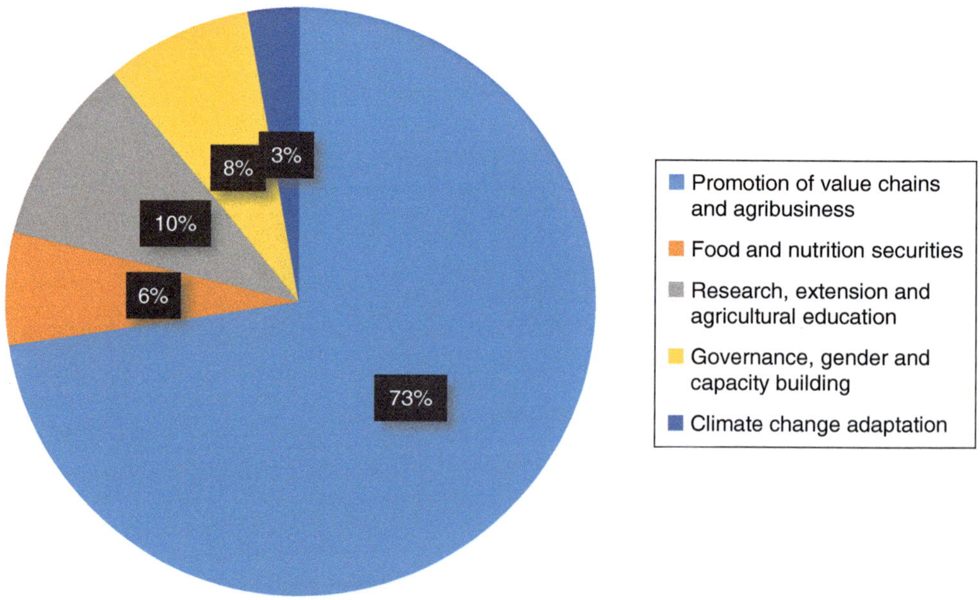

Fig. 3.2. Funds available for National Agricultural Investment Plan, by sub-program. From MINAGRI (2013).

Integrated Agricultural Research for Development (IAR4D) and IPs

The Lake Kivu region, capturing parts of eastern DRC, Rwanda and Uganda, is one of three project sites of the sub-Saharan Africa Challenge Program (SSA-CP), coordinated by the Forum for Agricultural Research in Africa (FARA). SSA-CP aims to develop technologies for sustainably intensifying subsistence-oriented farming systems; develop smallholder production systems that are compatible with sound natural resource management; improve the accessibility and efficiency of markets for smallholder and pastoral products; and catalyze the formulation and adoption of policies that would encourage innovation to improve the livelihoods of smallholders and pastoralists. It started in 2004 with the aim of introducing a new approach to promoting innovation and diffusion of innovations in African agriculture. IAR4D is based on the innovation systems perspective and creates coalitions of stakeholders to identify and address local bottlenecks to agricultural development. Through this approach, the program aims to promote agricultural innovations by utilizing farmers' indigenous knowledge through a participatory framework and interactions between different stakeholders. The innovation systems perspective focuses on recognition of a wider, differentiated set of innovation suppliers; demand responsiveness and better connectivity of agricultural research with a wider range of innovation actors beyond extension agents and farmers; and an expanded definition of the innovations being developed to include both economic and social applications (World Bank, 2007, 2011).

A central concept of this approach is IPs, which are decentralized local innovation systems. IPs are vehicles to bring stakeholders together. Each IP serves a group of villages and theoretically chooses representatives from different stakeholders via a participatory process. These representatives of farmers' associations, traders, researchers, extension workers, NGOs and government policy makers regularly meet at the platforms, articulate their views and negotiate joint strategies for action. To provide 'proof of concept', the IAR4D program and its IPs were rolled out as a large experiment whereby some communities received IPs and others did not. In addition, baseline and midline surveys were conducted. These processes ensure rigorous monitoring and evaluation of the concept. Available studies show mixed outcomes of these IPs (Pamuk *et al.*, 2012, 2014). Even within eastern DRC, results are mixed. The IPs in Kituva resulted in higher food security and reduced poverty, while those in Rubare led to lower food security and a higher poverty count. There were no significant impacts of IPs on poverty in Bweremana and Rumangabo (Pamuk *et al.*, 2012). The mixed outcomes of IPs in the context of the IAR4D program and SSA-CP are consistent with results obtained by other authors in other contexts (Kilelu *et al.*, 2013; van Paassen *et al.*, 2014).

Integrated multipronged communication approaches

Eastern DRC is one of the project areas for the Consortium for Improving Agriculture-Based Livelihoods in Central Africa (CIALCA), a research-for-development consortium led by the Tropical Soil Biology and Fertility Research Area of the International Center for Tropical Agriculture (TSBF-CIAT), the International Institute of Tropical Agriculture (IITA) and Bioversity International. It aims to improve the livelihoods of rural households in Central Africa through the identification, evaluation and promotion of technological options to enhance the productivity of banana-, maize-, cassava- and legume-based systems, and to create an enabling environment for their adoption. CIALCA promoted complete integrated soil fertility management (ISFM) comprising the use of improved germplasm, mineral fertilizer, appropriate organic resource management and local adaptation.

CIALCA follows an integrated, multipronged communication approach using a mix of tools to disseminate and promote ISFM at a large scale. The first steps towards ISFM were fertilizer and improved varieties. An essential condition for their adoption is access to farm inputs, markets and financial resources. CIALCA has worked on dissemination strategies, including ways to facilitate access to the required inputs, simple information fliers spread through extension networks and knowledge on how to avoid less responsive soils. CIALCA implemented campaigns that addressed farmers' constraints by offering them information, technology demonstrations, product exhibits, financial incentives and opportunities to develop their skills within their own farms. As efforts to promote the seed and fertilizer strategy were underway, activities such as FFSs were initiated to guide farming communities towards complete ISFM.

The CIALCA Knowledge Resource Centre was established in the African Great Lakes region to identify and leverage new impact pathways for ISFM technologies. By working closely with extension agents and outreach partners, targeted information tools can be developed to support adoption of practices by farmers in specific settings. A particular challenge was to develop innovative knowledge products that take into account the low rates of adult literacy and formal education prevalent in the region. Rural radio, one tool that was used, offered a wide reach and was very useful for raising awareness around a particular issue. However, it is less suitable as a training tool, particularly as knowledge complexity increases. With increasing complexity of knowledge, CIALCA focuses on rigorous and in-depth farmer facilitation and training.

CIALCA has intervened in markets by working with farmers' organizations to achieve a marketable production scale. Capacity building on collaborative action, marketing and business planning skills and management of credit and finances have ensured that farmers are now able to bulk their produce, wait for better prices and earn higher incomes from their produce. In addition to farmers, training also targets the institutions and organizations that support farmers' organizations, such as NGOs and national research staff, to ensure post-project sustainability. Farmers in South Kivu were able to raise their sales revenues by 50% through strategic storage facilitated by inventory credit schemes (*warrantage*): farmers did not have to sell immediately after harvest but were able to store their produce collectively, awaiting better prices for their products. Through group efforts, farmers were also able to acquire credit for their ISFM-based farming activities and, because they had targeted production to key markets, were able for the first time to borrow funds without collateral. In addition, farmers working in groups have been able to initiate mutual savings schemes that supplement other sources of finance, particularly for investment in new technologies.

Other approaches promoted by external partners

Various approaches to agricultural extension are being implemented by donor projects, including FFSs, rural radio and value chain approaches. The World Food Programme promotes rural radio as a powerful tool for broader dissemination of information on agricultural extension in rural areas. Since 2002 new approaches are being tested through various projects implemented with FAO technical assistance. These approaches include EU-, Technical Cooperation Program- and UNDP-funded projects, and use the FFS methodology developed by FAO. At present, 83 FFSs are operating and results are promising (MINAGRI, 2009). The aim is to extend the methodology to other crops and sites in other provinces.

The value chain approach has been used by major programs supporting the agricultural sector and technology dissemination in the DRC (e.g. ongoing projects of FAO, USAID and the World Bank), demonstrating that a range of services is needed to build sustainable value chains that effectively link farmers to markets. These services include:

- Organizing farmers into business-oriented entities.
- Offering training and advice for individual farmers, groups of farmers, farmer organizations, cooperatives and other agribusinesses along the value chain.

- Disseminating information about technologies, new research, markets, inputs and financial services.
- Providing participatory demonstrations and practical adaptation of new technologies and practices on-farm.
- Developing business management and negotiation skills among smallholder farmers and other local entrepreneurs.
- Facilitating linkages among market actors (including financial and non-financial inputs, processing, trading).
- Facilitating compliance with quality standards and traceability requirements of markets.

More projects and programs have been implemented in eastern regions due to the greater challenges they face resulting from past and current conflict. For example, FAO projects in the eastern regions particularly included water resources management among their main components, specifically to address post-conflict issues pertaining to water access and use policies among communities affected by civil unrest and local conflicts within the respective countries. The project also embedded gender and social inclusion safeguards during project implementation to ensure that reconstruction efforts and value chain development are able to reflect diverse voices and contributions.

For several regions in the country, in consideration of non-functional markets and lost planting materials due to past conflict, efforts to rebuild extension services have often been accompanied by provision of planting materials and other inputs in the short term, and rebuilding the markets and developing value chains in the longer term. Infrastructure development and basic services are also provided alongside agricultural information provision.

Key Lessons from Efforts to Rebuild the AES

The following seven points can be highlighted as major challenges and lessons for rebuilding the AES in the DRC.

Sustained government funding

It is critical to rebuild the lost capacity and livelihoods of rural communities. Therefore, it is crucial to mobilize substantive funds and commitment from the government. As discussed above, the Congolese government has often defaulted on its commitments in the implementation of key policies and programs. The process of decentralizing agricultural service provision in the DRC has been slow. Funds are not decentralized, leaving the local levels with fragile situations regarding financial, institutional and technical capacities.

No amount or degree of institutional reforms or approaches will work without sustained funding and commitment from government. In countries with major reforms in their AES, such as Brazil, Senegal and Uganda, state governments provide a pool of resources and advertise a call for providers (from any source: public, private, NGO or RPO) through a competitive process (Suresh Babu [April 2016] and Jean Charles Faye [November 2014], International Food Policy Research Institute, pers. comm.). Funding should be stable rather than relying on ad hoc projects. Governments must invest in the extension system in order to transform the agricultural sector.

In the absence of government funding, external funding has been crucial for the functioning of the extension system. External funders such as donor agencies and international NGOs also tend to place conditionality and emphasis on monitoring of adoption and impact, and this has implications for their support for a transitioning economy such as the DRC. However, as proved during the past decades of neglect of extension in the DRC, short-term support from external actors is extremely important but will need to be supported and continued by the government if sustainability is to be achieved.

Clearly defined objectives

Clear direction and vision, coupled with measurable targets, are extremely important. The huge number of extension agents and technicians in the agricultural inspection system within the MINAGRI (estimated at 11,245) offers an opportunity and asset to be utilized for extension delivery. To date, their mandate covers mainly data collection and 'monitoring' of farmers' activities. With a clear mandate, clear definition and communication of roles and responsibilities, and a corresponding performance assessment and incentive system, this inactive and defunct inspection system could be transformed to successfully support knowledge dissemination and technology transfer in rural areas.

The rationale for maintaining such an extensive field staff without a mandate to fulfil the 'public good' functions of agricultural extension remains unclear. Obviously, data on agricultural production could be collected more effectively by conducting sample surveys rather than maintaining an extensive network of field staff aiming to monitor every farmer.

Another function for the system of inspectors, agronomists, veterinarians and monitors, apart from data collection, is the prevention of crop and livestock diseases (Ragasa *et al.*, 2013b). While this is a public sector function, it appears more efficient to pursue this function together with, rather than separately from, the promotion of new technologies and farming practices.

Work with public institutions

Public sector extension remains crucial even in nations such as the DRC with very weak government institutions. As widely observed in the DRC and other fragile countries, many donor- and NGO-led projects tend to bypass government institutions, not involving them in their project design, implementation and capacity-strengthening activities. An important strategy therefore is to include public sector extension agents and subject matter specialists in capacity-strengthening and learning programs, instead of bypassing them and focusing only on NGOs, as is the practice for many development organizations. It will be crucial to work with government counterparts (agriculture ministries, extension systems, research institutes, seed inspection systems) in agricultural and rural projects. These efforts should occur even where there is some distrust of and perceived inefficiency in the government institutions, which continue to have human resources scattered throughout the country. These institutions should therefore form part of any capacity-strengthening efforts related to extension activities.

In the case of this pluralistic system, focusing on coordination, quality control and regulation to avoid conflicting extension messages and duplication of efforts will be a priority for the government, equal to or even more important than providing extension services themselves. While there remains pressure to continue service provision by the public sector, there have to be major decisions on how to invest limited resources more effectively. The public sector could focus more on coordination and regulation of extension services, thereby freeing more resources from actual service provision, which could be done mainly by, or contracted out to, the private sector and NGOs.

Demand- and supply-side strategies

Many government officials and stakeholders have high expectations that CARGs can offer the solution to the ineffective AES in the DRC. However, these expectations are unrealistic and stem from a lack of understanding and consideration of both the demand and supply sides of extension services. CARGs are starting to be used as a demand-side strategy, particularly as a platform for demand articulation and as a bridging institution between demand and supply of extension services. However, they are unlikely to perform the role of service providers. Supply-side strategies, including capacity building and coordination of pluralistic extension service providers, will need to be implemented in addition to strengthening CARGs. Especially in the case of the DRC, where almost a quarter of agents have only a primary school education and the majority of government agents have no university degree, regular training is extremely important and needs to be institutionalized rather than being provided as an ad hoc activity. Part of the supply-side strategies is to rebuild the agricultural education and training institutes so that they can be effective in training future field staff and service providers to work with rural communities. The AES relies on the strength of the agricultural education and training system, so any efforts to revitalize the former will need complementary investments in the latter.

CARGs would need to focus on facilitating consultation among stakeholders on agricultural extension policies and strategies, and on monitoring and evaluating the implementation of decisions that result from such consultation. To play such a complex role, CARGs need capacity building for their members and leadership on multi-stakeholder processes; management of multi-stakeholder platforms; social interaction and strategies for reaching a consensus; monitoring and evaluation of agricultural extension and agricultural production; and post-harvest methods. CARGS also need to implement appropriate self-financing policies which will allow them to collect substantial contributions from their organization members who have the financial capacity to support them, while setting financial targets for their other organization members with limited financial means.

Multipronged, targeted communication approaches

The experience with the T&V approach yielded limited results. Top-down approaches were used without taking sufficient account of farmers' needs and the variability of ecosystems. At the same time, the system was extremely costly and was unsustainable once donor financing ended. The experience with CIALCA highlights the need for integrated and multipronged communication approaches, and shows that the need for intensive farmer facilitation and training increases rapidly with increasing complexity of knowledge. By working closely with extension agents and outreach partners, targeted information and dissemination tools can be developed to support adoption of practices by farmers in specific settings. For example, a mix of input access facilitation, simple information fliers spread through extension networks, technology demonstrations, product exhibits and knowledge resource centers can be used based on the specific local setting. A particular challenge is to develop innovative knowledge products that take into account the low rates of adult literacy and formal education prevalent in the region. Rural radio is one tool that offers a wide reach and is very useful for raising awareness around a particular issue. However, it is less suitable as a training tool, particularly as knowledge complexity increases. Experiences with IPs are also mixed, so implementing them will require special consideration of the set-up process, focus on tangible outcomes, need for complementary skills and operational capacity needed to manage the IP process.

Holistic approaches to addressing farmers' constraints

The experience with CIALCA shows the need to look at other inputs, services, markets and policies. This is also the central theme of the African Development Bank's program of 'Centers for Integrated Development', designed as hubs of socio-economic development at the local level. This is also related to the value chain approach beginning to be adopted by various partners including FAO, USAID and the World Bank.

Weak policy and low investments in complementary inputs and services also limit the effectiveness of extension. Access to credit, inputs, markets, land, equipment and tools are the most common and most consistently mentioned constraints among farmers, based on the perspectives of agents, organization heads and farmers themselves. Training and extension services are mentioned as only a part of the problem, and therefore are only a part of the solution. Increasing the productivity and incomes of rural communities will not require changes in the extension system alone, but rather a holistic approach to address constraints in the input distribution system and technology adoption. The results warrant complementary review and reform in the policies and investments governing these inputs and services.

Recruiting female field staff

Only 5% of the field staff surveyed are women. Of all the extension organization heads interviewed, only 7% are women. There was no female supervisor or head in the public sector agencies.

In the DRC there is a limited dataset on female and male farmers' differentiated access to extension services, but evidence from the literature in many different countries suggests a strong correlation between the sex of the agent and the gender differential in access to extension services. For example, a 1993 FAO study of 24 extension programs in Africa, Asia and Latin America suggests that the presence of female extension agents was an important factor in female farmers' participation in extension activities. Female extension workers serve a higher proportion of female farmers than do male agents. The average ratio of women to men farmers is 1.3 for female agents and 0.53 for male agents (World Bank and IFPRI, 2009). This may suggest that extension services from female extension agents are better targeted to female farmers. In Tanzania, 40% of women farmers preferred to work with female extension agents, 26% preferred male extension agents and the remaining 34% had no preference (Due *et al.*, 1997). Female farmers stated that they preferred female extension agents because they were freer to discuss problems with them, and female agents were better able to accommodate their time preferences for meetings than were male agents. Gender-based constraints, such as social norms that limit women's school attendance or mobility, also limit their opportunities and willingness to work as agricultural extension agents (Ragasa *et al.*, 2012). It may be difficult at a practical level for a married woman to work in a rural area away from her husband and family or to find housing and schooling appropriate for her children.

The analysis by Ragasa *et al.* (2015b) suggests a strong correlation between the presence of female agents in extension organizations and performance in service provision, which confirms results of past studies (Ragasa *et al.*, 2013a). In places where female heads and agents are few, as in the case of the DRC, focusing on girls' education and encouraging more women students and graduates to work on agriculture and on AES will be an important strategy.

Conclusions and Recommendations

Although the post-conflict or emergency period is rife with difficulties, it is also a time of great opportunity. Particularly in post-conflict settings, long-standing structural issues can be addressed. Opportunities also are present for promoting economic and agricultural diversification and attaining sustainable livelihoods.

The ambitious strategy adopted by the Congolese government as part of the NAIP consists of concentrating its interventions in growth poles or corridors in which the necessary infrastructure (energy, water, roads) could be established to enable privately managed agricultural enterprises and farmers' cooperatives to focus on food crop and livestock production, processing and marketing in a coordinated manner in the context of an agro-industrial park (Ulimwengu *et al.*, 2015). In addition to providing infrastructure, the park could also host services. This is a good starting point towards more concerted efforts to rebuild the agricultural sector and provide the needed productive and social services to the rural population.

Rebuilding the agricultural sector is crucial for food security, restoring livelihoods and the economic recovery of the DRC. Revitalizing the AES will be extremely important in this regard. There is a clear need and opportunity to reach out and draw on the resources and expertise of both the public and private sectors. Community-based resources must also be harnessed. The extent of damage requires a full complement of labor, and the different skills and capacities of various actors.

Strategies to rebuild the AES in the DRC must look at both demand-side and supply-side constraints and should follow a defined sequence. In the short term, a first priority would be a unified policy and clearer strategy for agricultural extension. This will require designing and communicating a clear mission and mandate of the 11,000+ extension agents scattered throughout the country. If they could be utilized to focus on advisory services and technology transfer, rather than focusing on conducting censuses and monitoring farmers' activities, this could go a long way towards reaching out to the rural population with knowledge and technologies. Clear mandates, functions and performance targets for extension staff may trigger a change in perception and trust among farmers that extension agents are no longer 'monitors' but rather knowledge brokers and technical advisors.

The national workshop on agricultural extension and productivity held in June 2012 offered a good start, and the government can continue to bring the workshop's recommendations into its agricultural strategy planning and policy formulation.

A second priority in the short term is institutional coordination of extension services, especially in a pluralistic extension system. Greater coordination, quality control and regulation are critical to communicate consistent extension messages across many different extension service providers. This will also involve encouraging development organizations to work with local MINAGRI offices, INERA, SENASEM and SNV to avoid inconsistencies and duplication of efforts, as well as to ensure capacity building and continuity of activities. The capacity of SNV to play its role of coordination and technical backstopping should be strengthened.

A third priority in the short term consists of streamlining the number of extension staff by retiring the older and least efficient staff and hiring and retraining younger and more dynamic individuals. This can streamline the salary structure and free up valuable resources that can be shifted to much-needed operating funds (for actual extension service provision) and capital funds (e.g. buildings, vehicles and equipment). Implementation has already begun and the government should facilitate this for more immediate impacts.

A fourth priority addressing the demand side is to strengthen the role of farmers' organizations at the village level, and the role of CARGs at the sector and territory levels, in demand articulation and the capacity for planning and monitoring service provision. The role of CARGs in the policy making process should be strengthened so that they can advocate effectively for greater investment and commitment to agricultural extension.

In the medium term, complementary investment in reforming the agricultural education and training institutes will be a priority as they are the critical institutions that train and nurture the new wave of agricultural extension agents and officers. This will also include a review and reform of extension curriculum. And, given the serious food and nutrition insecurity in the DRC, inclusion of nutrition messages in agricultural extension services will be an important strategy. The scarcity of female extension workers requires investing in and supporting girls' education and mobilizing more female agents in the medium term. Given that extension services are just one of the factors determining productivity growth, complementary investments and policy reforms to facilitate access to markets, and affordability of inputs in order to complement the knowledge and extension provision, will also be important strategies in the medium term.

In the long term, there will be a need for better advocacy and communication for more stable funding (especially for operating and capital components) from the government, rather than persistent dependence on ad hoc and short-term projects. Institutional reform cannot succeed without sustained funding and commitment from government. The NAIP implemented in 2013, with detailed reforms of the AES, is encouraging and is in need of bold and firm commitment from the government.

References

AfDB (2004) Agricultural and Rural Sector Rehabilitation Support Project (PARSAR)—appraisal report. Available at: http://www.afdb.org/fileadmin/uploads/afdb/Documents/Project-and-Operations/Congo_-_Agricultural_and_Rural_Sector_Rehabilitation_Support_Project_in_Bas-Congo_and_Bandundu_Provinces_PARSAR_-_Appraisal_Report.pdf (accessed 18 January 2017).

AfDB (2012) Rehabilitation of the agricultural and rural sector in the Democratic Republic of Congo. Available at: https://www.afdb.org/en/news-and-events/rehabilitation-of-the-agricultural-and-rural-sector-in-the-democratic-republic-of-congo-9534/ (accessed 18 January 2017).

AfDB (2013) Democratic Republic of Congo—agricultural and rural sector rehabilitation project in Katanga, Kasai-Oriental and Kasai-Occidental provinces (Presar)—project completion report. Available at: http://www.afdb.org/fileadmin/uploads/afdb/Documents/Project-and-Operations/DRC_-_Agricutural_and_Rural_Sector_Rehabilitation_Project_in_Katanga-_Kasai-Oriental_and_Kasai-Occidental_Provinces__PRESAR__-_Project_Completion_Report.pdf (accessed 18 January 2017).

Akresh, R. and De Walque, D. (2010) Armed conflict and schooling: evidence from the 1994 Rwandan genocide. Working paper. Households in Conflict Network, Institute of Development Studies, University of Sussex, Brighton, UK. Available at: http://ideas.repec.org/p/hic/wpaper/47.html (accessed 18 January 2017).

Akresh, R., Verwimp, P. and Bundervoet, T. (2011) Civil war, crop failure, and child stunting in Rwanda. *Economic Development and Cultural Change* 59(4), 777–810.

Akresh, R., Bhalotra, S., Leone, M. and Osili, U. (2012) War and stature: growing up during the Nigerian Civil War. *American Economic Review: Papers and Proceedings* 102(3), 273–277.

Ali, R., Barra, A.F., Berg, C.N., Damania, R. and Nash, J.D. (2015) Infrastructure in conflict-prone and fragile environments: evidence from the Democratic Republic of Congo. Policy Research Working Paper 7273. World Bank, Washington, DC.

Anderson, J.R. and Feder, G. (2004) Agricultural extension: good intentions and hard realities. *World Bank Research Observer* 19(1), 41–60.

Badibanga, T., Ragasa, C. and Ulimwengu, J. (2013) Assessing the effectiveness of multistakeholder platforms: agricultural and rural management councils in the Democratic Republic of the Congo. Discussion Paper 01258. International Food Policy Research Institute, Washington, DC.

Bannon, I. and Collier, P. (eds) (2003) *Natural Resources and Violent Conflict: Options and Actions.* World Bank, Washington, DC.

Béguin, H. (1965) Espoirs, bilans et leçons d'un 'paysannat' au Congo. *Tiers-Monde* 6 (24), 891–913.

Bellows, J. and Miguel, E. (2006) War and institutions: new evidence from Sierra Leone. *American Economic Review* 96(2), 394–399.

Bellows, J. and Miguel, E. (2009) War and collective action in Sierra Leone. *Journal of Public Economics* 93(11/12), 1144–1157.

Blattman, C. (2009) From violence to voting: war and political participation in Uganda. *American Political Science Review* 103(2), 231–247.

Bundervoet, T., Verwimp, P. and Akresh, R. (2009) Health and civil war in rural Burundi. *Journal of Human Resources* 44(2), 536–563.

Chamarbagwala, R. and Morán, H.E. (2011) The human capital consequences of civil war: evidence from Guatemala. *Journal of Development Economics* 94(1), 41–61.

Congdon Fors, H. and Olsson, O. (2004) Congo: the prize of predation. *Journal of Peace Research* 41, 321–336.

Davis, K., Swanson, B., Amudavi, D., Ayalew Mekonnen, D., Flohrs, A., Riese, J., Lamb, C. and Zerfu, E. (2010) In-depth assessment of the public agricultural extension system of Ethiopia and recommendations for improvement. Discussion Paper 01041. International Food Policy Research Institute, Washington, DC.

Due, J.M., Magayane, F. and Temu, A. (1997) Gender again—views of female agricultural extension officers by smallholder farmers in Tanzania. *World Development* 25(5), 713–725.

FAO (2005) *Restructuring Plan for the Ministry of Agriculture, Livestock, and Fisheries (MINAGRI).* Project TCP/DRC/2904. DRC: Food and Agriculture Organization, Rome.

Gambino, T. (2011) Democratic Republic of Congo. Background case study for World Development Report 2011: Conflict, Security and Development. World Bank, Washington, DC.

GoM (2015) *Agriculture Joint Sector Performance Report.* Ministry of Agriculture, Irrigation, and Water Development, Government of Malawi, Lilongwe.

IDMC (2011) Democratic Republic of Congo: IDPs need further assistance in context of continued attacks and insecurity. Overview, 14 September. Internal Displacement Monitoring Centre, Geneva. Available at: www.internal-displacement.org/sub-saharan-africa/democratic-republic-of-the-congo/2011/idps-need-further-assistance-in-context-of-continued-attacks-and-insecurity (accessed 18 January 2017).

Inter-Congolese Dialogue (2002) Global and inclusive agreement on transition in the Democratic Republic of Congo (Pretoria Agreement). Available at: http://peacemaker.un.org/drc-agreementontransition2002 (accessed 18 January 2017).

International Alert (2010) *The Role of Exploitation of Natural Resources in Fuelling and Prolonging Crises in the Eastern DRC.* International Alert, London.

IRC (2007) *Mortality in the Democratic Republic of Congo: An Ongoing Crisis.* International Rescue Committee, New York.

Jewsiewicki, B. (1979) Le colonat agricole Européen au Congo-Belge, 1910–1960: questions politiques et economiques. *The Journal of African History* 20(4), 559–571.

Jewsiewicki, B. (1980) Political consciousness among African peasants in the Belgian Congo. *Review of African Political Economy* 19, 23–32.

Journal Officiel (1988) *Décret portant création du service national de vulgarisation (SNV).* République du Zaïre, Kinshasa.

Journal Officiel (1993) *Décret portant restructuration du service national de vulgarisation (SNV).* République du Zaïre, Kinshasa.

Justino, P. (2006) Collective action and economic development. *World Development* 34(7), 1254–1270.

Kakule, C.K. (2006) *Finance populaire et developpement durable en Afrique au Sud du Sahara. Application à la region Nord-Est de la Republique Démocratique du Congo.* Presse Universitaire de Louvain, Louvain, Belgium, pp. 314.

Kamau, M. (2012) Human resource development in staff training, adult learning and field methods. Presentation at Modernizing Extension and Advisory Services Global Learning Exchange on Best Fit Approaches in Extension and Advisory Services, Washington, DC, 8 June 2012.

Kaunda, E. (2011) National case study on extension and advisory services, Malawi. Innovation in Extension and Advisory Services: International Conference Proceedings, Nairobi, Kenya, 15–18 November 2011. CTA, Wageningen, the Netherlands.

Kilelu, C.W., Klerkx, L. and Leeuwis, C. (2013) Unravelling the role of innovation platforms in supporting co-evolution of innovation: contributions and tensions in a smallholder dairy development programme. *Agricultural Systems* 118, 65–77.

Kondylis, F. (2010) Conflict displacement and labor market outcomes in post-war Bosnia and Herzegovina. *Journal of Development Economics* 93(2), 235–248.

Kyle, J. (2010) The political spoils of war: the impact of conflict on political participation in Aceh, Indonesia. Working paper.

Le Plae, E. (1939) Native agricultural policy and European agriculture in the Belgian Congo. *Journal of the Royal African Society* 38(152), 357–369.

Maystadt, J.-F., De Luca, G., Petros, G., Sekeris, P.G. and Ulimwengu, J. (2014) Mineral resources and conflicts in DRC: a case of ecological fallacy? *Oxford Economic Papers* 66(3), 721–749.

Mills, W. (n.d.) Belgian colonial policy. Unpublished notes. Wallace G. Mills, St Mary's University, Halifax, Canada. Available at: www.saylor.org/site/wp-content/uploads/2011/12/HIST252-Belgian-Colonial-Policy.pdf (accessed 18 January 2017).

MINAGRI (2009) *Étude du Secteur Agricole: Rapport Bilan diagnostic et Note d'orientation.* Ministry of Agriculture, Livestock and Fisheries, République Démocratique du Congo, Kinshasa.

MINAGRI (2013) *Programme National d'Investissement Agricole.* Ministry of Agriculture, Livestock and Fisheries, République Démocratique du Congo, Kinshasa.

Minoiu, C. and Shemyakina, O.N. (2012) Armed conflict, household victimization, and child health in Côte d'Ivoire. ECINEQ Working Paper 2012-245. Society for the Study of Economic Inequality, Verona, Italy.

Mokili Danga Kassa, J. (1998) *Politiques agricoles et promotion rurale au Congo-Zaïre (1885–1997).* l'Harmattan, Paris, pp. 654.

Nest, M., Grignon, F. and Kisangani, E.F. (2006) *The Democratic Republic of the Congo: Economic Dimensions of War and Peace.* International Peace Academy Occasional Paper Series. Lynne Rienner Publishers, Boulder, Colorado.

Northrup, D. (2003) Coercion and exploitation: lever in the Belgian Congo. *The Journal of African History* 44(1), 167–168.

Pamuk, H., Bulte, E., Adekunle, A. and Diagne, A. (2012) Decentralized innovation systems and poverty reduction—quasi experimental evidence from Central Africa. Working Paper, Tilburg University, Wageningen University and FARA. Available at: https://www.tilburguniversity.edu/upload/ff29f5db-efc3-4693-aa49-dbc5b589b6bd_pamuk.pdf (accessed 18 January 2017).

Pamuk, H., Bulte, E. and Adekunle, A. (2014) Do decentralized innovation systems promote agricultural technology adoption? Experimental evidence from Africa. *Food Policy* 44, 227–236.

Peace Direct (2014) DR Congo: conflict profile. *Insight on Conflict* June. Available at: www.insightonconflict.org/conflicts/dr-congo/conflict-profile (accessed 18 January 2017).

Pellillo, A. (2012) Conflict and development: evidence from the Democratic Republic of the Congo. Unpublished manuscript, West Virginia University, Morgantown, West Virginia. Available at: https://iweb.cerge-ei.cz/pdf/events/papers/120402_t.pdf (accessed 19 January 2017).

Prunier, G. (2009) *Africa's World Wars: Congo, the Rwandan Genocide, and the Making of a Continental Catastrophe.* Oxford University Press, Oxford, UK.

Ragasa, C., Kinwa-Muzinga, A. and Ulimwengu, J. (2012) Gender Assessment in the agricultural sector in the Democratic Republic of Congo. Discussion Paper 01201. International Food Policy Research Institute, Washington, DC.

Ragasa, C., Berhane, G., Tadesse, F. and Taffesse, A.S. (2013a) Gender differences in agricultural extension, technology adoption and productivity: evidence from Ethiopia. *Journal of Agricultural Extension and Education* 19(5), 437–468.

Ragasa, C., Ulimwengu, J., Randriamamonjy, J. and Badibanga, T. (2013b) Assessment of the capacity, incentives, and performance of agricultural extension agents in western Democratic Republic of Congo. Discussion Paper 01283. International Food Policy Research Institute, Washington, DC.

Ragasa, C., Mazunda, J. and Kadzamira, M. (2015a) *The National Extension Policy of Malawi—Lessons from Implementation.* MSSP Policy Note 23. International Food Policy Research Institute, Lilongwe.

Ragasa, C., Ulimwengu, J., Randriamamonjy, J. and Badibanga, T. (2015b) Factors affecting performance of agricultural extension: evidence from Democratic Republic of Congo. *Journal of Agricultural Education and Extension* 9 April. DOI:10.1080/1389224X.2015.1026363

Shemyakina, O. (2011) The effect of armed conflict on accumulation of schooling: results from Tajikistan. *Journal of Development Economics* 95(2), 186–200.

Stearns, J. (2011) *Dancing in the Glory of Monsters: The Collapse of the Congo and the Great War of Africa.* Public Affairs, New York.

Tollens, E. (2004) Les défis: Sécurité alimentaire et cultures de rente pour l'exportation: principales orientations et avantages comparatives de l'agriculture en R.D. Congo. Presented at the Round Table on Agriculture in Democratic Republic of Congo, Kinshasa, 19–20 March.

Turner, T. (2007) *The Congo Wars: Conflict, Myth and Reality.* Zed Books, London.

Ulimwengu, J., Funes, J., Headey, D. and You, L. (2009) Paving the way for development? The impact of transport infrastructure on agricultural production and poverty reduction in the Democratic Republic of Congo. Discussion Paper 00944. International Food Policy Research Institute, Washington, DC.

Ulimwengu, J., Odunlami, N. and Bokanga, M. (2015) Special economic zones for agricultural production and processing: the case of Nigeria's staple crop processing zones and DR Congo's agricultural business parks. Background paper for the African Development Bank Conference on Unlocking Africa's Agricultural Potentials for Transformation to Scale, Dakar, 21–23 October 2015. Available at: https://www.afdb.org/fileadmin/uploads/afdb/Documents/Events/DakAgri2015/Special_Economic_Zones_for_Agricultural_Production_and_Processing.pdf (accessed 19 January 2017).

UNOCHA (2012) *Plan d'Action Humanitaire pour la République Démocratique du Congo 2013.* UN Office for the Coordination of Humanitarian Affairs, New York and Geneva.

Van Hoof, F. (2011) *Changer l'Agriculture Congolaise en Faveur des Familles Paysannes: des Dynamiques Paysannes dans les Différentes Provinces de la RDC.* Alliance AGRICONGO, Middelbeers, the Netherlands.

Vanlauwe, B., Pypers, P., Birachi, E., Nyagaya, M., van Schagen, B., Huising, J., Ouma, E., Blomme, G. and van Asten, P. (2012) Integrated soil fertility management in Central Africa: experiences of the Consortium for Improving Agriculture-based Livelihoods in Central Africa (CIALCA). In: Hershey, C.H. (ed.) *Eco-Efficiency: From Vision to Reality.* Centro Internacional de Agricultura Tropical (CIAT), Cali, Colombia.

Van Paassen, A., Klerkx, L., Adu-Acheampong, R. and Adjei-Nsiah, S. (2014) Agricultural innovation platforms in West Africa: how does strategic institutional entrepreneurship unfold in different value chain contexts? *Outlook on Agriculture* 43(3), 193–200.
Vlassenroot, K. and Raeymaekers, T. (2004) The politics of rebellion and intervention in Ituri: the emergence of a new political complex? *African Affairs* 103, 385–412.
WFP (2011) *Republique Democratique du Congo: Executive Summary: Results of the In-Depth Emergency Food Security Assessment (EFSA) of Displaced, Returnee and Host Households in the Orientale Province: September 2011*. World Food Programme, Rome. Available at: http://documents.wfp.org/stellent/groups/public/documents/ena/wfp242295.pdf (accessed 19 January 2017).
World Bank (2007) *Enhancing Agricultural Innovation: How to Go Beyond the Strengthening of Research Systems*. World Bank, Washington, DC.
World Bank (2010) *Agricultural Rehabilitation and Recovery Support—Project Appraisal Document*. World Bank, Washington, DC.
World Bank (2011) *Agricultural Innovation Systems Sourcebook*. World Bank, Washington, DC.
World Bank (2013) *Western Growth Poles—Project Appraisal Document*. World Bank, Washington, DC.
World Bank (2016) *Eastern Recovery Project—Project Appraisal Document*. World Bank, Washington, DC.
World Bank and IFPRI (2009) *Gender and Governance in Rural Services*. World Bank and International Food Policy Research Institute, Washington, DC.

4 Enhancing Food Security in the World's Youngest Nation: A Case Study of Agricultural Extension in South Sudan

Robert Strong Jr.,* Kirk Edney and Roger Hanagriff
Texas A&M University, College Station, Texas, USA

Background

History of conflict in South Sudan

Formed from the ten southernmost states of Sudan, South Sudan is a land of expansive grassland, swamps and tropical rainforest straddling both banks of the White Nile (Natsios, 2012). It is highly diverse ethnically and linguistically. Among the largest ethnic groups are the Dinka, Nuer and Shilluk. Unlike the predominantly Muslim population of Sudan, the South Sudanese follow traditional religions, while a minority is Christian (Jok, 2011).

As Sudan prepared to gain independence from joint British and Egyptian rule in 1956, southern leaders accused the new authorities in Khartoum of backing out of promises to create a federal system and trying to impose an Islamic and Arabic identity on the south (Schomerus and Allen, 2010). This tension led to a series of armed conflicts. Collins (2007) indicated that the longest, most destructive and violent of Sudan's five civil wars were the two between the north and south. The first occurred in 1955 when southern army officers mutinied, igniting a civil war between the south—led by the Anya Nya guerrilla movement—and the Sudanese government. A second conflict—termed the Anya Nya civil war—occurred from 1963 to 1972 and only ended when the Addis Ababa peace agreement of 1972 accorded the south a measure of autonomy.

In 1983 the south again rose in rebellion—led by the Sudan People's Liberation Movement (SPLM) and its armed wing, the Sudan People's Liberation Army (SPLA)—when the Sudanese government cancelled the autonomy arrangements (Belloni, 2011). This was the start of the second southern civil war, which took place between 1983 and 2006. During the second southern civil war, and specifically from 1991 to 1996, an internal 'civil war within a civil war' took place between the Dinka and Nuer tribes and caused massive casualties and human suffering before the SPLA could establish its authority (Collins, 2007: 1783).

The conflict officially came to an end with the signing of the 2005 Comprehensive Peace Agreement. The south was granted regional autonomy and guaranteed representation in a power-sharing government (Dagne, 2011). The agreement included a referendum in the south on independence, to be held in 2011. The people of the south voted 99% to

* Corresponding author e-mail: r-strong@tamu.edu

1% in favor of splitting from Sudan. The prospect of independence excited the South Sudanese, who were looking for a fresh start, a new country and lasting peace (Christopher, 2011). Consequently, South Sudan broke away from Sudan and became an independent nation in 2011. SPLA officers remained in military leadership positions as governmental reorganization proceeded.

This process was not without its costs. Hanzich (2011) reported that at least 1.5 million people are thought to have lost their lives and more than 4 million were displaced in the 22 years of guerrilla warfare. Food insecurity became a pressing issue as the war escalated (Keen and Lee, 2007). Massive numbers of people fled the conflict either northwards or to adjoining nations and have not returned. Thousands who fled took what livestock they could with them.

In December 2013, the young state again plunged into crisis amid a power struggle between the president and his deputy, whom he had dismissed. South Sudan's president is a member of the Dinka tribe and the former deputy is a Nuer tribe member (Jok, 2011). Fighting between government troops and rebel factions erupted, largely through tribal conflict between the Southern Sudan Defense Forces (SSDF) and the SPLA, which is controlled by the president.

Within weeks the conflict had killed thousands and prompted more than 800,000 to flee their homes (Arnold and LeRiche, 2013). Aljazeera America (2014) reported that nearly 100,000 South Sudanese civilians had been given refuge in UN camps in South Sudan. These civilians fled to escape killings and massacres, and continue to fear returning home. The large populations in the UN camps have given rise to specific challenges related to food security due to problems of access to food, distribution of food aid and the dietary needs of refugees with diverse age ranges.

A series of cattle raids between the Nuer and Dinka tribes in December 2013 and January 2014 caused conflict to spread once again. Hundreds of people were killed and thousands were displaced from their homes in retaliation (British Broadcasting Corporation, 2014). These raids are a long-running cultural practice that exists outside of national issues. Several cease-fire agreements have been proposed recently, but these have broken down since the summer of 2015.

The post-conflict context

South Sudan, like many fragile states, has not yet fully emerged from conflict. Therefore, terming it a post-conflict state is somewhat misleading. Nonetheless, South Sudan has many characteristics that are consistent with other countries facing sporadic and recurring violence following larger conflicts. These characteristics have important impacts on the roles of agriculture and agricultural extension.

South Sudan remains a resource-rich nation with poor infrastructure and continuing tribal conflicts that hinder agriculture and the conservation and responsible use of natural resources. Efforts to develop agriculture have been negatively impacted by the conditions surrounding the conflict. For example, much of the direction of economic activity in South Sudan is managed by generals and other leaders of the former SPLA. Given their backgrounds and the importance placed on security, approximately 40% of the budget of South Sudan is allocated to maintaining military infrastructure. This reduces the budget available for agriculture and other activities. In addition, military leaders have a de facto role in identifying and encouraging various endeavors and enterprises important to South Sudan, in terms of both economic viability for the nation and logistical considerations (e.g. food supply) for the military. Agriculture therefore has a unique role in this context of militarization and priority-setting.

Funding for agriculture is additionally compromised by the composition of the South Sudanese economy and reliance on oil revenues. Years of civil war have prevented the development of any effective infrastructure, and crude oil represents the only reliable revenue stream for the government.

Personal property taxes or income taxes are not a possible source of governmental funding, and the nation is not yet stable enough to allow for any significant income from tourism. However, the sporadic and ongoing conflicts, together with the current low price of oil, have led to a dramatic fall in oil production in South Sudan. The budget assigned to agriculture has decreased considerably as a result of the lowered oil revenues. This reduction has had a ripple effect, influencing the resources—human and other—that would typically have been devoted to food security.

Access to and the use of natural resources are also factors that affect agriculture in South Sudan. As Sudan—and later South Sudan—endured multiple civil wars between 1956 and the present, these conflicts reduced the country's access to natural resources and consequently its production of agricultural commodities (Collins, 2007). For example, restricted access to and the strain on water supplies during the civil wars led to decline in agricultural productivity (Fegley, 2009). In addition, managing South Sudan's water supply in the current period remains important in the fight against food insecurity (Rai *et al.*, 2012). Water conservation programs in South Sudan are paramount for agricultural production, due to domestic pressures on the Nile River and the needs of other countries that rely on the Nile. Meanwhile, the swamps in the southern part of the nation are in danger of being drained for cultivation, leading to a loss of biodiversity and soil fertility (Salman, 2011). Gorsevski *et al.* (2012) have also indicated that the conflicts have had a negative effect on forests and other natural resources.

Agricultural Extension Providers and Actors

The system of agricultural extension in South Sudan is largely disorganized and includes a range of actors with varied roles and operational statuses. Public sector programs and institutions from both the South Sudanese government and its military provide extension services. In addition, due to the youth of the nation and the corresponding lack of institutional capacity to serve farmers, a wide variety of non-state entities work to assist South Sudan with agricultural development. All provide crucial services and fill unique roles.

SPLA Agricultural Extension Battalion

Following independence in 2011, the SPLA was faced with the question of how to engage its fighters in productive activities that would provide food for the military but also serve the development goals of the country, while providing employment to a group of military officers lacking many marketable skills. At the same time, South Sudan's population was heavily reliant on agriculture as a livelihood and food production strategy. Given that many SPLA fighters came from agrarian backgrounds, an early concept of a post-conflict extension system for South Sudan was that a battalion from the SPLA would serve as extension officers. Efforts were made to create this agricultural battalion, which does provide limited services to farmers. However, the system was not fully developed due to the further uprising in December 2013. Nonetheless, the SPLA agricultural battalion is still a proposed model for agricultural extension services in South Sudan.

Ministry of Agriculture, Forestry, Tourism, Animal Resources and Fisheries

The Ministry of Agriculture, Forestry, Tourism, Animal Resources and Fisheries (MoAFTARF) also plays a major and central role in South Sudan's extension system, particularly as related to the development of the agricultural sector. The MoAFTARF has a broad mandate that includes efforts to:

- Establish and manage an effective agricultural extension service.
- Formulate legislation, policies, standards and plans for the development of agriculture and forestry in South Sudan and ensure adequate food availability.

- Promote and, where necessary, regulate the efficient production and marketing of agricultural and forest products.
- Conduct demand-driven research and collect data on production and its socio-economic impacts on incomes and well-being.
- Rehabilitate and expand training and research institutions.
- Develop the capacities of farmers and other stakeholders in the fields of agriculture and forestry, especially related to the development and adaptation of appropriate technologies.
- Provide technical assistance and training to state and local governments to build their capacity to assume responsibilities for agriculture and forestry.

Within the context of this mandate, the priority objectives of the Ministry are: (i) to establish and maintain an effective agricultural extension service; (ii) to develop the human resources of stakeholders working in agriculture and forestry; and (iii) to rehabilitate and expand training and research institutions. To accomplish these objectives, the Ministry is supported by the USA, the EU and the Ministry of Agriculture of the People's Republic of China. However, to date the MoAFTARF does not support or provide training for any type of extension system in South Sudan. Only the SPLA's agricultural battalion performs this role.

These three Ministry objectives align with US Department of State and SPLA agricultural battalion goals for agricultural development and extension in South Sudan. However, historical and political barriers—such as perpetual tribal and political conflicts—keep the leaders of the SPLA's agricultural battalion and the Ministry from working together. At the time of this case study, the Ministry had no working relationship with the extension efforts provided by the SPLA. Over time, international organizations may be able to help improve working relationships between the Ministry and the SPLA for the benefit of the rural poor, but more effective collaboration is not expected in the short term due to internal socio-political conflict and reduced involvement from foreign governments. Instead, other organizations are anticipated to provide agricultural extension services in South Sudan to address institutional weaknesses and gaps due to poor coordination between the MoAFTARF and the SPLA.

Agricultural non-governmental organizations (NGOs)

Several NGOs provide extension services in South Sudan. The specific objectives and approaches of different non-state actors are both similar and dissimilar to governmental efforts, although they share the common mission to increase food security in South Sudan.

Catholic Relief Services (CRS)

CRS has been working in the southern part of Sudan—and later South Sudan—since 1983. After independence in 2011, CRS supported development and recovery in parts of the country where post-conflict remedial services were otherwise entirely lacking. Against a background of political and economic turmoil resulting from the conflict that flared up in December 2013, the CRS website for South Sudan announced: 'The presence of Catholic Relief Services in South Sudan is needed now more than ever to strengthen these relationships and promote healing, while continuing to deliver lifesaving emergency and development assistance' (Pozniak, 2015).

CRS South Sudan offers programs in agriculture, civil society and governance, disaster response, microfinance, peace building, and water and sanitation. The immediate goal of the CRS South Sudan agriculture program is to improve family well-being through agro-economic development and environmental stewardship. The agency's long-term goal is to strengthen the capacity of local agencies and farming communities to take control of their own development. The formal objectives of the agriculture program are: (i) to engage with vulnerable communities to meet their long-term food and livelihood needs; (ii) to foster sustainable socio-economic development; and (iii) to promote agricultural practices that link production to conservation, using local sources where available.

World Vision International (WVI)

WVI has been involved in South Sudan for over 25 years. It has provided numerous agricultural extension programs during this time and in 2014 worked with approximately 128,000 people. World Vision's South Sudan country program reported in 2015 that their organization was the only entity combating food insecurity in the northernmost state, the Upper Nile.

World Vision South Sudan provides training on improved agricultural practices such as soil conservation techniques, no-till farming techniques, native tree regeneration and the development of small seed-saving banks (Africa CSA Alliance, 2014); post-harvest handling and packaging; agricultural marketing; and financial management. Extension programs focus on commodities that include sesame, pineapple, fish, tomato and maize, among many others.

Besides teaching farmers, the organization also provides inputs such as hoes, seeds and fishing equipment. The organization donated approximately 174 oxen to farmers in Warrap State to increase food production (World Vision–South Sudan, 2013). In South Sudan, World Vision has had an impact on farmers' yields, income and well-being as a result of these efforts (Oxfam, 2014).

African Centre for the Constructive Resolution of Disputes (ACCORD)

ACCORD is one of several agricultural NGOs that are still operational despite past and current conflicts in South Sudan. It has been involved in South Sudanese agricultural extension work for approximately 30 years. ACCORD operates in the Central and Eastern Equatoria States.

ACCORD offers extension programs related to conflict resolution, leadership development, fishery production, animal management and water conservation. The organization maintains collaborative relationships with evangelical Christian groups, the United States Agency for International Development (USAID), the Konrad Adenauer Foundation and the Norwegian Institute for International Affairs.

United Methodist Committee on Relief (UMCOR)

UMCOR currently conducts up to 69 projects in South Sudan (2015–2016) and has collaborative relationships with USAID, the EU, the US State Department and various UN agencies. Sustainable agriculture and food security are a programmatic focus of UMCOR (2015). Under its food security project, 16 fish farms have been established and over 200 farmers in the Central Equatoria state of South Sudan have been trained in aquaculture and received fish management inputs. An additional 2400 farmers have been trained in cassava intercropping. UMCOR also provides training and advice on water quality for rural people living in the Upper Nile State and Bahr el Gazal/Darfur region (UMCOR, 2015), and estimated that over 40,000 individuals will be impacted by this project.

Agricultural colleges

JOHN GARANG MEMORIAL UNIVERSITY OF SCIENCE (JGMUST). John Garang Memorial University of Science in Bor, the capital of Jonglei State, provides agricultural services and education in the agricultural sciences. The university, in collaboration with the US State Department and the Norman Borlaug Institute for International Agriculture of Texas A&M University, designed the Consortium for Development project to improve agricultural education; create institutional linkages to improve productivity; encourage agricultural development and resource conservation; and develop programs to address youth entrepreneurship, conflict resolution and gender-based education issues. Based on JG-MUST faculty input, a livestock production course was developed and is currently being offered. Curriculum for the course was delivered by three methods: (i) traditional pencil and paper; (ii) separate student and instructor course CDs; and (iii) a basic web-based format. Challenges relative to maintaining the infrastructure at JG-MUST make it difficult to determine the effectiveness of any of these delivery methods.

Agricultural Extension Projects and Activities

Various efforts have been made by state and non-state actors to provide and strengthen agricultural extension in South Sudan. However, given the influential role of the military, many efforts have sought to engage the SPLA directly. USAID, in particular, has created a number of projects to assist the SPLA in improving food security, working closely with the US State Department and Texas A&M University's Norman Borlaug Institute for International Agriculture to design and deliver these projects.

Enhancing Agricultural Infrastructure for the Military of South Sudan Project

A major collaborative project—funded by the US Department of State—was the Enhancing Agricultural Infrastructure for the Military of South Sudan Project. The project aimed to support the Republic of South Sudan in its efforts to transform its military from a largely guerrilla force to a national army under the newly formed government. Part of the project involved the development of agricultural infrastructure, specifically the building of roads linking farmers to markets, and the development of water points for people and livestock.

Food security, and particularly a sustainable food supply for the SPLA, was another key objective. The SPLA was spending upwards of US$100 million a year to import food for its troops. Through a grant, the US Department of State worked with the South Sudan Ministry of Defense to establish an agricultural support and training program, which was implemented through the Borlaug Institute for International Agriculture and Texas A&M AgriLife Research. The objective was to allow the SPLA to develop sustainable food sources for soldiers and their families while stationed at bases or training camps and while deployed on mission. The project designed an approach that engaged military personal, employed local people in food production activities that contributed to rural livelihoods and utilized available resources to generate food locally for consumption by the SPLA.

In addition, the project focused heavily on designing and delivering basic and expanded agricultural training curricula, with the understanding that the curricula would function as infrastructure by creating a system for knowledge to be transferred from trainer to learner (Knowles *et al.*, 2005). In curricula design, the project team from Texas A&M and the US State Department assessed the needs of participants, developed and provided curricula to meet those needs and evaluated the change in knowledge gained due to the curricula and instruction. Ultimately, the project contributed to improved food supplies for both the SPLA and rural people.

Extended Agricultural Training (EAT) project

The EAT project—conducted with SPLA—was derived from the Enhancing Agriculture Infrastructure for the Military of South Sudan Project. The project was originally a class designed to train members of the SPLA to serve as extension officers and to operate as farmers themselves. Upon graduating from EAT, army officers were provided with a farm and were expected to teach modern farming practices to local farmers through demonstrations and individual consultations, effectively employing army personnel as extension officers through the SPLA's agricultural battalion. Trainees learned the skills needed to teach local farmers the sustainable management and production strategies to improve their food security but also developed the competencies to themselves manage the army's farms, to increase the SPLA's food production and reduce food costs for the army.

The implementation of the EAT project occurred in stages. First, the project's Training

Impact Specialist—under the direction of the Project Training Coordinator—conducted an observational needs assessment of future participants' training needs and determined whether curricula needed to be developed from scratch, sourced from instructional resources or transformed from existing materials.

The assessment showed that the majority of future participants needed more training on how to teach adults. In technical areas, 67% needed more training in small animal production; 80% wanted to be more knowledgeable in identifying crop pests and diseases; 73% of the participants believed they needed more farm management knowledge to help farmers become more successful; and 93% wanted to better understand evaluation techniques to be able to measure the impact of their teaching. Future participants also requested more visual materials due to their target audience being illiterate (67%); materials that would help them teach soil testing procedures and pest and disease identification (73%); and dictionaries to help them better understand English and thus work more effectively with US and European agricultural development organizations (47%).

Following the assessment, the EAT Training Content Specialist developed new training materials—largely in pictorial formats—to fill gaps in existing curricula and allow better access to information for trainees with varying levels of literacy. A complete training manual was produced for use with farmers to help them understand budgets, markets, options for improving soil fertility and other relevant topics. The first manual influenced subsequent attempts to develop materials for SPLA extension officers and farmers, many of whom were illiterate in Arabic as well as English.

The EAT project then reviewed its current training methodologies as compared to principles of adult education (Knowles *et al.*, 2005). Recommendations for additional adult training methodologies were developed for use in conjunction with the improved training materials, and resources were sourced from program partners to support their use.

Finally, the EAT project sought to identify environmental factors that could potentially influence the adoption and diffusion of program objectives and educational resources using a political, economic, social, technological, environmental and legal (PESTEL) analysis (Table 4.1). The PESTEL analysis was selected because of its suitability in changing environments and was therefore a more valuable assessment of external factors of influence than a strengths, weaknesses, opportunities and threats (SWOT) analysis. The analysis was completed through interviews and qualitative measurement of extension officers involved in the EAT project.

Ultimately, these themes were exemplified in the actual outcomes of the EAT

Table 4.1. Barriers identified through the PESTEL analysis.

Political (P)	Land ownership disputes; lack of trust in the government and governmental programs; limited education of farmers and officers; difficulties for outsiders in understanding the issues faced by farmers
Economic (E)	Limited incomes of farmers; high food prices; inadequate housing; lack of sustained funding and support from the government for basic needs
Social (S)	Separation from families to attend the training conducted at SPLA headquarters outside of Juba; lack of English skills; health problems; inadequate training facilities; short duration of the training program
Technological (T)	Lack of electricity; insufficient equipment and computers; poor quality of equipment; lack of textbooks/manuals; shortage of livestock inputs
Environmental (E)	Climate; poor living conditions at the school; high transportation costs; lack of medical care
Legal (L)	Inadequate follow-through on commitments from the local government, outside governments and non-governmental organizations

SPLA, Sudan People's Liberation Army.

project, which were affected by political influences and sustainability issues related to ongoing conflict in regions affected by the civil war. However, PESTEL factors are useful not only for assessing the EAT project but also for providing insight into the larger context of agricultural extension services in South Sudan.

Conclusions and Lessons Learned

The examples described above provide a range of themes and lessons learned from South Sudan that may have larger importance for other post-conflict countries.

Safety and security

The ongoing conflict between the Dinka and Nuer tribes makes the rebuilding of agricultural extension in South Sudan challenging, to say the least. The safety of trainers, evaluators, curriculum designers, farmers and local government officials involved in agricultural extension projects is a serious worry, both for the individuals concerned and for the organizations that employ them.

Extension actors struggle with the question of whether or not to approve activities in environments where safety cannot be evaluated because the situation changes from one day to the next. Ruiz-Postigo *et al.* (2012) reported that safety concerns prevented NGO healthcare workers from screening South Sudanese citizens for human African trypanosomiasis. Likewise, Spencer *et al.* (2013) found that security issues prevented a health screening team from testing children for the debilitating and potentially fatal nodding syndrome. Similar limitations are found in agriculture. Agricultural development workers with good intentions towards development and farmers' empowerment are often deterred by safety concerns, resulting in needs going unaddressed and depriving people of the hope that conditions can improve.

Security concerns also hinder the growth and development of South Sudan at the level of farmers. Branch and Mampilly (2005) suggested that until national conflicts are reduced or eliminated, farmers are unlikely to return home to plant and harvest their crops, causing food insecurity to persist. Farmers are also reluctant to participate in group-based programs or meetings commonly employed by extension providers. Only by reducing the number and scale of conflicts in South Sudan will local people feel safer in attending public educational events (Themnér and Wallensteen, 2012).

The safety of both extension workers and farmers must be improved. Unless this need can be met, the challenges of rebuilding agricultural extension in South Sudan will be hard to overcome. However, effective agricultural development can actually mitigate some of these same safety and security concerns. Agricultural extension programs that provide the rural population with the means to improve their own production can help the nation achieve food security and reduce poverty, ultimately lessening conflict and promoting social cohesion. This should, in turn, raise the chances of sustaining peace.

Agricultural training for extension providers

Pre- and in-service training

The lack of a formal extension system in South Sudan underscores the need for consistent and ongoing training for extension officers. Individuals acting as extension officers are often unprepared for the demands of the job. Graduates of JG-MUST, for example, lack the technical and process skills needed to work with farmers. Data from the EAT project also pointed to a number of unmet or inadequately met training needs of army officers preparing to serve as extension officers. The training competencies most needed were in developing and evaluating educational programs for adults. Strengthening these training programs and the institutions that prepare extension officers is therefore crucial.

Ongoing training in teaching, learning and disseminating information to farmers is also needed. EAT graduates will continue to require professional development and resources to be successful in their roles as extension officers. In-service training is particularly needed on topics related to crop and small animal production; pest and disease control; farm management and agribusiness; and agricultural marketing. Like other post-conflict nations, South Sudan needs the support of international organizations to address its extension training gaps and create competent and excellent agricultural extension programs.

Needs assessment and evaluation for development organizations and extension actors

Experiences in South Sudan—and specifically with the EAT project—illustrate the importance of proper needs assessment and evaluation when training extension professionals. Evaluating existing materials and approaches against learner needs is essential for creating training programs that address knowledge and skill gaps and better prepare graduates for work with farmers.

Using a PESTEL analysis at the onset of developing training programs to determine barriers to learning and the adoption of new techniques can contribute to greater successes and sustainable outcomes. The results of a PESTEL analysis can offer insight into needs likely to continue as issues for extension workers, such as land ownership, living conditions, family needs and insufficient infrastructure. Understanding these factors can help extension programs make strategic decisions and improve services. Consideration of experiences in other conflict regions suggests that identifying these results or external factors may offer insight into issues that may lead to program failure if not properly managed.

Agricultural extension curricula

Effective and appropriate curricula are also crucial to training future extension officers. For example, program leaders and SPLA generals described their training needs in a meeting held at the outset of the EAT project. At the same time, existing curricula and training materials were predominately written in English and at a ninth-grade (age 14) reading level. However, effective training was confounded by two major factors: (i) literacy levels in the workforce are low due to 50 years of conflict; and (ii) those who were literate mostly read Arabic (over 60% of the population are Arabic speakers) rather than the official language of English.

It was concluded that training materials needed to be bilingual (English/Arabic) to encourage learning in both languages, more pictorial than text-based and with content delivered at an appropriate level of understanding. Materials were supported by teaching through demonstrations as much as possible. Texas A&M University's Instructional Materials Service provided a variety of reference materials and resources to the EAT project, either as PDFs or CDs for instructors and students. The materials were better suited to the audience and helped trainees learn to make decisions in many areas of farm management, including: (i) site location for crop production; (ii) crop production designs, including patterns and rotations; (iii) water management; (iv) tillage and land preparation; (v) the use of inputs such as fertilizer and lime to control soil acidity; and (vi) harvest and post-harvest operations. Extension programs operating in post-conflict settings are most effective when using a similar participatory and needs-driven process for curriculum design.

Agricultural training to farmers

Programmatic targeting

Agricultural extension programs that identify and understand the characteristics and desired outcomes of their target audience are more likely to be successful. There are several examples of extension efforts that have had a positive impact on agriculture in South Sudan. Successful NGO projects working on food security are particularly evident. The common denominator for these successful projects is that they targeted a specific audience or commodity and the organization has a long, positive history in

South Sudan and a strong understanding of the local context.

In contrast, some projects observed a lack of these characteristics. For example, the methodology espoused by EAT extension personnel was not congruent with the context of resource-poor and risk-averse small farmers. The project focused on national-level issues and did not adequately consider the needs of local farmers. The EAT project was further compromised because it was instituted by an internal organization and by the military, which was not 'well-perceived' by a large percentage of the South Sudanese.

Alternative approaches to reaching farmers

Given the safety and security conditions that often limit the ability of extension officers to interact directly with farmers, there are useful alternative means of reaching farmers in South Sudan. Many South Sudanese—including EAT officers and farmers—have cell phones, which can be charged using solar power. This indicates that information and communications technologies (ICTs) could be an avenue for delivering educational content to farmers throughout the conflict-torn nation. Cell phones could be used to get information to farmers in remote locations, as well as offering professional development to extension officers, thereby saving on travel time and expenses. The benefit of ICT use for agricultural development efforts in Africa, Latin America and the Caribbean is well documented. This opportunity needs further investigation in the context of South Sudan, but it appears that ICTs could help extension personnel deliver agricultural content to farmers more effectively than current face-to-face meetings, especially when conflict and safety are also factors.

Service provision to women farmers

There is great need and potential to better serve women farmers in South Sudan. Globally, the diversity and number of females in food production is increasing as women become more involved in agriculture (Charlton, 1984), yet extension programs regularly struggle to reach and benefit women farmers (Meinzen-Dick *et al.*, 2011). Due to the lack of training and educational opportunities for women in production agriculture, women farmers are largely excluded from some business aspects of farming (Kabeer, 2012). However, when women are equally engaged in extension and have equal access to resources they see greater improvements in agricultural production than men (Davis *et al.*, 2012), suggesting greater emphasis on serving women through extension could be beneficial to South Sudan's agricultural sector. Benefits are greater when women farmers are taught by female extension officers, especially in contexts and communities with strict social and gender norms.

However, the EAT project consisted only of male participants, since women were not involved in the agricultural battalion. This raises the question of how women farmers will react to male instructors, whether women will participate in extension programs and if EAT graduates will be able to serve women farmers to the same degree they serve males. This suggests that the benefits of the EAT project to women will be minimal.

In contrast, WVI has effective and well-documented experience serving women farmers in South Sudan. Unlike the SPLA, WVI specifically targets women farmers with their programs and develops programs strictly for women. The primary lessons learned by WVI were the critical importance of identifying the target audience, conducting a needs assessment with that audience and developing educational programs based on those identified needs. These needs assessments can highlight the aspects of agricultural education training most needed by women. In the case of WVI, evaluations were conducted during and after program implementation to help coordinators monitor and improve programs for women farmers.

Agricultural extension organizations—whether governmental or non-governmental—should examine the experiences of WVI in serving women farmers in South Sudan. The organization's approach has proved effective and replication of certain elements may be possible. This would benefit all extension providers, but more importantly it would benefit women farmers. Increasing their inclusion in extension programming will help improve the food security of South Sudan, assist a new democracy and provide citizens of both genders with a sense of ownership in the nation's future.

Coordination and collaboration

Coordination between providers is an area of concern for agricultural extension in South Sudan, as in many post-conflict countries. As discussed above, MoAFTARF did not collaborate with the SPLA on extension. There could be numerous reasons why these breakdowns in collaboration exist (e.g. tribal differences, socio-political power struggles), although in this case there is no clear single reason.

Regardless of the cause, a lack of coordination between state actors certainly compromises the quality of the services provided to farmers. Building better collaboration within the public sector would enable better solutions and more focused efforts in addressing food security across the young nation. Collaboration could also foster more national unity and promote peace, improving the lives of all South Sudanese and their descendants.

In addition, despite actively providing agricultural and extension services, many NGOs operate in isolation from state actors. At the time of this case study, collaboration between the SPLA and agencies such as ACCORD, CRS, UMCOR and WVI seems to be minimal. Nevertheless, collaboration between NGOs, the SPLA and MoAFTARF is needed to more effectively train extension staff to train farmers across South Sudan. All extension actors could strengthen each other, creating synergistic relationships in which the impact of the whole system is greater than the sum of its parts. Opportunities for collaboration readily exist, and external and internal NGOs should seek positive collaboration with the government to implement projects that align with respective organizational objectives to help South Sudan become a food secure and peaceful nation.

References

Africa CSA Alliance (2014) *The Africa Climate Smart-Agriculture Alliance*. CSA Alliance, Johannesburg, South Africa.

Aljazeera America (2014) South Sudan rivals sign new cease-fire deal. Available at: http://america.aljazeera.com/articles/2014/8/25/south-sudan-peace.html (accessed 8 February 2016).

Arnold, M. and LeRiche, M. (2013) *South Sudan from Revolution to Independence*. Oxford University Press, New York.

Belloni, R. (2011) The birth of South Sudan and the challenges of state-building. *Ethnopolitics (Formerly Global Review of Ethnopolitics)* 10(3/4), 411–429.

Branch, A. and Mampilly, Z.C. (2005) Winning the war, but losing the peace? The dilemma of SPLM/A civil administration and the tasks ahead. *Journal of Modern African Studies* 43(1), 1–20.

British Broadcasting Corporation (2014) South Sudan: what is the fighting about? Available at: www.bbc.com/news/world-africa-25427965 (accessed 19 January 2017).

Charlton, S.E. (1984) *Women in Third World Development*. Westview Press, London.

Christopher, A.J. (2011) Secession and South Sudan: an African precedent for the future? *South African Geographical Journal* 93(2), 125–132.

Collins, R. (2007) Civil wars in the Sudan. *History Compass* 5(6), 1778–1805.

Dagne, T. (2011) The Republic of South Sudan: opportunities and challenges for Africa's newest country. *Congressional Research Service* R41900, 1–24.

Davis, K., Nkonya, E., Kato, E., Mekonnen, D.A., Odendo, M., Miiro, R. and Nkuba, J. (2012) Impact of farmer field schools on agricultural productivity and poverty in East Africa. *World Development* 40(2), 402–413.

Fegley, R. (2009) Local needs and agency conflict: a case study of Kajo Keji County, Sudan. *African Studies Quarterly* 11(1), 25–56.

Gorsevski, V., Kasischke, E., Dempewolf, J., Loboda, T. and Grossman, F. (2012) Analysis of the impacts of armed conflict on the Eastern Afromontane forest region on the South Sudan–Uganda border using multi-temporal Landsat imagery. *Remote Sensing of Environment* 118, 10–20.

Hanzich, R. (2011) Struggles in South Sudan: five months to resolve 55 years of structural violence. *Harvard International Review* 33(1), 38.

Jok, J.M. (2011) Diversity, unity, and nation building in South Sudan. United States Institute of Peace Special Report. Available at: http://www.usip.org/sites/default/files/resources/Diversity,%20Unity,%20and%20Nation%20Building%20in%20South%20Sudan%20(Jok).pdf (accessed 19 January 2017).

Kabeer, N. (2012) Women's economic empowerment and inclusive growth: labour markets and enterprise development. International Development Research Centre. Available at: https://www.idrc.ca/sites/default/files/sp/Documents%20EN/NK-WEE-Concept-Paper.pdf (accessed 19 January 2017).

Keen, D. and Lee, V. (2007) Conflict, trade, and the medium-term future of food security in Sudan. *Disasters* 31(s1), s9–s24.

Knowles, M., Holton, E. and Swanson, R. (2005) *The Adult Learner: The Definitive Classic in Adult Education and Human Resource Development*, 7th edn. Butterworth Heinemann, Elsevier, Oxford, UK.

Meinzen-Dick, R., Quisumbing, A., Behrman, J., Biermayr-Jenzano, P., Wilde, V., *et al.* (2011) Engendering agricultural research, development, and extension. Available at: www.ifpri.org/publication/engendering-agricultural-research-development-and-extension (accessed 19 January 2017).

Natsios, A.S. (2012) *Sudan, South Sudan, and Darfur: What Everyone Needs to Know*. Oxford University Press, New York.

Oxfam (2014) Aid effort to avert South Sudan famine in jeopardy. Available at: https://www.oxfam.org/en/pressroom/pressreleases/2014-07-09/aid-effort-avert-south-sudan-famine-jeopardy (accessed 19 January 2017).

Pozniak, K. (2015) Conflict and hunger in South Sudan: CRS and the Church provided assistance. Catholic Relief Services. Available at: www.crs.org/our-work-overseas/where-we-work/south-sudan (accessed 19 January 2017).

Rai, R.K., Ramadhan, A.A. and Tulchinsky, T.H. (2012) Prioritizing maternal and child health in independent South Sudan. *Maternal and Child Health Journal* 16(6), 1139–1142.

Ruiz-Postigo, J.A., Franco, J.R., Lado, M. and Simarro, P.P. (2012) Human African trypanosomiasis in South Sudan: how can we prevent a new epidemic? *Neglected Tropical Diseases* 6(5), 1–7.

Salman, S.M.A. (2011) The new state of South Sudan and the hydro-politics of the Nile Basin. *Water International* 36(2), 154–166.

Schomerus, M. and Allen, T. (2010) Southern Sudan at odds with itself: dynamics of conflict and predicaments of peace. Development Studies Institute, London School of Economics and Political Science, London.

Spencer, P.S., Vandemaele, K., Richer, M., Palmer, V.S., Chungong, S., *et al.* (2013) Nodding syndrome in Mundri County, South Sudan: environmental, nutritional, and infectious factors. *African Health Science* 13(2), 183–204.

Themnér, L. and Wallensteen, P. (2012) Armed conflicts, 1946–2011. *Journal of Peace Research* 49(4), 565–575.

UMCOR (2015) *Sudan Development*. United Methodist Committee on Relief, General Board of Global Ministries of the United Methodist Church. Available at: www.umcor.org/UMCOR/Programs/Country-Offices/South-Sudan (accessed 19 January 2017).

World Vision–South Sudan (2013) World Vision South Sudan helps increase food production. Available at: http://www.wvi.org/south-sudan/article/world-vision-south-sudan-helps-increase-food-production (accessed 19 January 2017).

5 Mozambique: Rebuilding Agricultural Extension in the Post-Conflict Period (1993–2015)

Hélder R. Gêmo*
Consultant, Ministry of Agriculture and Food Security, Maputo, Mozambique

Introduction

The economies of most sub-Saharan African countries are agriculture based with the majority of the population living in rural areas and depending on farming for their livelihood (World Bank, 2008). Under these circumstances, the outbreak of protracted armed conflicts, which often particularly affect rural areas, can result in the decline and stagnation of agriculture, including the provision of key agricultural services such as extension.

Mozambique celebrated its national independence in June 1975, following a 10-year war of liberation against the former colonial power, Portugal. However, in 1977 the country was back at war, this time with two rival factions—the government and the 'rebels'—pitted against each other in a struggle that lasted approximately 16 years and resulted in the loss of many lives and the widespread destruction of social and economic infrastructure. A combination of external and internal factors, related to regional politics as well as government policies, is thought to have fuelled the conflict (Baden, 1997). Governments in former Rhodesia (now Zimbabwe) and in former apartheid South Africa opposed the socialist policies embraced by Mozambique's post-independence government (Unruh *et al.*, 2003).

The war was fought mainly in rural areas, affected different areas to different degrees and reached its peak in the late 1980s (Baden, 1997). Foremost among its negative legacies was poverty. By the early 1990s, an estimated 74% of the population lived below the World Bank's poverty threshold (less than US$1 per day), and Mozambique was considered one of the world's poorest countries (Fauvet, 2000; Brück and van den Broeck, 2006).

The Peace Agreement was signed in October 1992. By that time most of the vital rural infrastructure (roads, bridges, energy and water supply) was either badly damaged or completely destroyed. The agricultural sector was virtually stagnant in the wake of the displacement of millions of people from the rural areas due to widespread insecurity. There was little provision of key public agricultural services and very limited investment by either the public or the private sector.

* Corresponding author e-mail: gemohelder@gmail.com

Farming can play a major role in countries recovering from armed conflict through the provision of food for the rest of the country and the pursuit of export opportunities (Unruh, 1995; Unruh *et al.*, 2003). During the post-war period, reviving agriculture was a top priority for the Government of Mozambique and high on the agenda of key stakeholders (MAP, 1995; Lindau, 1998; Hester and Harrison, 2005). Rebuilding extension was seen as one of the most critical tasks to revive agriculture. Extension was an immediate need for thousands of smallholders returning to their place of origin to resume farming activities (Gêmo *et al.*, 2005).

This chapter looks at how the government and key stakeholders performed in rebuilding extension. It outlines the main challenges faced, the strategies adopted and the key factors that contributed to the establishment of a pluralistic extension system, from the early post-conflict period (1993–1998) to the present (2016). The chapter also identifies some continuing challenges and highlights the issues that need to be addressed to further enhance agricultural extension.

Very few studies (e.g. Gêmo and Rivera, 2001; Gêmo *et al.*, 2005; Gêmo *et al.*, 2013) have discussed the rebuilding of Mozambique's pluralistic post-conflict extension system. This study provides information on what was, and was not, achieved; why; and how.

Mozambique's Agriculture and Extension System

Mozambique's total population in 2016 is estimated at 26.4 million, about 68% of whom live in the rural areas and rely directly or indirectly on agriculture (INE [National Institute of Statistics], 2012). Mozambique's agriculture can be disaggregated into the smallholder subsector, dominated by households cultivating relatively small plots of land principally for home consumption, and the large-scale commercial subsector (Benson *et al.*, 2014). Small and medium-sized farms are the most common, numbering 3.9 million, 99.3% of the total number of farms (INE, 2010).

The country consists of ten agro-ecological regions, each having several production systems. These regions indicate agricultural potential, which is determined mainly by the predominant soil types and the length of the growing period for rain-fed and irrigated agriculture.

The different agro-ecological regions, particularly the high- and moderate-potential regions, are suitable for a wide range of annual and perennial crops and also for livestock. The main food crops are cassava, sweet potato, maize, rice, sorghum, pearl millet and pulses. Maize and cassava are the two most commonly produced food crops. The main cash crops include cotton, tobacco and banana, and perennial crops such as cashew, coconut, citrus and mango. Livestock are another important component across the different agro-ecological regions, comprising mainly cattle, goats, pigs and poultry (MADER, 2002; MADER, 2003; MINAG, 2005; MINAG, 2007a; MINAG, 2008).

Public extension

Mozambique's extension system has been pluralistic since the 1990s, with government, private sector and mainly international non-governmental organization (NGO) extension providers. The public extension service was the first to be established, in 1987. This was followed by growing involvement of the private sector in the early 1990s, and of NGOs, particularly beginning in the 1993–1994 growing season.

Hosted at the then Ministry of Agriculture and implemented through the former National Directorate of Rural Extension (DNDR) from 1987 to 1992, the public extension service limited its activities to fewer than 30 districts out of the then 128 rural districts due to insecurity in many rural areas coupled with limited financial, human and other resources. The 30 districts covered were mainly those close to provincial capitals, which were considered to be relatively safe. The service went through two main expansion periods: from 1993 to 1998, when at least 21 new rural districts were included; and from 2005 to 2008, when a further 59 districts were added. In 2015 the public extension system, renamed the National Directorate of Agrarian Extension (DNEA) in 2006, covered geographic parts of 141 districts and 26 peri-urban areas (MASA, 2015). MASA (Ministério da Agricultura e Segurança Alimenta, the Ministry of Agriculture and Food Security) was established in January 2015 as a transformation of the previous Ministry of Agriculture (MINAG).

Since its establishment, public extension has used various extension models, including the Training and Visit (T&V) system, the farmer-to-farmer approach and Farmer Field Schools (FFSs). Funding since 1987 has been provided by bilateral and multilateral development partners and by the government. Since 2007 four areas have been recognized as critical—for institutional development of the public extension, in particular—within the scope of implementation of the National Extension Program (PRONEA): human capital development; planning, monitoring and evaluation (PM&E); adequate logistic support; and research and extension linkages (MINAG, 2007b; IFAD, 2012).

Private extension

Private sector extension activities started in the early 1990s following the privatization of former large state farms, particularly those engaged in cotton production. Privatization marked the shift from large-scale public investment in large farms based on capital-intensive mechanization to private investment through labor-intensive out-grower schemes involving thousands of smallholders. Private extension was based on 'cotton concessions', whereby the government authorized local and foreign investors to promote cotton production and marketing in agro-ecologically suitable areas, mainly in districts in the northern and central regions of the country (Gêmo *et al.*, 2005). Particularly since 1996, private extension has been extended through private sector investments in tobacco production and marketing, and also through out-grower schemes.

NGO-based extension

The role of NGOs in extension was limited until the Peace Agreement of 1992, when most of the international NGOs present in the country were involved in emergency activities. These activities included the free distribution of food, second-hand clothes, blankets and agricultural inputs (seeds and hand tools), targeted at the rural poor. Starting in the 1993–1994 growing season—the first following the Peace Agreement—many international NGOs decided to engage in development activities alongside their emergency activities, including the provision of extension. These NGOs assisted thousands of economically and socially debilitated smallholder households with agricultural advice and inputs, particularly from 1993 to 1998. From 2000 to 2010, most international NGOs remained widely present in rural areas throughout the country, although some restructured their interventions to focus on one or two provinces rather than having geographically scattered interventions (Gêmo *et al.*, 2005; Gêmo and Chilonda, 2013). In the past 5 years some international NGOs have downsized their interventions in extension in provinces such as Zambezia and Nampula, probably due to the limited funding received from bilateral development programs. However, international NGOs, together with the few national NGOs working in extension, are still important providers contributing to enhanced access to extension by smallholder farmers.

Extension coverage at the national level and some issues with the pluralistic system

Despite being remarkably pluralistic, extension coverage at the national level has decreased since the early 2000s according to surveys conducted by MASA in collaboration with the National Institute of Statistics, as shown in Fig. 5.1.

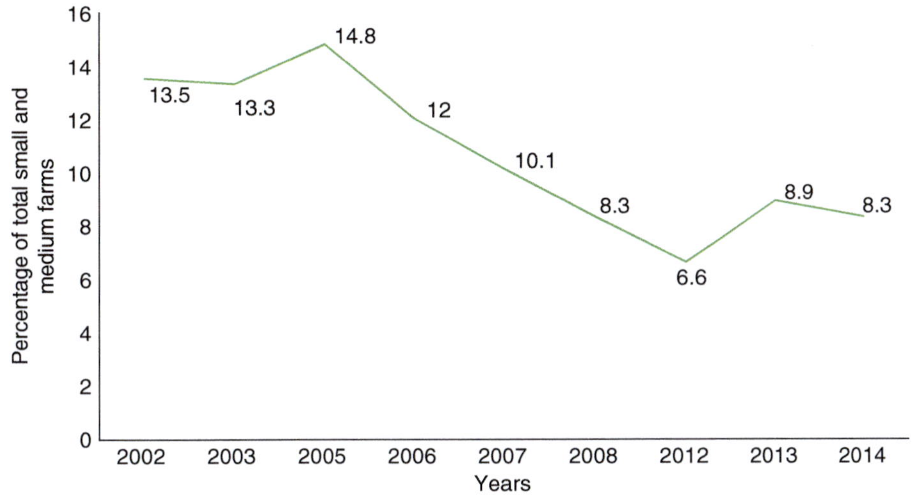

Fig. 5.1. Proportion of small and medium-sized farms with access to extension. From MASA (2002–2014).

The decline in extension coverage appears to be related to population growth, which stands at 2.8% per annum, as estimated in 2014 (World Bank, 2016). This has driven an increase in the number of new farmers and farms, while there has been little or no corresponding increase in extension coverage by all extension providers. The reasons behind the brief upward blip in 2012–2013 were not clearly identified during the writing of this chapter.

Extension as a whole (public, private and NGO) has been targeted mainly at smallholders. Private extension is commodity oriented, focusing mainly on non-food export crops (cotton and tobacco) and operating in areas of high or moderate agro-ecological potential. Public and NGO-based extension focuses primarily on food crops (cereals, roots and tubers, legumes and vegetables), selected livestock species (mainly chicken and goats, but also cattle) and agro-forestry activities (beekeeping). These activities span the full range of agro-ecological zones from high-potential areas to those with marginal agro-ecological potential.

Other important providers in Mozambique's national extension effort include farmers' organizations; the National Union of Peasant Farmers (União Nacional de Camponeses, UNAC); donor and development agencies which fund extension; agricultural colleges and university faculties; and input suppliers, although the latter are still very limited in terms of number, penetration into rural areas and volumes of inputs traded. Despite the considerable diversity of extension providers, there has been little informed public debate intended to enhance the role of extension and create a more enabling policy environment for agriculture.

The legacy of war and the rebuilding of extension

The war created an exceptionally challenging environment in which to rebuild extension, particularly during the immediate post-war years (1992–1998), for the following reasons:

- The rural population had suffered widescale displacement, the loss of family members and the near total disappearance of rural social networks. Mozambique's civil war created 1.5–1.7 million external refugees and 3.0–4.3 million internally displaced people out of a total population of 13.2 million (Schwartz, 2010, citing Aird *et al.*, 1997).
- The country's rural areas were split between 'rebel'-held areas and those held by the government. This meant that

government interventions, including public extension, were restricted to the areas they could reach. Attempts to deliver services in 'rebel'-held areas were likely to meet with mistrust at the very least and often with some outright hostility.

- Markets for agricultural inputs and outputs had virtually collapsed in rural areas. Key rural infrastructure (roads, bridges, warehouses, water supplies and energy) was either seriously damaged or completely destroyed (Hoeffler, 1999; IFAD, 2014). The poor condition of most rural roads was a special challenge to rebuilding extension because of the difficulties in accessing smallholders in many rural areas.
- Thousands of landmines in rural areas posed a threat to reviving agriculture and rebuilding extension at a time when many smallholders were returning and opening new farming plots in areas that had been the theater of war for many years.
- Only two colleges provided training at diploma level in agriculture and livestock production: the Agrarian Institute of Chimoio (IAC) and the Agrarian Institute of Boane (IAB). These schools trained only a few dozen diploma-level technicians per year during the war (Gêmo, 2004). The lack of qualified extension workers was a severe challenge at a time when more field staff were needed.

Strategies adopted in rebuilding extension

A number of strategies were used by the government to rebuild agricultural extension in the post-conflict years. While some strategies were implemented mainly by the government, most involved collaboration with other stakeholders in the agriculture sector. Box 5.1 summarizes these strategies, together with the period of their implementation.

More detailed discussion of these strategies follows.

1993–1998

EMERGENCY FOOD DISTRIBUTION FOLLOWING THE PEACE AGREEMENT. Free or highly subsidized food distribution was critical during the second half of the 1980s and until 1992, when agriculture was virtually stagnant. During this period almost 80% of the country's food requirements were supplied from abroad, mainly as international food aid (Barnes, 1998; Selvester and Castro, 2003). The main organizations involved in distribution were, on the government side, the former Department for the Prevention and Combat of Natural Disasters (DPCCN), which worked closely with the World Food Program (WFP) and the United Nations Children's Fund (UNICEF), together with a number of international NGOs and some religious organizations. The continuation of food distribution after the Peace Agreement, particularly until 1996, allowed a smooth transition from

Box 5.1. Strategies adopted to rebuild extension in the post-conflict period.

1993–1998 (the first six years after the 1992 Peace Agreement)

Continuation of substantial emergency food distribution, particularly during the first three years following the Peace Agreement
Enhancement of the emergency agricultural input distribution program among smallholder farmers across the country
Use of field extension workers with low levels of education

1993–1998 to present

Rehabilitation, maintenance and construction/expansion of rural infrastructure and basic social services
Mobilization of resources from donor and development agencies to support agriculture in general, and extension in particular
Pilot dissemination of new technologies
Design and implementation of medium-term investment programs for public extension

widespread hunger and complete dependency on aid in 1992 to a productive and largely self-reliant smallholder sector by 1998 (MAP, 1999). Millions of smallholders benefited from this transition.

ENHANCING THE EMERGENCY AGRICULTURE INPUT DISTRIBUTION PROGRAM. As noted by Brück (2001), considerable household assets, including agricultural inputs and tools, were lost during Mozambique's conflict. Restoring households' assets and tools was critical to increasing agricultural production and also in buffering households against shortfalls in income during the early post-conflict years. Against this background, the government, bilateral development partners and various international NGOs made a tremendous effort to distribute improved seeds and tools, including hoes, machetes, axes and sickles. Thousands of smallholders negatively affected by war and drought benefited, particularly from 1992–1993 until 1995–1996 (MAP, 1996). This intervention was made under the Emergency Program for Seeds and Hand Tools (PESU), a program implemented from the late 1980s onwards. The Swedish International Development Agency (SIDA) and later the Japanese International Cooperation Agency (JICA) were the main organizations involved in the distribution of hand tools, while the African Development Bank (AfDB) worked directly with the government/MINAG to boost the dissemination of inputs, particularly seeds. Improved seeds were provided free of charge for poorer people and highly subsidized (50–75%) for those who were somewhat better off. During the 1993–1994 growing season about 33,000 tonnes (t) of improved seed, mainly maize, rice, sorghum, groundnut and cowpea, were distributed to smallholders (MINAG, 1994). While substantial quantities of maize and rice seed were produced locally (100% of rice) by the national Seed Company (SEMOC, Sementes de Moçambique), some maize hybrids, sorghum, groundnuts and cowpea seeds were imported from the southern African region, mainly Zimbabwe (Seed CO) and South Africa (Pannar Seed) (MINAG, 1994).

By providing basic inputs to thousands of smallholders who were starting to interact with extension (public sector and NGO-based) at the time when input markets simply did not exist in most rural areas, PESU provided an opportunity to demonstrate to farmers how good agronomic practices (planting in rows, two seeds per hole, timely weeding, water harvesting, etc.), combined with improved varieties, could increase crop yields. The large-scale distribution of inputs through PESU ended in 1996, when the government decided that sufficient progress had been made to allow a transition to normal market mechanisms for the supply of inputs.

USE OF FIELD EXTENSION WORKERS WITH LOW LEVELS OF EDUCATION. Given the limited number of qualified field extension workers (FEWs) in the early 1990s, the government allowed field staff with low educational levels to do extension work. In the Mozambique context, qualified FEWs are mainly diploma holders with 3 years of training in agriculture, livestock or forestry following 10 years of primary and secondary education. A limited number of BSc holders, with 4 years of post-secondary education, are also employed in extension. FEWs possessing lower educational levels were also common, including a mix of those with a basic education level (7 years primary and pre-secondary schooling plus 3 years in agricultural training) and some FEWs with only elementary level education (5 years primary schooling plus 2 years in agricultural training).

The government's strategy of suspending the required qualification levels for hiring extension staff during the early post-conflict period (1993–1998) allowed extension providers to hire many technicians despite low educational levels. The need to improve the educational level of FEWs began to receive increasing emphasis from 1999 onwards, as the first Extension Master Plan (EMP) was implemented (Gêmo, 2004; Gêmo *et al.*, 2005).

1993–present

PRIORITIZING REHABILITATION AND MAINTENANCE OF THE RURAL INFRASTRUCTURE. As noted by Brück (2006), post-war reconstruction policies should support the re-establishment of markets and the expansion of household production to meet demand, as these are unlikely to occur spontaneously or with any rapidity when households are weak and their resources depleted by war. In Mozambique, road rehabilitation, vital for linking smallholders to markets, was a major post-conflict priority. The country benefited from a multi-donor Roads and Coastal Shipping (ROCS) project, which rehabilitated 3800 km of roads and improved road maintenance (OECD and AfDB, 2002). In rural areas the government and key stakeholders made efforts to ensure annual rehabilitation of selected unpaved rural roads at district level, some of them linking different districts. The ROCS project was instrumental in enabling the expansion of extension as well as the restoration of market linkages. However, rural road rehabilitation and maintenance were limited in technical scope and restricted in geographic focus compared with the overall needs across the country. Even rural roads that are annually maintained are still often impassable during the rainy season from October to April.

GOVERNMENT SUPPORT IN MOBILIZING RESOURCES TO REBUILD EXTENSION. The government was able to mobilize resources to rebuild extension, particularly from international NGOs and development partners who provided funding and political support to the reconstruction effort. This was particularly critical from 1993 to 1998, when there were huge tasks and resource requirements for rebuilding extension as part of the national reconstruction effort. Various funding mechanisms were used by the government and development partners to channel resources, as shown in Box 5.2.

As noted above, the private sector has been an important extension player since the early 1990s, when commodity-oriented extension began to expand through out-grower schemes focusing at first on cotton and later (1996) on tobacco. These schemes involved thousands of smallholders. Local private investment (mainly for cotton subcontracted production) and direct foreign investment (mainly for tobacco, such as Mozambique Leaf Tobacco Co. Lda) are the two sources of funding for private sector extension.

In summary, while the government was able to secure the resources needed from development partners to rebuild and restart public extension services, political support and external funding from international NGOs and private sector companies were also needed to accelerate the expansion of extension services. In turn, the government provided political support to private extension to build the early relationships between private extension providers and smallholder farmers involved in the out-grower schemes. For instance, the government played a significant role in negotiations and in reaching agreement on the annual market prices of different grades of cotton. The government also joined forces with a few large private extension providers in joint venture companies for cotton production and marketing (Gêmo *et al.*, 2005).

Piloting and Disseminating Improved Technologies

In the second half of the 1990s, MINAG made notable efforts to pilot the use of vaccines against Newcastle disease in chickens kept by smallholders in rural areas. Most rural households rear chickens (MADER, 2002; MADER, 2003; MINAG, 2005; MINAG, 2007a; MINAG, 2008), mainly with low use of improved inputs. However, Newcastle disease has seriously constrained production and its control is seen as crucial for poverty alleviation at the household level (Perttula, 2009; Tomo, 2009). Analysis and modeling have indicated that control of Newcastle disease alone has the potential to increase household income from chickens by 42% (Woolcock *et al.*, 2004).

The piloting of vaccinations against the disease in selected areas of Mozambique was important because it added 'hard content'—or a physical technology—to extension, rather than just advice. In general, this has been well accepted and has contributed to the relevance of extension in the eyes of

Box 5.2. Funding mechanisms used by the government and development programs to rebuild extension.

1993–1998

Public extension funding through large projects

The World Bank provided resources for public extension through two large projects: the Agricultural Services Rehabilitation and Development Project (ASRDP) for Nampula and Cabo-Delgado provinces in the north of the country (1992) and the Agricultural Rehabilitation and Development Project (ARDP) (1990) for Inhambane and Gaza provinces in the south (see Lindau, 1998).

The two projects were vital in contributing to the expansion of public extension to at least 21 new districts between 1993 and 1998, with adequate logistical support (Gêmo *et al.*, 2005). This included human capital development (contracting new qualified extension workers and providing in-service staff training) and purchasing means of transport (cars, motorcycles and bicycles) and equipment for fieldwork. Such logistical support is crucial in the retention of qualified field staff when salaries are low, as they traditionally have been in most extension systems in the developing world.

Funding from bilateral development partners for NGOs

The European Union (EU), the United States Agency for International Development (USAID) and, to a lesser extent, the UK's Department for International Development (DFID) were the main donor agencies involved in funding the activities of international NGOs. ACDI/VOCA, ActionAid, Care International, the Cooperative League of the USA (CLUSA) and World Vision were the main NGOs involved. They contributed to strengthening extension by employing hundreds of agriculture/livestock technicians holding diplomas and even some holding a BSc in agriculture.

1999–2016

Public extension funding through the National Agricultural Development Program (PROAGRI) (1999–2011)

This joint initiative by the government and development partners aimed to support the annual budget of the Ministry of Agriculture and Fisheries, implemented from 1999 to 2006 (Phase I) and from 2007 to 2011 (Phase II). Extension was one of eight PROAGRI components, together with crop production, irrigation, livestock, forestry, research, land management and capacity development.

Under PROAGRI, public extension became underfunded (MINAG, 2007c; Gêmo and Chilonda, 2013), resulting in limited logistical support for field operations, in-service training and monitoring and evaluation (M&E) at central, provincial and district levels. The effectiveness of public extension was compromised (MINAG and IFAD, 2010; Gêmo and Chilonda, 2013). The funding shortfall is thought to have occurred because extension was simply not seen as a priority by PROAGRI decision makers.

Public extension funding through the government and the International Fund for Agricultural Development (IFAD) and the EU (2012 to the present)

Since 2012 the government, IFAD and the EU have been funding public extension in 42 districts through the Pronea Support Project (PSP), which aims to strengthen implementation of the National Agrarian Extension Program (PRONEA), launched in 2007 and due to run until 2016. However, the timely availability of the government's contribution (17% in most of the costs incurred by IFAD), as well as delays in the disbursements of the approved PSP annual budget, have constrained project implementation across the 42 districts (DNEA, 2014). In order to identify feasible solutions, these issues were recently addressed by the government and IFAD during the mid-term review (MTR) of the DNEA/PSP held in the second half of 2015.

Public extension funding through the government and several bilateral development partners (2012 to the present)

In parts of the other 99 rural districts and the 26 peri-urban areas where public extension was operating in 2015 (MASA, 2015), the government and a few bilateral donors have been the main sources of funding from 2012 to the present. However, delays in releasing funds and erratic disbursements (of amounts below the approved budget) have negatively affected implementation in these areas (MINAG, 2014).

Continued

Box 5.2. Continued.

Funding from several bilateral development partners for NGOs (1999 to the present)

Some bilateral donors continue to provide funds to international NGOs and to some national NGOs. This funding mechanism has been crucial in keeping the pluralistic extension system alive. Funding for NGO-based extension efforts has declined over the past 5–10 years. International NGOs that had large-scale programs in the 1990s, such as Care International and World Vision (Gêmo *et al.*, 2005), have to some extent downsized their extension activities in the most densely populated provinces of the country, including Nampula and Zambezia provinces.

smallholders. Vaccination against Newcastle disease is currently the major livestock extension activity under the public Unified Extension System (SUE) adopted in 1998–1999. However, limited quantities and delayed distribution of vaccines at the district level by the veterinary/livestock services of MASA, together with other logistical issues, have prevented a major success for extension in this area (MINAG, 2014).

Another initiative was the piloting of technology packages aimed at intensifying the production of food crops, mainly maize and rice, by smallholders. This initiative was implemented by public extension in collaboration with Sasakawa Global 2000 (SG 2000), an international NGO active in Mozambique from 1995 to 2004. SG 2000 promotes intensification of smallholder production in a number of African countries, often through partnerships with public extension (Fertilizer Toolkit, n.d.). Launched during the 1995–1996 growing season, the piloting of technology packages was particularly strong from the 1998–1999 to the 2002–2003 growing season. The packages consisted of free on-farm demonstrations of herbicides, mainly to support the introduction of no-till practices; inorganic fertilizers (mainly NPK: 12-24-12 and urea 46%), applied using blanket/uniform application levels for different locations; and improved seeds (mainly open-pollinated varieties of rice and some medium- and long-season maize hybrids). By 2002–2003, selected areas in 51 districts had been covered by this initiative. About 1110–1200 on-farm demonstrations (with sizes varying from 100 m^2 to 5000 m^2) were conducted per year, particularly from 1998 to 2003. Participating farmers were able to obtain yields varying from 2.5 to 6.0 t/ha of maize and 2.5 to 5.0 t/ha of rice, depending on how accurately the packages were applied. Average rain-fed maize and rice yields without the use of inorganic fertilizer in high- to moderate-potential agro-ecological zones in Mozambique are estimated at approximately 1.0 t/ha for maize (MacauHub, 2013) and for rice (FAO, 2014). The majority of extension workers involved in the initiative also received free inputs during two successive cropping seasons in order to farm 0.5 ha 'model demonstrations' on their own plots and for home consumption. The main objectives of extending access to inputs by the field staff was to expand demo plots and also to promote major technical ability among extension workers in using the technological packages then under dissemination.

The technology packages initiative was important not just in raising yields, but also as a means of boosting smallholders' access to, and appetite for, public extension. The initiative was discontinued after 2004, when SG 2000 funding ended, and the public extension service was not financially prepared to continue with the field demonstrations and the related training needed for both FEWs and farmers in the newly covered districts.

Designing and Implementing Public Extension Investment Programs

The then National Directorate of Rural Extension (Direcção Nacional de Extensão Rural, DNER) implemented the first EMP from 1999 to 2006. It identified key institutional milestones, main activities, the stakeholders that should be involved within and outside the then MINAG and the estimated total budget to be spent within the 5-year time frame (1999–2004, subsequently extended to

2006). In 2007, a second EMP was approved and began to be implemented, with a time frame of 10 years (2007–2016). In summary, the two EMPs can be characterized as shown in Table 5.1.

The two EMPs were important in rebuilding extension because they: (i) provided a more strategic means of pursuing the development of public extension by specifying the target beneficiaries (smallholders), the implementation pillars, the intended geographic coverage and, since 2005, projections of the total number of farmers to be covered in each of the 5-year periods; and (ii) were a means of mobilizing medium-term funding from the government and development partners. For instance, the first EMP was budgeted at US$24 million over 5 years and by 2006 it had been funded at about US$20 million in total under PROAGRI (MINAG, 2007c). The second EMP was implemented through PRONEA, whose initial budget was estimated at US$50 million over 10 years (2007–2016). IFAD allocated US$20 million to DNEA, while other development partners and the government were expected to contribute the balance of US$30 million.

Put on hold in 2010, pending redesign due to poor performance (MINAG and IFAD, 2010; Gêmo and Chilonda, 2013), IFAD financial support to PRONEA resumed in late 2012, with a budget of US$20 million and at least US$1.08 million from the EU, in addition to the governmental contribution consisting of 17% of PRONEA expenses (a few category costs such as daily allowances and staff travel do not require a government contribution). These funding developments show how the design and implementation of the medium-term EMPs have been important in mobilizing resources.

However, a weakness in implementing the EMPs has been the poor definition of medium-term indicators or intermediate outcomes. In livestock, for instance, the annual targets for vaccination against Newcastle disease have only recently (in the past 2–3 years) begun to be defined by the National Directorate of Veterinary Services (DNSV) at MASA. This is a critical intervention area for DNEA that deserves proper planning in terms of setting medium-term targets and identifying the required logistics—planning that should have been conducted jointly by DNSV, DNEA and MASA research services. The latter are responsible for producing one of the most widely used vaccines (I-2). I-2 was produced by Mozambique's Agrarian Research Institute (IIAM, from Instituto de Investigação Agrária de Moçambique), Animal Science Directorate (DSA, Direcção de Ciências Animais). The ACIAR project AS1/1995/040 developed seed cultures of the I-2 vaccine suitable for vaccine production in developing countries. The I-2 master seed culture is kept at the University of Queensland and is provided to developing countries free of charge (Australian Center for International Agricultural Research (see ACIAR, 2014).

Learning from the Rebuilding Process

This section highlights the key issues encountered in rebuilding extension since the 1992 Peace Agreement. Four issues are outlined:

1. Key supporting factors that contributed to rebuilding extension.
2. Challenges in rebuilding a properly coordinated national extension system (Sistema Nacional de Extensão, SISNE).
3. Challenges in linking research and extension.
4. Institutional concerns related to public extension.

Key factors that contributed to the post-conflict rebuild

Various political, social and economic factors were instrumental in contributing to the post-conflict rebuilding of extension. The political stability achieved by creating democratic institutions has been critical. This is epitomized by the multi-party National Parliament, which sat for the first time in 1995. Despite disputes between the two

Table 5.1. Summary characteristics of the first and second Mozambique Extension Master Plans (EMPs).

The two EMPs	Implementation pillars				
	Decentralization	Human capital development	PM&E	Implementing approaches	Contribution to pluralistic extension
First EMP (1999 to 2004–2006)	Decentralization of extension management to provinces	Stated targets in improving qualifications of FEWs	Need to strengthen PM&E within scope of SUE under decentralized management of public extension	• Adoption of SUE • Openness to piloting/adopting new extension models • Predominantly supply-driven extension • Outsourcing of pilot initiatives	Need to be a leading institution (DNEA) in building the national extension system
Second EMP (2007–2016)	Enhancement of decentralization to district level	Stated emphasis on continuing human capital development but without quantified targets	Need to strengthen PM&E under SUE, decentralized management of public extension and expected scaling up of extension outsourcing at provincial and district levels	• Strengthening SUE • Openness to piloting/adopting new extension models and intended scaling up of FFS model • Emphasis on need to develop demand-driven extension • Need to expand outsourcing initiatives	Need to strengthen national extension system, including through outsourcing initiatives and empowerment of farmer organizations and other community-based organizations

FEW, field extension worker; FFS, farmer field school; DNEA, National Directorate of Agrarian Extension; PM&E, Planning, monitoring and evaluation; SUE, Unified Extension System.

main parties, which recently spilled over into localized armed confrontations in 2013–2014, the country has enjoyed a largely peaceful coexistence among its political parties since the 1992 Peace Agreement.

The government's support for reviving agriculture and rebuilding a pluralistic extension system immediately after the 1992 Peace Agreement was of paramount importance in keeping the peace and restoring prosperity. The financial support provided by development partners to agriculture, and particularly to the rebuilding of extension, was and still is crucial. Finally, after 16 years of devastating war, thousands of vulnerable smallholders were brave enough to return to their destroyed villages and farms, risking their own safety to clear land riddled with landmines and to resume agriculture, especially between 1993 and 1996. Since then the smallholders have interacted well with extension wherever it operates across the country, and they increasingly demand the inputs and services they need to improve yields and raise family farm incomes. These factors contribute to the country's economic and social recovery. Figure 5.2 summarizes the supporting factors that were critical to the post-conflict rebuilding of agricultural extension.

Even in the tragic aftermath of the war, Mozambique found ways to facilitate peaceful recovery by making intelligent use of the contributions of both local and foreign partners. The country was able to rebuild its extension system, creating a pluralistic model that became a reality in less than 10 years (Gêmo and Rivera, 2001; Gêmo *et al.*, 2005).

Challenges in rebuilding the national extension system

The first and second EMPs both refer to the need to develop a truly national extension system with coordination among all service providers (a public, private and NGO-based SISNE) (MAP, 1998; MINAG, 2007b). This includes the exchange of information on planned interventions at the local level to

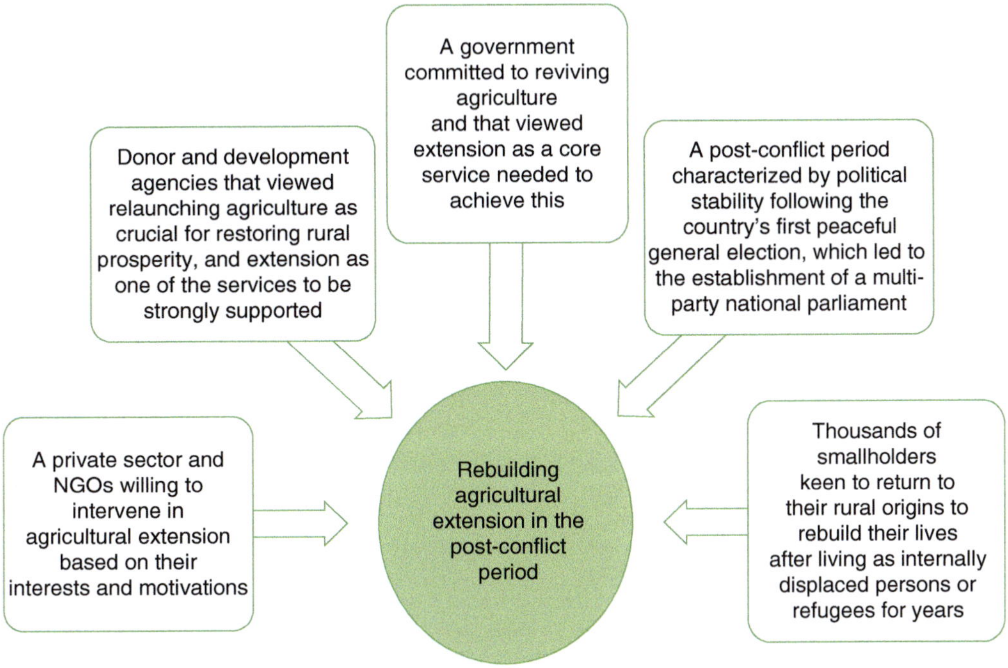

Fig. 5.2. Five supporting factors in rebuilding extension in the post-conflict period.

avoid conflicting or overlapping activities; collaboration in the field and in other core activities, such as in-service training for staff and field days for demonstrating technologies; and the sharing of experiences and lessons in delivering extension, such as environmental concerns, social and gender equity issues, difficulties in achieving quality standards and the effectiveness of different extension models, among other matters.

However, ensuring an effective SISNE has been challenging (MAP, 1998; Gêmo *et al.*, 2005; Gêmo and Chilonda, 2013). SISNE has been inadequately achieved because many NGOs are mainly accountable to their donors rather than to local authorities, partners and beneficiaries; private extension interacts more with related MASA institutions (e.g. Mozambique's Cotton Institute rather than the DNEA); and the DNEA itself has made little effort to act as the lead institution in generating and sharing evidence-based knowledge that can make extension more effective and efficient. The DNEA has also been weak in promoting policy debate on extension. For instance, the DNEA has only occasionally discussed how to involve private extension in SISNE to increase food production, particularly with smallholders participating in cotton and tobacco outgrower schemes. Very little progress, if any, has been made on this issue at MASA.

In conclusion, developing SISNE requires holistic vision and dynamic implementation, led by the DNEA and MASA but involving all the different extension actors and other relevant stakeholders in the pursuit of shared objectives.

Challenges in linking research and extension

Research and extension linkages have been emphasized as crucial in Mozambique, as in nearly all developing countries (MAP, 1998; MINAG, 2007b; MINAG, 2010). In Mozambique, research and extension linkages refer to collaboration between the institutions (agricultural universities and research institutes) and private sector companies that generate knowledge, and the public and NGO-based extension service providers. Agricultural research is almost all public with the exception of a few cases of adaptive research for cash crops (sugar cane and tobacco), carried out by large commercial companies.

Effective research and extension linkages in Mozambique means three things: (i) a substantial number of functioning linkage mechanisms and activities at the district, provincial and central/national levels and across different agricultural subsectors; (ii) linkages that generate expected outputs and outcomes among target farmers and farming systems; and (iii) successful cases of linkages that can be replicated across the country, where applicable.

Since the 1990s, various mechanisms and activities have been implemented at district, provincial and central levels aimed at improving research and extension linkages. But despite progress in a few areas, effective linkages are still far from being accomplished (Gêmo, 2007; Gêmo, 2013). Evidence of strong research and extension linkages is scanty. This suggests that, with a few exceptions (Low *et al.*, 2000; HarvestPlus, 2012; Low *et al.*, 2013), linkages between research services and extension have been weak.

Achieving effective public sector linkages has been challenging for various reasons:

1. There has been weak demand for them from the government/MASA and from key stakeholders (farmers' organizations, the private sector and development partners). Consequently, demand for information on the status and performance of linkages has also been weak to non-existent. Debate on research and extension linkages has been sporadic and has very seldom led to action.

2. Resources allocated to public sector research and extension, specifically to the Agricultural Research Institute of Mozambique (Instituto de Investigação Agrária de Moçambique, IIAM) and the DNEA have been very limited, even for their respective core business activities and critical investments (Flaherty *et al.*, 2010; Gêmo and Chilonda, 2013), let alone for outreach or linkage activities such as joint field activities with extension workers or farmers.

3. As most linkage mechanisms and activities have been funded by development partners, their duration is often limited to

2 or 3 years while project funding lasts, with limited likelihood of bringing about permanent change.

4. Little has been done by either research or extension to analyze and document the few linkage mechanisms and activities that have been implemented, as a source of learning on how to create more effective linkages.

In summary, despite their role in contributing to research and extension effectiveness, linkage problems have not been properly addressed, and this is an issue on which MASA will need to perform better in future.

Institutional concerns related to public extension

Three issues are discussed below, specifically geographic expansion and its implications; PM&E; and the implementation of different extension models.

Geographic expansion and institutional preparedness

Public sector extension went through two important phases of expansion, the first from 1993 to 1998 (with at least 21 new districts covered) and the second from 2005 to 2008 (from 69 districts covered in 2005 to 128 in 2008). Comparing the first and second phases reveals a number of issues, as summarized in Table 5.2.

In summary, during and after the 2005–2008 accelerated expansion, public extension was substantially understaffed, underequipped and undertrained. For instance, eight extension workers per district was the minimum number targeted by public extension since its establishment in 1987 (Gêmo *et al.*, 2005; Gêmo and Chilonda, 2013). However, until 2012 two to four extension workers was the norm in at least 40 districts, well below the critical number needed per district. Under these circumstances it is unlikely that public extension could be effective in providing services to farmers. However, since 2010, when there were 770 extension workers, the number of DNEA extension workers has risen to 1361 (in June 2015). Despite this increase, more effort is needed in hiring new staff.

In conclusion, failure to address the main factors critical to institutional preparedness seriously constrained the DNEA's capacity to achieve the second geographic expansion with the desired level of effectiveness.

Planning, monitoring and evaluation (PM&E)

PM&E in public extension has been decentralized since the late 1990s in line with the requirements of the first EMP, which had decentralization as a key milestone (MAP, 1998). Since that time the districts and provinces have been primarily responsible for ensuring a sound bottom-up approach to PM&E. The organizations responsible have been the District Services of Economic Activities (SDAEs), in which agriculture/extension services have been hosted since 2006, and the Provincial Services of Rural Extension (SPERs), which are hosted at the Provincial Directorates of Agriculture and Food Security (DPASAs). However, developing effective PM&E in public extension is still a challenge for the reasons outlined in Table 5.3.

Despite the weaknesses described in Table 5.3, PM&E staff have been able to prepare the annual plans that have been the basis for budget allocations for all agriculture services within MASA, including extension. They have also ensured regular reporting consisting of quarterly, half-yearly and annual progress reports at district, provincial and national levels. Reports mainly focus on output indicators and less on input and outcome indicators, a tendency that masks the persistent underinvestment in PM&E. Major institutional efforts will be needed to develop the DNEA's PM&E capacity in the future.

Assessing the implementation of extension models

When public extension began in 1987, the T&V model developed and widely promoted by the World Bank was adopted. Until 1992, the model consisted of eight extension workers per district involved in

Table 5.2. Comparing variables between the two expansion phases of public extension. From Gêmo *et al.* (2005); MINAG and IFAD (2010); Gêmo and Chilonda (2013).

Variable	First expansion (1993–1998)	Second (accelerated) expansion (2005–2008)
Main objectives	To expand extension assistance to reach thousands of people returning to their rural areas after years of internal displacement or living as refugees in neighboring countries	To ensure 'extension for all' by expanding services to all rural districts, in response to the political directive of the then MINAG leadership
Staff recruitment	Emphasis on hiring new staff members to be allocated mainly to the new districts covered. The World Bank provided funding for new staff until they could be placed on the government payroll	Little emphasis on hiring a critical mass of new extension workers to be allocated to new districts, causing a huge deficit of FEWs until 2012. The need to hire more staff was not addressed before MINAG's political directive to accelerate the expansion
Human resources development	Availability of financial resources for in-service training of field extension staff but also supervisors, who were in turn regularly supervised at field level by provincial- and national-level M&E officers	Frequent lack of resources for in-service training at the provincial level. Most districts did not have extension supervisors as many were reassigned to new districts and some appointed to new activities within and outside the agriculture sector at the district level
Operational logistics	Availability of adequate logistical support for transport, field equipment and inputs for demonstrations, and housing in some cases	Serious limitations in logistical support, particularly means of transport and inputs for field demonstrations, with negative effects on field activities
Effectiveness of service provision	No studies were conducted on this issue. Despite the low educational levels of extension workers at that time, there was adequate logistical, technical and methodological support from district supervisors and extension officers at provincial and central levels. Public extension was effective in disseminating agronomic practices such as row planting, timely weeding, improved granaries and seed germination testing, and in introducing new varieties of food crops	The first phase of PRONEA was implemented from 2007 to 2010. The program was put on hold following a joint decision by MINAG and IFAD (the main development partner funding PRONEA). PRONEA's temporary suspension was caused by poor performance, indicating how weak service provision was during the 2005–2008 period and even in subsequent years. In 2015 the DNEA was involved in a comprehensive mid-term review of the PRONEA Support Project (PSP), which was implemented in 42 districts from June 2012 (IFAD, 2012). With a budget of at least US$21 million (2012 to 2017–2018), PSP aims to strengthen the implementation of PRONEA

DNEA, National Directorate of Agrarian Extension; FEW, field extension worker; M&E, monitoring and evaluation; MINAG, Ministry of Agriculture; PRONEA, National Extension Program.

the dissemination of improved technology and practices through contact farmers selected from among the more innovative farmers in different rural communities. In 1992 the T&V model was modified to focus on farmers' groups rather than on individual contact farmers, and to allow a more flexible agenda determined according to local conditions rather than the uniform extension agenda prescribed across the country. Over the 2000–2010 period a few provinces such as Manica and Sofala in the central region of the country also introduced the farmer-to-farmer model alongside T&V.

From 1998 to 2003, public extension and the international NGO Sasakawa Global also implemented a widespread technology-transfer model in at least 51 districts, seeking to promote the intensification of maize and rice production among smallholder farmers.

Table 5.3. Main weaknesses and challenges in improving planning, monitoring and evaluation (PM&E) in public extension.

Planning	Monitoring and evaluation
External	
Weak farmers' organizations at the national level that could demand and lobby for greater farmer participation in PM&E	Limited, weak and inconsistent debate on extension performance in general and M&E in particular by government and key stakeholders (farmers' organizations, input suppliers, donor agencies)
Limited critical mass of smallholder farmers' organizations across the rural areas, which could lead to an increasing lack of pressure in public extension for more demand-driven planning	Lack of evidence-based decision making in public agriculture services, especially with regard to strategic prioritization and investment
Annual plans made and approved 6 months before implementation, often under unknown financial ceilings and often subject to subsequent budget reduction and delayed allocation, depending on the speed of budget disbursements by relevant government bodies. This is a key issue as financial resources for field activities are often unavailable during the critical period of the agricultural season	
Internal	
Limited qualified and experienced staff working in planning at district, provincial and central levels. In mid-2014 newly recruited staff (two agronomists with BSc degrees) strengthened central DNEA planning and M&E. Before that the unit relied on just one experienced agronomist with a BSc and two other professionals trained in social sciences at the BSc level. This very small team has been responsible for overseeing all PM&E at the national level	Limited qualified staff allocated to M&E, particularly at the provincial level. Over time, only one and occasionally two M&E officers have been available at the provincial level, whereas at the central level there have been three or four people (also involved in planning)
Limited in-service training on planning over time	Limited investment in M&E in-service training at district, provincial and central levels. In 2014 the DNEA engaged some technical staff to help modernize the M&E system. The aim was to improve data recording and digital inputing at district level and to build a comprehensive database at the central DNEA. This was the only major investment in M&E over the past 10 years
Limited efforts to review and assess the planning process and difficulty in securing evidence-based action at district, provincial and central levels. DNEA promotes annual planning meetings for information sharing among provinces and with central government, but documented outcomes from these meetings are not available at MASA/DNEA	Limited review of M&E procedures and tools used for data recording, compilation and verification at district, provincial and central levels

DNEA, National Directorate of Agrarian Extension; MASA, Ministry of Agriculture and Food Security.

Additionally, between 2000 and 2010, public extension introduced the FFS model in collaboration with FAO. Since 2013 there have been efforts to scale up the FFS model as part of PRONEA implementation.

Little effort has been made by the DNEA to analyze and compare the performance of the different models as a source of learning. For instance, although FFS were piloted as long ago as 2000 (with modest expansion since then), little is known about their effectiveness and outcomes, the challenges faced in implementing this model and the lessons that can be drawn for the model's future adaptation, with the exception of a handful of studies (Djeddah *et al.*, 2006; Braun and Duveskog, 2008; Dzeco *et al.*, 2010). There is nothing wrong with importing different extension models, especially those successfully implemented in other African countries. However, a major effort is needed to assess their effectiveness and learn from experiences with them to improve the efficiency and effectiveness of public extension in Mozambique.

Conclusions and Recommendations

The rebuilding of extension in post-conflict Mozambique has been a multi-stakeholder process involving both domestic and international organizations. Domestic stakeholders include government institutions, in particular MASA, which now hosts the public extension program; farmers' organizations and a few farmers' unions; and input suppliers and output buyers. International organizations include donors and development partners, which have funded extension; international NGOs; and private sector companies involved in commodity production through out-grower schemes.

The first 6 years (1993–1998) after the 1992 Peace Agreement were of paramount importance in rebuilding extension and saw steady progress towards the current, relatively stable, pluralistic extension system. However, while public extension has been substantially funded by development partners and by the government, NGOs are funded solely by development partners. The extent to which these international NGOs will continue to access financial resources from development partners is an important issue for the future of the pluralistic system. Promoting the emergence of competent national NGOs is thus a challenge that should be considered by the country's bilateral development partners. This challenge seems not to have been addressed, in part because international NGOs are better at accessing foreign resources for extension than are the less well-resourced national NGOs.

Finally, more efforts are needed from different extension providers and other key stakeholders to create a better coordinated and more collaborative national extension system to replace the current rather fragmented model. This will include relevant information sharing, collaborative fieldwork (particularly in technology transfer), capacity building, farmer training and innovative partnerships to expand access to extension by farmers, particularly smallholders in many rural areas. Extension coverage at the national level in 2012 was estimated at 6.6% of the total number of small and medium-sized farms, which were estimated at 3.9 million in 2010. Expanding extension coverage while retaining its effectiveness is still a critical challenge for Mozambique's agricultural sector.

The DNEA has been a major provider of public extension services. Its presence in 129 rural districts and 23 peri-urban areas in 2014 illustrates the importance of its contribution to the pluralistic extension system. However, there are some critical issues that must be better addressed if the DNEA's operations and performance are to be enhanced. The second geographic expansion of DNEA's operations, from 2005 to 2008, when a total of 59 new districts were covered, was poorly executed as the DNEA's institutional preparedness was weak. As a result, the expanded service was underfunded and understaffed, and personnel were undertrained. Although the DNEA has since hired more extension workers (who nearly doubled in number from 770 in 2010 to 1360 in June 2015), there is a need for the DNEA to assess its current institutional capacity and increase its staffing levels still further, particularly in areas where extension coverage is currently poor.

Human capital development, PM&E, analysis and comparison of different extension models, research and extension linkages are areas of intervention considered to be vital for the future effectiveness of public extension. They should feature in the priorities for periodic performance assessments and studies at MASA/DNEA. This is particularly important because the DNEA has operated in selected areas of almost every Mozambican rural district, across different agro-ecological zones and in varying farming systems since 2010. Extension therefore serves smallholders who are socially, economically and culturally heterogeneous and face different production and marketing challenges and opportunities.

The DNEA should also think about preparing an evidence-based medium-term (5–10-year) institutional development plan. As the second EMP is due to end in 2016, now is the opportunity to think about a comprehensive institutional development plan as part of the design of the third EMP.

References

ACIAR (2014) Newcastle disease control in Africa. ACIAR impact assessment series 87. Available at: http://aciar.gov.au/files/ias_87-web.pdf (accessed 19 January 2017).

Baden, S. (1997) Post-conflict Mozambique: women's special situation, population issues and gender perspectives: to be integrated into skills training and employment promotion. Institute of Development Studies, Brighton, UK. Available at: www.bridge.ids.ac.uk/sites/bridge.ids.ac.uk/files/reports/re44c.pdf (accessed 19 January 2017).

Barnes, S. (1998) Humanitarian aid coordination during war and peace in Mozambique: 1985–1995. Studies on emergencies and disaster relief, Report no. 7. Nordiska Afrikainstitutet (Nordic Africa Institute) and Swedish International Development Cooperation Agency, Uppsala, Sweden. Available at: http://www.diva-portal.org/smash/get/diva2:272739/FULLTEXT01.pdf (accessed 19 January 2017).

Benson, T., Mogues, T. and Woldeyohanne, S. (2014) Assessing progress made toward shared agricultural transformation objectives in Mozambique. IFPRI Discussion Paper 01370. International Food Policy Research Institute, Washington, DC.

Braun, A. and Duveskog, D. (2008) The farmer field school approach—history, global assessment and success stories. IFAD Background Paper. International Fund for Agricultural Development, Rome. Available at: http://web.ifad.org/rural/rpr2008/chapter3/3.pdf (accessed 19 January 2017).

Brück, T. (2001) Determinants of rural poverty in post-war Mozambique: evidence from a household survey and policy implications for government and donor policy. QEH Working Paper 67. Queen Elizabeth House, University of Oxford. Available at: www3.qeh.ox.ac.uk/pdf/qehwp/qehwps67.pdf (accessed 19 January 2017).

Brück, T. (2006) War and reconstruction in northern Mozambique. *Economics of Peace and Security Journal* 1(1), 30–39.

Brück, T. and van den Broeck, K. (2006) Growth, employment and poverty in Mozambique. ILO Discussion Paper 21. International Labour Organization, Geneva, Switzerland. Available at: www.ilo.org/wcmsp5/groups/public/---ed_emp/documents/publication/wcms_120678.pdf (accessed 19 January 2017).

Djeddah, C., Mavanga, R. and Hendrickx, L. (2006) Junior farmer field and life schools: experience from Mozambique. In: Gillespie, S. (ed.) AIDS, poverty, and hunger: challenges and responses. International Food Policy Research Institute, Washington, DC, pp. 325–339. Available at: http://citeseerx.ist.psu.edu/viewdoc/download?doi=10.1.1.129.4868&rep=rep1&type=pdf (accessed 19 January 2017).

DNEA (Direcção Nacional de Extensão Agrária) (2014) Síntise do encontro entre Investigação e Extensão. Projecto de Apoio ao PRONEA (PSP). Bilene, Província de Gaza, Mozambique. Unpublished.

Dzeco, C., Amilai, C. and Cristóvão, A. (2010) Farm field schools and farmers' empowerment in Mozambique: a pilot study. Paper presented at the 9th European IFSA Symposium, 4–7 July 2010, Vienna, Austria. Available at: http://ifsa.boku.ac.at/cms/fileadmin/Proceeding2010/2010_WS1.1_Dzeco.pdf (accessed 19 January 2017).

FAO (2014) Análise de incentivos e desincentivos para o arroz em Moçambique. Série de notas técnicas por Loganemio, Dolito Lucas, Rome. Available at: http://www.fao.org/fileadmin/templates/mafap/documents/technical_notes/MOZAMBIQUE/2005-2013/MOZAMBIQUE_TN_ARROZ_web.pdf (accessed 19 January 2017).

Fauvet, P. (2000) Mozambique: growth with poverty. A difficult transition from prolonged war to peace and development. Africa Recovery, 12–19 October. Available at: www.un.org/en/africarenewal/subjindx/subpdfs/143moz1.pdf (accessed 19 January 2017).

Flaherty, K., Mazuze, F. and Mahanzule, R. (2010) Moçambique: últimos desenvolvimentos na investigação agrária. Agricultural Science and Technology Indicators. International Food Policy Research Institute, Washington, DC. Available at: www.asti.cgiar.org/pdf/Mozambique-Pt-Note.pdf (accessed 19 January 2017).

Gêmo, H. (2004) An assessment of human resources development program in Mozambique's public extension service. MSc thesis. University of Pretoria, South Africa. Available at: http://repository.up.ac.za/handle/2263/28391?Show=f (accessed 19 January 2017).

Gêmo, H. (2007) Ligação investigação—extensão como contributo para o fluxo de conhecimento e tecnologias: breves considerações sobre Moçambique. Documento Apresentado no Workshop Sobre Transferência de Tecnologias, Instituto de Investigação Agrária de Moçambique and USAID,

Nampula, Mozambique, 6–8 March 2007. Available at: http://www.iiam.gov.mz/documentos/dfdtt/wshop_namp_mar_07/gemo_tabela_mct_apresent_nampula_revCXD.pdf (accessed 19 January 2017).
Gêmo, H. (2013) *Ligação Investigação & Extensão em Moçambique: Aprendendo de Experiências do Passado Para Maior Efectividade no Futuro*. MINAG/Direcção Nacional de Extensão Agrária and Instituto de Investigação Agrária de Moçambique, Maputo, República de Moçambique.
Gêmo, H. and Chilonda, P. (2013) Why did Mozambique's public extension halt the implementation of the national extension program (PRONEA)? Working Paper 6. IFPRI-Maputo, Mozambique and Mozambique Strategy Support Program. Available at: http://citeseerx.ist.psu.edu/viewdoc/download?doi=10.1.1.309.5069&rep=rep1&type=pdf (accessed 19 January 2017).
Gêmo, H. and Rivera, W.M. (2001) Mozambique's move towards a pluralistic national system of rural extension. ODI, AgREN, Network Paper N. 110. Available at: http://www.odi.org/sites/odi.org.uk/files/odi-assets/publications-opinion-files/5114.pdf (accessed 19 January 2017).
Gêmo, H.R., Eicher, C. and Teclemariam, S. (2005) *Mozambique's Experience in Building a National Extension System*. Michigan State University Press, East Lansing, Michigan.
Gêmo, H., Stevens, J.B. and Chilonda, P. (2013) The role of a pluralistic extension system in enhancing agriculture productivity in Mozambique. Research paper, University of Pretoria, South Africa. Available at: http://repository.up.ac.za/handle/2263/37267 (accessed 19 January 2017).
HarvestPlus (2012) *Disseminating Orange-fleshed Sweet Potato: Findings from HarvestPlus Project in Mozambique and Uganda*. HarvestPlus, Washington, DC.
Hester, R.E. and Harrison, R.M. (eds) (2005) *Sustainability in Agriculture, Issues in Environmental Science and Technology*. Vol. 21. Royal Society of Chemistry, Cambridge, UK.
Hoeffler, A. (1999) Challenges of infrastructure rehabilitation and reconstruction in war-affected economies. Background paper for the African Development Bank Report. Centre for the Study of African Economies, University of Oxford. Available at: www.icafrica.org/fileadmin/documents/Knowledge/DFID/Challenges%20of%20Infrastructure%20Rehabilitation%20and%20Reconstruction%20in%20War%20Affected%20Economies.pdf (accessed 19 January 2017).
IFAD (2012) National Agricultural Extension Program: mid-term review. International Fund for Agricultural Development and Republic of Mozambique, Rome.
IFAD (2014) Investing in rural people in Mozambique. International Fund for Agricultural Development, Maputo, Mozambique. Available at: www.ifad.org/operations/projects/regions/Pf/factsheets/mozambique_e.pdf (accessed 19 January 2017).
INE (2010) *Censo Agro-Pecuário (CAP)*. Instituto Nacional de Estatística, Maputo, República de Moçambique.
INE (2012) *Projeções Populacionais*. Instituto Nacional de Estatística, Maputo, República de Moçambique.
Lindau, L. (1998) *Rebuilding the Mozambique Economy. Assessment of a Development Partnership*. World Bank, Washington, DC.
Low, J., Uaiene, R., Andrade, M.I. and Howard, J. (2000) Orange-flesh sweet potato: promising partnerships for assuring the integration of nutritional concerns into agricultural research and extension flash. Research results from the Department of Policy Analysis, Ministry of Agriculture and Rural Development, Directorate of Economics, Mozambique), 10 November. Available at: http://fsg.afre.msu.edu/mozambique/flash/flash20e.pdf (accessed 19 January 2017).
Low, J., Arimond, M., Labarta, R., Andrade, M. and Namanda, S. (2013) The introduction of orange-fleshed sweet potato in Mozambican diets: a marginal change to make a major difference. Case study 5. In: Fanzo, J., Hunter, D., Borelli, T. and Mattei, T. (eds) *Diversifying Food and Diets. Using Agricultural Biodiversity to Improve Nutrition and Health*. Routledge, Abingdon, UK, pp. 283–290. Available at: www.bioversityinternational.org/fileadmin/user_upload/online_library/publications/pdfs/CS5_The_introduction_of_orange_fleshed_sweet_potato_in_Mozambican_diets.pdf (accessed 19 January 2017).
MacauHub (2013) Produção de milho em Moçambique com produtividade das mais baixas da África Austral. Available at: http://www.macauhub.com.mo/pt/2013/09/11/portugues-producao-de-milho-em-mocambique-com-produtividade-das-mais-baixas-da-africa-austral/ (accessed 10 April 2017).
MADER (2002) *Trabalho de Inquérito Agrícola (TIA)*. Ministério da Agricultura e Desenvolvimento Rural, Maputo, República de Moçambique.
MADER (2003) *Trabalho de Inquérito Agrícola (TIA)*. Ministério da Agricultura e Desenvolvimento Rural, Maputo, República de Moçambique.
MAP (1995) *Política Agrária e Estratégia de Implementação*. Ministério da Agricultura e Pescas, Maputo, República de Moçambique.
MAP (1996) *Relatório Anual do PESU*. Ministério da Agricultura e Pescas, Maputo, República de Moçambique.

MAP (1998) *Plano Director de Extensão Rural*. Direcção Nacional de Extensão Rural (DNER), Ministério da Agricultura e Pescas, Maputo, República de Moçambique.

MAP (1999) *Relatório Anual da DINA*. Ministério da Agricultura e Pescas, Maputo, República de Moçambique.

MASA (Ministério da Agricultura e Segurança Alimentar) (2002–2014) Anuário de Estatísticas Agrárias (2002–2014). MASA, Maputo, República de Moçambique.

MASA (2015) *Reunião Anual Técnica de Extensão Agrária*. Direcção Nacional de Extensão Agrária (DNEA), Ministério da Agricultura da Agricultura e Segurança Alimentar. Namaacha, Província do Maputo, República de Moçambique.

MINAG (1994) *Relatório Anual do PESU*. DINA, Maputo, República de Moçambique.

MINAG (2005) *Trabalho de Inquérito Agrícola (TIA)*. Ministério da Agricultura, Maputo, República de Moçambique.

MINAG (2007a) *Trabalho de Inquérito Agrícola (TIA)*. Ministério da Agricultura, Maputo, República de Moçambique.

MINAG (2007b) *Plano Director de Extensão Agrária*. Direcção Nacional de Extensão Agrária (DNEA), Ministério da Agricultura, Maputo, República de Moçambique.

MINAG (2007c) *PROAGRI Phase 1 Final Evaluation*. Ministério da Agricultura, Maputo, República de Moçambique.

MINAG (2008) *Trabalho de Inquérito Agrícola (TIA)*. Ministério da Agricultura, Maputo, República de Moçambique.

MINAG (2010) *Plano Estratégico do Instituto de Investigação Agrária de Moçambique (IIAM) (2011–2015)*. Ministério da Agricultura, Maputo, República de Moçambique.

MINAG (2014) Reunião Anual de Planificação.Bilene, Província de Gaza. Direcção Nacional de Extensão Agrária (DNEA), Ministério da Agricultura, Maputo, República de Moçambique.

MINAG and IFAD (2010) *Missão de Supervisão ao PRONEA, Dezembro: Aid Memoire*. Maputo, República de Moçambique.

Nippon Foundation (n.d.) *Sasakawa Global 2000: Building Food Security in Africa Global 2000: Building Food Security in Africa*. The Nippon Foundation. Available at: http://www.nippon-foundation.or.jp/en/what/spotlight/food_shortage/overview/ (accessed 3 March 2017).

OECD and AfDB (2002) Mozambique. In: African Economic Outlook. Organisation for Economic Co-operation and Development and African Development Bank Group, Paris, France and Abidjan, Côte d'Ivoire, pp. 215–228. Available at: www.oecd.org/countries/mozambique/1825790.pdf (accessed 19 January 2017).

Perttula, L. (2009) Epidemiology and characterization of Newcastle disease in smallholder poultry in Mozambique. Swedish University of Agricultural Sciences, Uppsala. Available at: http://stud.epsilon.slu.se/794/1/perttula_l_100125.pdf (accessed 19 January 2017).

Schwartz, S. (2010) *Youth in Post-Conflict Reconstruction: Agents of Change*. United States Institute of Peace, Washington, DC.

Selvester, K. and Castro, M.A. (2003) Mozambique food security issues paper. Forum for Food Security in Southern Africa, Overseas Development Institute, London. Available at: www.odi.org/sites/odi.org.uk/files/odi-assets/publications-opinion-files/5617.pdf (accessed 19 January 2017).

Tomo, A.A. (2009) Economic impact of Newcastle disease control in village chickens: a case study in Mozambique. MSc thesis. Michigan State University, East Lansing, Michigan. Available at: http://citeseerx.ist.psu.edu/viewdoc/download?doi=10.1.1.156.6976&rep=rep1&type=pdf (accessed 19 January 2017).

Unruh, J.D. (1995) Post-conflict recovery of African agriculture: the role of 'critical resource' tenure. *Ambio* 24, 343–350.

Unruh, J.D., Heynen, N.C. and Hossler, P. (2003) The political ecology of recovering from armed conflict: the case of landmines in Mozambique. *Political Geography* 22, 841–861.

Woolcock, R.F., Harum, M. and Alders, R.G. (2004) The impact of Newcastle Disease in village chickens on the welfare of rural households in Mozambique. Presented at the Fourth Co-ordination Meeting of the FAO/IAEA, Vienna, Austria, 24–28 May 2004.

World Bank (2008) World Development Report 2008. World Bank, Washington, DC.

World Bank (2016) Data: Mozambique—Population growth (annual %). Available at: http://data.worldbank.org/indicator/SP.POP.GROW/countries/MZ?display=default (accessed 19 January 2017).

6 Beyond Enabling Livelihoods: Agricultural Extension in Post-War Rebuilding and Reconciliation in the North of Sri Lanka

M.W.A.P. Jayatilaka*
University of Peradeniya, Kandy, Sri Lanka

Introduction

This chapter focuses on Sri Lanka's recent ethnopolitical conflict, its effects on agricultural extension and attempts made to address farmers' needs in a post-war context. The evidence summarized in this chapter is drawn primarily from published sources and also from unpublished government reports, interviews with key figures and focus group discussions in five villages: Maharambeikulam and Samalankulam in Vavuniya district and Arasady, Manipay and Navaly in Jaffna district.

Background on Sri Lanka

Sri Lanka is a small island nation south of India. The island has been inhabited by travellers, traders and settlers from Arabia, Africa, the Far East and the West, although the island nation's 21 million inhabitants (as of 2012) descend predominantly from settlers from the Indian subcontinent. The island came under Western colonial rule, initially of the Portuguese in 1597 and then of the British in 1815, more than two centuries before gaining independence from the British colonial empire in 1947. Modern Sri Lanka has a land mass of 65,610 km^2, with 29.4% covered in forest (Table 6.1). The country has a gross domestic product (GDP) of US$59,421 million, a per capita gross national income of US$2761.30, an unemployment rate of 4%, a population growth rate of 8% and a rural annual population growth rate of 0.7% (World Bank Data—http://data.worldbank.org/country/sri-lanka). After independence from the British Empire, the country enjoyed a high quality of life, with a welfare economy drawing predominantly on plantation exports of tea, rubber and coconuts (Sandaratne, 2004). The state heavily subsidized key services including education, healthcare, transport and food. Sri Lanka has had a high Physical Quality of Life Index for many years, and on the basis of many social indicators the country is ranked high among developing countries (UN, 2014).

Sri Lanka is a diverse country (Table 6.2) where ethnic groups have unique cultures, histories and values; communities that are close knit; and memories going back over several generations. The majority of Sri Lanka's population (75%) speak the Sinhalese language and are of the Buddhist faith. The Tamil group is the single largest minority, and is split between Sri Lankan Tamils, who are predominantly Hindu, and Indian

* Corresponding author e-mail: wijayajayatilaka@yahoo.co.uk

Table 6.1. Land mass, population, percentage urban and average household size (2011). From Department of Census and Statistics (2011).

	Land (km^2)	Total population	Urban (%)	Average household size	Dead or untraceable persons	Resettled population
Northern province	8,884	997,754	16.7	3.98	71,201	388,517
Jaffna district	1,025	567,229	19.8	4.13	42,410	158,747
Mannar district	1,996	95,430	23.2	4.01	4,912	34,229
Vavuniya district	1,967	164,852	19.2	3.84	7,925	40,722
Mullaithivu district	2,617	66,526	–	3.75	6,346	58,114
Kilinochchi district	1,279	103,717	–	3.56	9,608	96,705

Table 6.2. Sri Lanka's diversity.

Ethnic breakdown	Percentage	Religious composition	Percentage
Sinhalese	74.9	Buddhist	70.2
Sri Lankan Tamils	11.2	Hindu	12.6
Indian Tamils	4.2	Islam	9.7
Moors	9.2	Christian	7.4
Others	0.5	Other	0.1

Tamils, who are predominantly Muslim. There are more than 20 lesser-known ethnic minorities in Sri Lanka (Galkanda, 2007), and many struggle to maintain their identity.

Background of the Conflict

Post-independence Sri Lanka suffered from ethnic divisions, and successive elected governments have failed to manage the nation's culturally diverse peoples in such a way as to ensure peace. Tension between the predominantly Sinhala state and the Tamil minorities started when the government disenfranchised Tamil plantation workers in 1949, depriving them of the right to vote. These workers had been brought to the country from India by the British colonial rulers to work in the tea plantations as laborers. In 1956, Solomon Bandaranaike was elected on a wave of Sinhalese nationalism as leader of a new Sri Lanka Freedom Party, which empowered the large Sinhalese rural population. The populist government addressed the strong feeling of deprivation and loss of opportunities felt by the Sinhala ethnic majority during British rule and made Sinhala the sole official language and Buddhism the state religion. These measures bolstered the popularity of successive governments but drove a deep division between the two main linguistic groups—Sinhalese and Tamil. Since these early policy interventions, there has been a sharp ethnic divide in the strategies of electoral politics in the country (de Silva, 2005; Bandarage, 2009).

These divides have led to several cycles of violent uprisings, which occurred at almost regular intervals in the 1970s, 1980s and 1990s (Uyangoda and Bastian, 2008). Feelings of disenfranchisement among educated rural youth in the predominantly Sinhala regions resulted in two major youth uprisings in the 1970s and 1980s, led by the Janatha Vimukthi Peramuna (JVP, or People's Liberation Front), a Marxist revolutionary political party (Alles, 1990; Gunaratne, 1990).

The Liberation Tigers of Tamil Eelam (LTTE) movement was formally established in 1976 (de Silva, 2012) with the purpose of founding a separate country for the Tamil-speaking ethnic group, carving out more than two-thirds of the island's land mass in the north and east. The first Eelam war broke out in 1983 (see Table 6.3). The uprising in the north, led by LTTE, escalated its violent campaign in the late 1980s and a full-scale

Table 6.3. Timeline of the conflict and related events.

Date	Event
1948	Ceylon gains independence from British colonial rulers
1956	Wave of Sinhalese nationalism, with Sinhala made the official language of Sri Lanka
1958	Anti-Tamil violence erupts
1971	First JVP uprising—a Sinhalese Marxist youth group launches a rebellion
1972	Ceylon becomes Sri Lanka; Buddhism and Sinhala language given prominence in new constitution
1976	Launch of LTTE
1977	Post-electoral violence leaves 100 Tamils dead
1983	In Black July riots, hundreds of Tamils are killed and more than 100,000 flee to India Outbreak of first Eelam war
1985	Peace talks fail; Tamil militants attack the city of Anuradhapura
1989	Operation Liberation and the Indo-Lankan Peace Accord
1990	Second Eelam war
1994	Peace talks
1995	Third Eelam war; conflict intensifies, thousands killed
2002	Norwegian-mediated permanent ceasefire; road link to the north reopens, flights resume to Jaffna; relief, resettlement and reconciliation efforts launched by the government
2003	LTTE withdrawal from peace talks; war escalates in the north and east
2004	Tsunami: about 35,000 people, both Sinhalese and Tamil, die and many thousands are displaced
2008	Conflict escalates; government withdraws from ceasefire agreement
2009	End of war with LTTE; people begin to return to war-affected areas
2010	Mahinda Rajapaksa re-elected; parliament approves a constitutional change allowing the president to seek an unlimited number of terms
2012	UN report on civil war crimes released
2013	Amnesty International and UN Human Rights Commissioner accuse government of eroding democracy and the rule of law
2015	Maithripala Sirisena, a liberal elected president at early elections, brings in sweeping changes

war broke out, with both sides using heavy arms from foreign suppliers. Prime Minister Rajiv Gandhi of India (in 1991) and Prime Minister R. Premadasa of Sri Lanka (in 1993) and large numbers of leading politicians in Sri Lanka were assassinated. Among the latter were nine cabinet ministers: Ranjan Wijeratne (in 1991); Lalith Athulathmudali (in 1993); and Gamini Dissanayake, Weerasinghe Mallimarachchi and G.M. Premachandra (in 1994); C.V. Gunaratne (in 2000); Lakshman Kadirgama (in 2005); D.M. Dasanayake and Jeyaraj Fernandopulle (in 2008); and 27 members of parliament from 1975 to 2008. Police officers and civilians, including women and children, were also killed. Fear was pervasive across the country. Sporadic attacks on civilians and public property, coupled with disruption of civil life, displacement of families, an outflow of refugees, the presence of land mines and coerced recruitment of fighters by LTTE, combined to make life extremely difficult for villagers caught between the warring sides.

The effects on agriculture and livelihoods during the conflict were severe and inflicted deep damage. Movement and access to farmland became difficult. The military set up numerous checkpoints and bunkers, large areas of land were taken over and declared high-security zones, and the military eventually occupied and controlled the northern and eastern regions (Fig. 6.1). Formerly these were mostly residential and farming areas, thus the residents were deprived of their livelihoods and homes. Markets for selling produce and purchasing inputs ceased functioning, severely affecting production and disrupting cash flows and livelihoods. Any material transported to the north, and any produce moved out of the north, was subjected to detailed and lengthy security checks by both the Sri Lanka military and LTTE, at the border between the regions controlled by each of the two warring factions (UN, 2010). These involved time and clearance certifications from designated authorities, and LTTE imposed a tax system

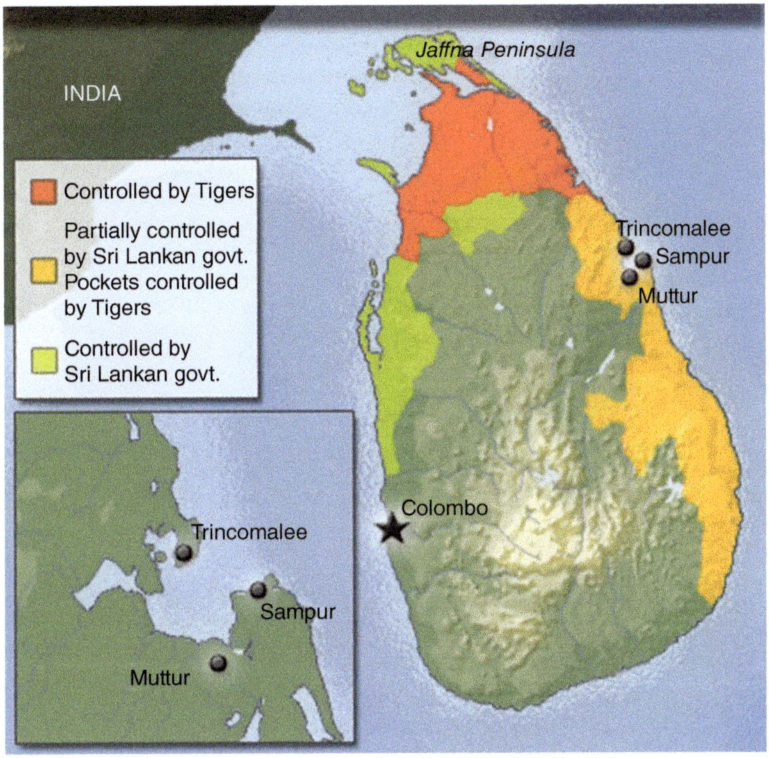

Fig. 6.1. Map showing the areas under Liberation Tigers of Tamil Eelam and government control during the height of the conflict. Source: Stratfor.

at the border crossings. This situation prevailed for nearly three decades and was highly visible, extremely intimidating and very disruptive.

Movement of people was also problematic. Irrigation systems no longer operated, since the officials controlling them were unable to work. Farmers' organizations, village development societies, women's village development societies and other forms of community-based organizations (CBOs) foundered as their members could not meet, a problem compounded by a general breakdown in social relationships. Many people migrated to safer areas and became refugees for decades (focus group discussions, 2014).

Agricultural extension, credit and insurance providers were unable to work in the conflict areas. Travel restrictions, security checks, disruptions of public transport and fear on the part of extension workers visiting clients outside of main towns all took their toll. Interactions between farming families and service personnel gradually disintegrated as the violence escalated. Regular systematic agricultural extension became impossible in the context of sporadic intense violence, security concerns, tensions and fear (focus group discussions in four villages in 2014).

Ultimately the civil war ended with the military defeat of LTTE in 2009, and Sri Lanka recently emerged from several decades of intense ethnopolitical violent conflict. The conflict-related damage was severe. The northern region was previously a major supplier of vital food crops and seafood, as well as human resources important to public administration and professional groups. Key agricultural, educational, industrial and cultural centers were located in this region. The war in the north disrupted the functioning of these institutions, destroyed much of the infrastructure, tore apart families and communities and resulted in enormous economic losses (NPC, 2001). Thousands of lives were lost, large numbers of women were widowed and children orphaned and many civilians were left physically disabled (de Silva *et al.*, 2011).

Over a million people were internally displaced and about 200,000 migrated overseas as refugees from the affected regions (NPC, 2001; Gomez, 2002; World Bank, 2003; UN, 2010; UNICEF, n.d.). Although thousands of people migrated out of these regions, many more remained and were displaced several times. Nearly 40% of the population in this part of the country were displaced once or several times (Department of Census and Statistics, 2011). Life in these regions has gradually returned to normal since the end of the war in May 2009, with over 400,000 persons moving back to the north of the country (see Table 6.1), assisted by the International Organization for Migration (IOM) and other international agencies and supported by family and friends.

Schools, places of worship, public transport and telecommunications systems are being restored. More importantly, the social fabric torn apart by 26 years of war and distrust is gradually being rebuilt. Large numbers of war widows, single-parent families and disabled persons are starting to return to a semblance of normal life, with support from the state, faith-based organizations, non-governmental organizations (NGOs) and numerous international agencies.

It has also been challenging to rebuild agricultural production. Much of the agricultural infrastructure in towns was destroyed by years of heavy shelling, as were roads and irrigation systems in rural areas. However, a large number of tanks previously used for irrigating rice cultivation and later either abandoned or destroyed by the war have since been reconstructed, and the irrigation canals have been renovated for re-use. Agricultural service infrastructure that was destroyed or abandoned is being rebuilt. Agricultural extension personnel, who had left their workstations due to the war, are returning, or new recruits are taking up duties.

Ethnic tensions contributed to the conflict; thus, any attempts to rebuild after a long and violent conflict must recognize the root causes of the conflict (Perera, 1999) and also the complex yet fragile social relationships that make communities function and meet the needs of individuals and families. Transforming fractured communities into cohesive and wholesome units will have a significant bearing on the outcome of efforts to rebuild the country, develop agriculture and make agricultural extension effective.

Agriculture and Extension in Sri Lanka

This section examines continuity and changes in the agricultural sector in the north, focusing mainly on agricultural extension. Following a brief history of agriculture in Sri Lanka, it describes how agrarian livelihoods continued during the war and what adjustments had to be made due to the disruption of institutions in the region; how the agricultural extension system functioned before, during and after the period of conflict; how agricultural extension is re-establishing operations in the post-war context; what key challenges and opportunities the state faces; and what should be the roles of key stakeholders in ensuring an effective and accountable agricultural extension service in the region.

Agriculture in Sri Lanka

Sri Lanka has always been, and remains, a predominantly agricultural country, even though recent years have seen contributions to the economy dominated by remittances, tourism and industrial outputs. Still, the country's culture and national ethos is deeply agrarian. The agriculture of Sri Lanka is diverse and provides employment and livelihoods to many people. It has had important historical antecedents. Sri Lanka's hydraulic civilization, dating from as early as 161 BC (Seneviratna, 1987), continues to be culturally and politically significant today.

The country's rice-based irrigation settlement systems date back as far as 200 AD, when there was a sophisticated agrarian management system for land preparation, irrigated rice cultivation, payment of taxes and sharing of labor (Hettiarachchi, 1982; Gunasinghe, 1985; de Silva, 2005). The ancient irrigation system consisted of a cascade of man-made reservoirs that stored excess water and were refilled during the rainy monsoon period, before being released as required for rice farming during the dry periods. This ancient agrarian system was reinforced to some extent by the development of tank-based irrigation systems introduced by British colonial rulers, who attempted to increase rice production in the country.

The system was also transformed by the introduction of plantation crops such as coffee, tea and rubber (de Silva, 1985a, 1985b). The plantations introduced by the colonial rulers transformed the predominantly agrarian economy into a ‘dual economy’, with plantations exporting coffee, cinnamon, spice, tea, rubber and coconut, and small-scale farmers engaged in rice and other field crop cultivation on smallholdings primarily for home consumption and for local markets (de Silva, 2005). Crop farming in Sri Lanka has continued to be dominated by subsistence smallholders (Abeysekera, 1985). They remain poor, since attempts to introduce high-value crops, market integration and commercialization have failed to produce improvements in incomes or lifestyles (Silva *et al.*, 1999).

Agriculture in Sri Lanka is an important sector, currently contributing about 3% to its GDP, compared with 6.1% and 12.6%, respectively, for the service and industrial sectors (Table 6.4; Central Bank of Sri Lanka, 2014). Although the relative contribution to GDP is low, the sector continues to be of social and economic importance to the country (Tables 6.4 and 6.5).

The highest proportion of employed people (33%) were engaged in agricultural activities in 2001 when the most recent census was conducted (Table 6.5). However, the proportion of people directly engaged in farming and the contribution that agriculture makes to the economy have continued to diminish over the past six decades. This trend directly relates to the conflict, as agricultural activities in the north were disrupted.

The unemployment rate is currently about 10%, and the agriculture sector has been unable to absorb the growing number of people entering the labor force each year. An increase in opportunities for access to

Table 6.4. Trends of contributions to GDP by agriculture and manufacturing sectors (US$ million). From Department of Census and Statistics (n.d.).

	1999		2000		2001		2002	
Sector	Value	%	Value	%	Value	%	Value	%
Agriculture	1422.2	17.3	1491.9	16	1592.3	15.4	1761.6	15.1
Manufacturing	1494.49	18.2	1723.4	18.5	1942.5	18.7	2210.4	19.0

Table 6.5. Percentage distribution of employed population by industry in selected sectors, 1953–2012. From Department of Census and Statistics (n.d.).

	Census				
Sector	1953	1963	1971	1981	2012
Agriculture, hunting, forestry and fishing	52.9	52.6	50.1	45.2	31.0
Manufacturing	9.7	9.2	9.3	10.1	17.7
Construction	1.9	2.7	2.2	3.9	8.4
Wholesale and retail sale	9.4	10.9	9.4	10.5	14.0
Transport, storage and communication	3.5	4.3	4.9	4.8	8.1
Finance, insurance, real estate and business services	2.2	0.5	0.7	1.1	3.5

free public sector education, from primary to university levels, has created a large cohort of highly educated but unemployed young people. These educated youth could not be absorbed into the agrarian economy and had no other avenues for social or economic mobility other than state sector employment. However, the introduction of a language policy favoring the Sinhala-speaking majority limited access to state sector employment for Tamil speakers, maintaining the feelings of marginalization and disempowerment that previously led to the violent conflict which afflicted the country for several decades (Perera, 1999).

Agriculture in Sri Lanka is dominated by smallholder farmers in both the food crop and the export-oriented plantation sectors (Table 6.6). A total number of 3,252,954 agricultural landholdings exists in Sri Lanka, covering a land area of 4,797,004 ha. Smallholdings (units of less than 0.1 ha) account for 45%, while only 2% of holdings are larger than 8.09 ha. This has important implications, such as production units being unable to generate sufficient surplus and agricultural incomes to ensure a middle-class lifestyle, and independent access to information and inputs being heavily dependent on state and collective efforts. State agriculture development policies and extension programs have to a large extent been shaped by this reality.

Agricultural extension

Agricultural extension has long been a part of the agricultural sector in Sri Lanka. Post-independence agriculture policies were highly sensitive to the needs of the rural peasantry (Sandaratne, 2004). Consecutive governments have attempted to improve conditions for smallholder subsistence farmers. A major strategy has been to revitalize the ancient rice-based irrigation systems in an attempt to increase food production and achieve self-sufficiency. Extension services have been a component of this strategy.

There are several key stakeholders in agricultural extension, especially in the post-war context of northern Sri Lanka (Fig. 6.2). Each plays a crucial role and can have a major impact on successful outcomes. They include farmers, their families and the communities within which they operate. They often work collectively as CBOs such as farmers' organizations or women's rural development societies. The individuals within these groups are direct beneficiaries of agricultural extension services.

Agricultural research and extension services for the various agriculture production systems are provided to farmers mainly by state agencies. The state operates two tiers of extension service providers, at national and provincial levels. These state agencies also act as the key regulators. The private sector and NGOs also play a role, but this is limited to specific geographical areas where their projects are implemented, or to marketing inputs or outputs (Sri Lanka Agricultural Extension Association, 1992). The non-governmental sector consists of local and international agencies. All NGOs offer resources, as well as strengthening local communities to help them engage meaningfully with the state so they can obtain services such as advice, information, agricultural inputs and access to markets. Farmers also provide extension information

Table 6.6. Number of holdings and extent of land under major crops as a percentage of total cultivable land (6.5 million ha). From Department of Census and Statistics (n.d.).

Sector	Number of holdings	Extent (ha)	Percentage of total cultivated land
Rice	493,382	897,076	13.7
Tea	264,758	212,716	3.2
Rubber	83,497	287,814	4.4
Coconut	71,347	394,836	6.0
Total agriculture/farm holdings	3,252,954	4,797,004	73.1

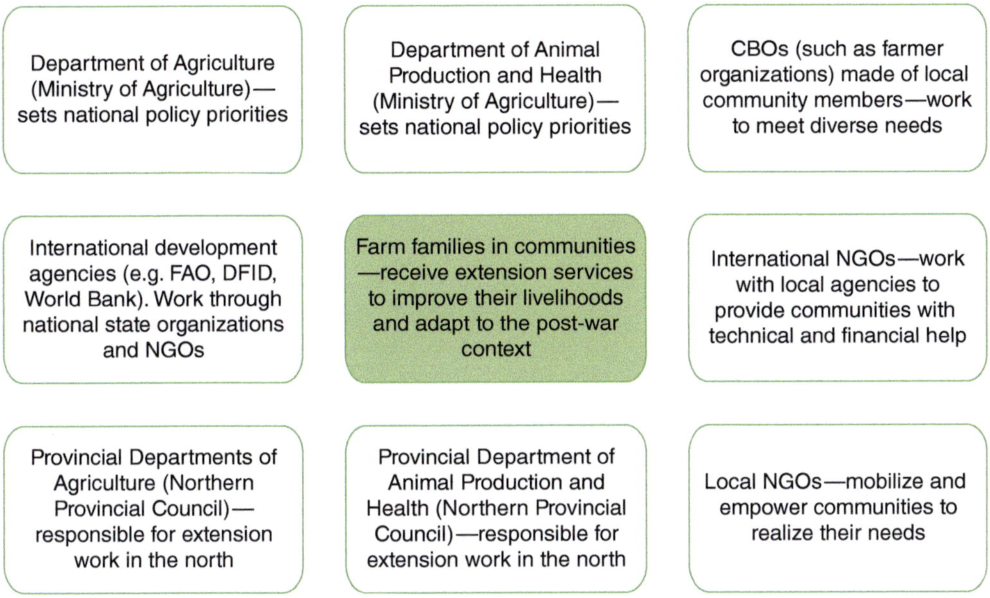

Fig. 6.2. Key stakeholders in agricultural extension in north Sri Lanka. CBO, community-based organization; DFID, Department for International Development (UK); FAO, Food and Agriculture Organization of the United Nations; NGO, non-governmental organization.

to each other, engaging in peer-to-peer extension of sorts.

Public sector extension

Public extension operates at many levels. In January 2015 the Ministries responsible for the agricultural sector were redefined, together with the various departments, institutes and divisions, as follows:

- Ministry of Mahaweli Development;
- Ministry of Plantation Infrastructure Development;
- Ministry of Irrigation and Water Resource Management;
- Ministry of Rural Economic Affairs;
- Ministry of Plantation Industries.

Different Ministries were assigned different subsectors of agricultural production (Table 6.7). Farmers may be provided with public sector extension services by a multitude of agencies depending on the combination of agricultural production they are involved in. Specialized agencies and extension officers exist at the field level to provide advice and guidance with technology and know-how. Thus, a multi-enterprise farmer who has an apiary, livestock, pond fisheries, coconuts, tea and rubber will be served by six different field-level extension personnel, each directed by a different state agency.

The Ministry of Mahaweli Development and Environment is unique. A multipurpose irrigation settlement scheme was launched in 1979, with a target of providing irrigation and land settlement by diverting the largest river in the country—the Mahaweli. The vision was to provide irrigation to 365,000 ha of land, covering 55% of the country's Dry Zone. Most settlers to this part of Sri Lanka were from the southern, Sinhala-speaking areas, and this has been a politically contentious issue (Perera, 1999). All lands and settlers that come under the Mahaweli irrigation system are managed by a specialized agency of the Ministry of Mahaweli Development and Environment, the Mahaweli Authority of Sri Lanka, which has its own administrative system including research, extension and community development (Mahaweli Authority of Sri Lanka, n.d.). The agricultural extension system within this

Table 6.7. Departments and institutes responsible for agricultural extension for the subsectors of agriculture. Compiled from respective Ministry websites.

	Responsible for agriculture extension	
Subsector	Department	Ministry
Rice, field food crops, horticulture, floriculture	Department of Agriculture Nine provincial departments of agriculture (provincial councils) that come under provincial councils	Ministry of Agriculture (http://www.agrimin.gov.lk/web/index.php/en)
Spices, other export crops	Department of Export Agriculture	Ministry of Primary Industries (http://mpi.gov.lk)
Tea	Tea Research Institute. Tea Small Holdings Development Authority	Ministry of Plantation Industries (http://www.kppk.gov.my/mpic/index.php/en)
Rubber	Rubber Development Department	Ministry of Plantation Industries
Coconut	Coconut Research Institute	Ministry of Plantation Industries
Sugar	Sugar Research Institute	Ministry of Plantation Industries
Livestock	Department of Animal Production and Health, National Livestock Development Board	Ministry of Rural Economic Affairs

vast project has its own management hierarchy and field presence in operational areas.

In the crop and livestock sectors, extension services are decentralized and are handled by each of the nine provincial governments, known as provincial councils, each with a Provincial Director of Agriculture. The provincial agricultural extension services have their own recruitment, training and funding mechanisms. Their efforts are coordinated for effective management and administration at provincial, district and divisional levels.

In each district, agricultural extension work is managed by a District Director of extension, who oversees agricultural officers and instructors responsible for various agriculture extension activities. These different field officers provide extension services to farmers, as listed below:

- Rice and food crops: agriculture instructor of the Provincial Department of Agriculture.
- Livestock and veterinary services: livestock development instructors, livestock officers and veterinary surgeons of the Department of Animal Production and Health.
- Tea: tea inspectors of the Tea Small Holdings Development Authority or extension officers of the Advisory Division of the Tea Research Institute.
- Rubber: rubber inspectors of the Rubber Development Department.
- Coconut: coconut inspectors and research officers of the Coconut Research Institute.
- Sugarcane: extension officers of the Sugar Research Institute.

In the fisheries sector, for both inland and coastal enterprises, fisheries development officers also undertake extension work.

In addition, the Extension and Training Division of the Department of Agriculture within the Ministry of Agriculture also provides extension services to farmers. It has several divisions that deliver specialized services (MoA, n.d.):

- In-Service Training Institutes;
- District Agriculture Training Centres;
- Sri Lanka School of Agriculture, providing 2-year professional training to Agriculture Instructors;
- Farm Mechanization Training Centre;
- Bee Development Unit;
- Agro-enterprises Development and Information Service;
- Farm Women Agriculture Extension;
- Horticultural Training and Development Institute.

The plethora of agencies operating at national and local government levels—attempting to service farmers and rural families with meager resources—can be a source of considerable confusion with regard to responsibilities and accountability.

The public sector fills the main role in rebuilding post-war agriculture and providing agricultural extension. The main efforts made to rebuild agricultural extension services fall into several categories.

Policy and planning are central to post-conflict extension. The Department of Agriculture of the Government of Sri Lanka, within the Ministry of Agriculture, remains the key agency responsible for agricultural development, particularly for the food crop sector. A master plan developed for the agricultural sector envisages comprehensive rebuilding and restoration of production, but also ensuring that farmers can contribute to the country's agricultural GDP (Northern Provincial Council, 2010). The planners also recognize the need for public sector extension to enable farmers to increase their efficiency, competitiveness and ability to compete in a global market. Faced with strong competition, producers need a range of information and skills to be able to respond to market signals and constantly evolving markets. These conditions require extension programs to provide more diverse services, but also respond to changes in extension approaches that promote sustainable production (Northern Provincial Council, 2010).

The effort to restart and/or resume institutional operations is another post-conflict extension approach used in Sri Lanka. The restoration of the governance and public administration system to cover the war-affected areas, where these were paralyzed, is a crucial first step. The previous mix of systems is being brought under one domain now that state power has been restored. In the past, open conflict meant that there were regions where no state administrative system was present and operational. These systems needed to be re-established, re-staffed and supported to function again, and developing institutional infrastructures remains a key challenge in these areas (Marikar, 2012; IFAD, 2014).

Resuming institutional operations includes the provision of public extension services. This institutional system was severely hampered by the war and is now being restored through monthly meetings and implementation of extension and development work programs. The Food and Agriculture Organization of the United Nations (FAO) has provided support to restore agricultural extension work in the war-affected areas of the north, initially for a 4-year period (FAO, 2009). This effort, coordinated by the provincial directors of agriculture, livestock and fisheries, consists of support for:

- the recovery and rehabilitation of crop production;
- the recovery and rehabilitation of livestock production;
- the recovery and rehabilitation of fisheries production;
- water management systems.

These support programs include strengthening community-level participation; strengthening the recruitment and training of extension personnel; providing relief as well as production inputs to all three sectors; assistance to rebuild required infrastructure; and improving extension training capacity. Livelihood support in the form of relief, as well as inputs to strengthen or relaunch agricultural, livestock or fisheries ventures, is also being provided by FAO, with the relevant Ministries coordinating efforts with regional entities such as provincial councils.

The public sector is also charged with coordination of other development actors. National and international NGOs are key partners in development interventions. Enabling responsible and accountable engagement in post-war development efforts in the former war zone is a strategic requirement that the state should nurture and support. The NGOs bring essential human, technical and financial resources to the region. These must be judiciously used through effective coordination if optimum benefits are to be achieved. As agriculture is the main source of revenue for the majority of families in the north, the support the NGOs provide is critical. This assistance must be coordinated with overall development efforts in the north, particularly in the case of making agricultural extension efforts more effective.

For these reasons, early post-war development work—during the previous government—was coordinated and strictly controlled by a committee under the Ministry of Economic Development. This was the Presidential Task Force for the Development of the North, which approves all projects in the northern region of Sri Lanka, including any conceived by NGOs or international development agencies. These proposals, as well as the organizations submitting them, were subjected to intense intelligence scrutiny by the Ministry of Defence. All NGOs were required to be registered with the NGO Secretariat or other prescribed state administrative agencies in the regions. All international staff working for such organizations had to obtain clearance from the Ministry of Defence to work, travel to the north or obtain visas. Many were turned down, and some organizations, such as the Berghof Foundation and Friedrich-Ebert-Stiftung, were forced to terminate their operations in Sri Lanka. The current government has changed these dynamics somewhat.

NGO-based extension

NGOs have operated in all parts of Sri Lanka for many years, actively engaging in development efforts before the war and continuing to serve farmers in conflict areas even during the conflict period. NGOs are also directly engaged in present post-war development efforts (Box 6.1) and will continue to work towards development and agricultural extension in the future.

Many NGOs have contributed. Five key organizations that have played a critical role in the northern and eastern regions are CARE International, Deutsche Gesellschaft für Internationale Zusammenarbeit (GIZ), the United Nations Development Programme (UNDP), World University Service of Canada and ZOA. Their efforts are usually carried out in collaboration with Sri Lankan NGOs operating at national, provincial, district or local village levels.

During the war, these organizations and other NGOs played a key role in relief and humanitarian work. Many engaged closely in strengthening livelihoods and attempted to fill the void left by the withdrawal or ineffectiveness of state institutions (Box 6.2). NGOs have also been strong proponents of human rights and have earned the wrath of the government and the defence establishment for these efforts. Some NGOs have had limited engagements, focusing instead on developing and transferring their capacities to local organizations.

After the conflict, NGOs' work in the northern and eastern regions focused primarily on issues regarding internally displaced

Box 6.1. Opinion of a senior NGO officer in the north, January 2015.

> We have always worked in the north. The war did not deter us. At that time, the requirement was humanitarian relief, to help those affected directly by the war. At present, the need is to restore livelihoods, heal the wounds of war and help people to get on their feet. We need a human touch in the field, not officialdom. NGOs have an ethos of non-bureaucratic flexibility to address human needs. We find it difficult to work when the government officials and the military are suspicious of us and do not trust us. But there are some reasonable officers, and we manage to continue to do a little bit of work. Many have given up and left.

Box 6.2. Respondent in focus group discussion, Vavuniya, January 2014.

> We survived because of the food, shelter and medicine that the NGOs provided. Government agencies were practically non-existent during the long years of war. But often the NGOs gave us the same things in duplicate. We took them because it was free; we could use them later. We did not throw away or waste anything.

persons and those directly affected by the war. This involved relief efforts, clearing landmines, and investigating disappearances and human rights violations. Interventions have since changed in nature and scope. The work undertaken by NGOs in the formerly war-affected areas now falls into three categories:

Category A: Undertaken by most NGOs, these tasks are low risk and aim to meet the basic needs of families

- Helping returnees to resettle and rebuild their livelihoods, homes and families from within Sri Lanka and overseas.
- Assisting with rebuilding local common infrastructure, such as village schools, dispensaries, irrigation canals, water wells, roads, primary health clinics and village markets.
- Restoring irrigation water reservoirs and distribution canals.
- Strengthening livelihoods and economic support to agriculture, industry enterprises and marketing.

Category B: Undertaken by some international NGOs, with collaboration and approval from the state

- Obtaining land and homes acquired or encroached upon illegally, especially where the original legal documents have been destroyed.
- Tracing lost family members.
- Providing care and support to the marginalized, especially war widows, the physically disabled, orphaned children and single-parent families.
- Rehabilitating and mainstreaming ex-combatants.
- Restoring rightful ownership of livestock abandoned prior to the outbreak of war.
- Psychosocial support for those traumatized by war, including providing counseling and psychiatric or institutional care.
- Reconciliation with past violators of human rights and the root causes of the war by restoring social relationships with different groups at local, regional and national levels.

Category C: Implemented by the state with technical and financial support from international NGOs

- De-mining lands to make them safe.

In the short term, NGOs will have a positive impact on the lives of people directly affected by the war. Their livelihoods will be improved, the local infrastructure developed and institutional support to villagers restored. However, the long-term needs of reconciliation, healing and peace-building, as well as conflict prevention efforts, may need more concerted, professional and well-planned intervention of high strategic relevance.

Furthermore, the work of NGOs was severely restricted by the government in the early post-conflict period through direct bans as well as by indirect means, even though these organizations have been instrumental in providing relief and rehabilitation as well as in providing inputs for agriculture, livestock and fisheries development. NGOs have worked closely with the provincial administration but were not allowed to work independently in local communities and were required to be under the control and constant monitoring of the chief government administrative officers of the province and districts, although these dynamics have changed somewhat in recent years.

Restrictions imposed by the government on NGOs and international development agencies severely hampered development efforts and the restoration of agriculture in the conflict-affected areas of Sri Lanka (Saravanathan, 2003). The agencies' ability to function is affected by conditions created by the state, and the potential to achieve agricultural extension outcomes in the post-war context is seriously limited.

Community-based and peer-to-peer extension

Other approaches to more informal agricultural extension include CBOs and knowledge sharing among farmers. CBOs, including rural development societies, women's rural development societies, farmers' organizations, death donation societies, rotation credit groups and labor exchange systems, all have long histories in Sri Lanka. These grassroots organizations, some of which are formal while others are informal, provide the social basis for collective action and solidarity in agriculture, but also more broadly to communities. Community-level social relationships have a strong bearing on the effective functioning of these structures. In most war-affected villages these organizations, which had been developed over long periods of time by generations of families providing mutual support, were dissipated. Rebuilding them, and strengthening their democratic and participatory ethos, is essential to addressing the deep-rooted causes of Sri Lanka's ethnopolitical conflict. Proactive action is required in all sectors, including agricultural extension.

Successes and Challenges in Rebuilding Agricultural Extension for a Conducive Agrarian Environment

More than 25 years of war and tensions in the post-war period were not conducive to agricultural production, extension service delivery, marketing activities and agricultural value chains. The damage done by the conflict and the conditions that followed it brought many challenges in rebuilding the sector and re-establishing rural livelihoods, particularly for displaced people. However, the post-conflict period also offered a unique opportunity for agricultural development and extension, supported by political will and external support. Sri Lanka has been successful, to some degree, in capitalizing on this opportunity:

1. Since the beginning of the accelerated resettlement process, led by the Government of Sri Lanka from October 2009 to December 2012, more than 413,000 people have returned to their homes in their district of origin or are staying with host families in the Northern province. This effort, coupled with the removal of landmines, allowed agricultural activities to resume in large numbers in former conflict areas, significantly affecting production and rural livelihoods. The emergence of retail marketing agencies for direct purchase of agricultural produce has also provided a boost to farmers in the region, as supermarkets have started opening up in the north, providing goods to consumers and a market for fresh produce. The opening up of highways and modern retail markets in cities enables farmers to sell their produce directly to private retail companies.

2. Despite large-scale disruption to services provided by key extension agencies of the Ministries of Agriculture and Livestock Development, where there was no serious fighting many farming families continued to grow crops, rear animals and sell produce in nearby markets. With the end of the war, the government extension services quickly recommenced their activities and connected with these remaining farmers. Extension activities, including the distribution of seeds and fertilizer, field visits and farmer training, were delivered by the Department of Agriculture and the Department of Livestock Development without much delay after the fighting had ended.

Another major boost to creating a conducive environment for the development of agricultural extension came in the shape and appropriateness of projects launched by the government in the conflict regions. These include:

- Provincial Accelerated and Medium-Term Development Programs;
- 'Vadakkin Wasantham' District 180-Day Northern Development Programs;
- World Bank-funded Re-Awakening Project.

These projects were successful because they targeted small-scale producers seeking to relaunch or step up activities in the agriculture and livestock sectors, but also offered support to the fisheries sector. These projects also identified and provided crucial components to rebuild agricultural production by providing inputs; repairing irrigation systems, roads and markets; building infrastructure; helping to strengthen transport; and also providing capital grants and credit to farmers (FAO, 2009). Specific project efforts are listed in Table 6.8.

Table 6.8. Government project efforts to rebuild agriculture and fisheries, by sector.

Sector	Project efforts
Food crops	Revitalization of paddy and other field crop production
	Promotion of carbonic fertilizers
	Revival of home gardens
	Community-based produce marketing
	Provision of water pumps
	Renovation of office-*cum*-training centers
	Revival/development of seed production
	Revival/development of fruit crops, vegetable crops, household farm forests
	Promotion of organic food production
	Promotion of sprinkler and drip irrigation systems
	Revival of social institutions
	Development of agro-based enterprises
	Revival of agricultural extension systems
	Renovation of district agricultural training centers and district farms
Livestock	Rounding up of loose cattle
	Establishment of dairy villages and animal banks
	Strengthening/reorganizing livestock breeders' cooperative societies
	Establishment/renovation of milk sales outlets
	Re-establishment of milk collection networks
	Setting up chilling centers, mini-milk coolers, liquid milk processing facilities
	Goat rearing and restocking
	Backyard poultry restocking
	Immunization program
	Training of livestock farmers and artificial insemination technicians (including youth and ex-combatants)
	Renovation of veterinary offices
	Establishment of nucleus herds
	Providing vehicles for livestock transportation
	Setting up cattle and buffalo breeding farms
	Establishment of goat breeding farms
	Establishment/strengthening of range veterinary offices and quarters
	Establishment/strengthening of regional and district animal production and health training centers
Fisheries	Provision of fishing craft, gear, fish boxes and bicycles (for marine fishing and aquaculture)
	Construction of shore facilities and block ice plants
	Establishment of fish purchasing centers
	Restocking of fingerlings in tanks
	Re-establishment/establishment of Ministry of Fisheries and Aquatic Resources (MOFAR), National Aquaculture Development Authority (NADA), Ceylon Fisheries Corporation (CFC) and National Aquatic Resources Research and Development Agency (NARA) district offices
	Skills development for fishermen and fisherwomen
	Re-establishment/strengthening of fisheries cooperative societies
	Establishment of fish selling centers and development of market linkages
	Undertaking a fisheries census and fishing boat registration
	Fisheries resources survey
	Zonal planning for shrimp farming
	Rehabilitation of fisheries harbors
	Provision of multi-day fishing boats
	Establishment of aquaculture development centers
	Construction of Department of Fisheries and Aquatic resources (DFAR) district offices, NARA regional centers and College of Fisheries and Nautical Engineering
	Restoration of mangroves and salt marshes

Continued

Table 6.8. Continued.

Sector	Project efforts
Irrigation	Rehabilitation of agriculture and irrigation infrastructure facilities to enable abandoned land to be cultivated Promotion of new irrigation schemes Construction of tube-wells Improved water management practices

Prospects for positive change to rebuild the agricultural sector, including agricultural extension, received a further boost after the 8 January 2015 presidential elections, when the incumbent president was replaced by a liberal social reformer. Sweeping changes towards reconciliation and restoration of public life and confidence are underway, creating conditions conducive to community-level action and strengthening of agricultural extension activities.

Post-war development in the north, and agricultural restoration in general, also received a major boost from the concerted and coordinated efforts of UN agencies through their Common Humanitarian Action Plans (CHAPs). The 2009 CHAPs, administered through the Office for the Coordination of Humanitarian Affairs (OCHA), operated 12 clusters/sectors led by specialized UN agencies. FAO led the agriculture sector, with OCHA coordinating the economic recovery and infrastructure development efforts made by UNDP; food aid was provided by the World Food Programme; education was provided by UNICEF and Save the Children Sri Lanka; health issues were addressed by the World Health Organization; nutrition, water, sanitation and hygiene were supported by UNICEF; shelter, non-food relief items, camp management and protection issues were overseen by the United Nations High Commissioner for Refugees; and security was provided by the United Nations Department of Safety and Security. This comprehensive approach provided a sound basis for families to benefit from the recovery and rebuilding processes (FAO, 2009). Several international NGOs (e.g. ACTED, Asia Onlus, FORUT International, International Relief and Development, Oxfam and ZOA) also collaborated in these efforts. Additional NGOs provided agricultural extension services, as discussed above.

Despite the successes, in the post-war period several key challenges can also be identified, which affect Sri Lanka's efforts to restore normality in the conflict-affected areas, especially in agriculture and agricultural extension:

1. A major contributing factor has been the government's historic and consistent failure to adequately support agricultural extension as a priority. The deterioration of agricultural extension services due to low investment by the government has continued to take its toll on the presence and efficacy of extension personnel (Panabokke, 2002; Fernando, 2014). Reduction in recruitment levels for the extension services, and minimal financial allocations for extension programs, have also had a negative impact and make extension services inaccessible to many farmers and of low quality. The current weak accountability and poor links between research outputs and extension services must also be strengthened as part of an overall approach to making country-wide improvements to agricultural extension services. Special focus should be placed on the war-affected areas.

2. Development efforts in the affected areas soon after the end of the war were heavily dependent on donor support due to paucity of state funding. Often the financial commitments made by donors were not followed up. Underfunding has had serious effects on the fragile efforts to rebuild these regions. Attempts to restore agricultural extension have been adversely impacted as well (FAO, 2009). A related factor is that the donor funding flow is linked to identification of cycles of needs, with an implementation phase that is often limited to 3 years. This time frame is completely insufficient to enable a socially transformative process to take place at the community level. Also, donor-implemented projects are often inadequately and ineffectively monitored and evaluated, and lessons learned are not used or shared to develop a body of knowledge, as stipulated in the Paris Declaration on Aid Effectiveness (OECD, 2005).

Excessive state control over post-war development efforts and intense security arrangements combined to make the conditions extremely difficult for development agencies and NGOs to function at ground level. The work of the Presidential Task Force for the Development of the North was severely criticized for its arbitrary control of any external NGO- or donor-led development interventions. The processes of scrutiny were not transparent and the assessment criteria were not made known to the public. There were no effective mechanisms for appeal. Despite the severe need for agricultural development and efforts to restore farming and improve agriculture, governmental restrictions on activities by NGOs and international agencies have slowed any progress made by these actors.

3. The development of agricultural extension should be placed in the proper political and social context and be addressed as part of the restoration, rebuilding and reconciliation process underway in post-war situations. Such a broad perspective is essential to address root causes of conflict and as a long-term conflict-prevention measure. Agricultural extension should therefore be integrated with conflict transformation and peace-building (CTPB) principles. This will require a paradigm shift away from the notion that the purpose of agricultural extension is to increase productivity and enhance farm incomes, towards an approach that addresses the social needs of communities. Such needs are human security, peace, strengthening democratic space and citizenship for effective participation and inclusive approaches to governance and development. Agricultural extension must therefore undergo a substantial change in Sri Lanka and other post-war settings.

Recommendations and Opportunities for Improvement

There are several areas where agricultural extension in post-conflict Sri Lanka can be improved.

Planning and policy

The Lessons Learnt and Reconciliation Commission (LLRC) report by the Government of Sri Lanka explicitly recognized that any efforts to improve agricultural extension must be placed in the perspective of the ethnopolitical conflict and governance failure (de Silva *et al.*, 2011). Agricultural extension program planning and human resource development must therefore be undertaken in this context. The LLRC report is considered a fundamental guide for crafting a peaceful Sri Lanka, in which all stakeholders take responsibility for building a new future. The principles of 'do no harm' as advocated by former Irish President Mary Robinson, along with inclusion, good citizenship and social integration, should be woven into a more holistic framework to develop a broader perspective for agricultural extension in Sri Lanka (Ministry of National Languages and Social Integration, 2014).

However, the master plan for the north, as developed by the Provincial Government, does not fully recognize the importance of post-war reconciliation, the need to strengthen community trust in social institutions and governance systems, or the need to invest in democratic citizenship, as recognized by the LLRC (LLRC, 2011). Yet at the community level, many agricultural activities and efforts are collective and require cordiality and trust if they are to be effective. Such qualitative transformation of communities, leadership and institutions must be planned for with a proper understanding of the past and present situations, as well as what is to be hoped for in the future.

Coordination

The efforts of the many different development and agricultural extension services must be consolidated so that the system is more efficient and accountable to stakeholders (Somaratne, 2002). In the short and medium term, there needs to be close coordination among the different agencies. In the long term, the organizational architecture of agricultural extension must be critically assessed and a more rational system developed. Initiatives launched by the NGO sector should be well planned and coordinated with governmental authorities at national and regional levels so that beneficiary needs are met efficiently and effectively.

Strengthening social relationships

Agriculture remains an intensely social and collective activity. It involves people coming together to work in the fields at the community level, with farming families dependent on each other to meet many of their needs (Saravanathan, 2003). These include land preparation, preparing the irrigation canal networks for water supply, obtaining inputs and marketing produce, and planning the seasonal cropping calendar.

Farmers must coordinate collective efforts that involve men and women bound by different social and economic relationships. Traditions, social values and rituals color the multitude of transactions that take place in agriculture among the different people engaged in the sector. Agriculture therefore offers unique promise and the potential to address deeply ingrained social divisions and the root causes of conflicts, and to heal and restore social harmony.

However, this outcome can be achieved only if agriculture and farming are recognized as a social activity, with people placed on center stage. Such an approach requires that the stakeholders, as opposed to markets and technology, are given priority and are kept informed, engaged and in control, with all people able to assert themselves and to be treated as equals.

Stronger social relationships will also require efforts to rebuild trust. At community level, farmers are trying to restore working relationships with each other and to continue to farm, rear animals and sell their produce to markets. However, some repercussions from the war culture continue to exert a negative influence. Fear, mistrust and hostility have long-term consequences, and social interactions and collective action that could be of mutual benefit to communities may be compromised. There is a need to rebuild the subtle and often fragile social interpersonal relationships as social capital for mutual benefit. Many agrarian families suffer from the effects of events during the war, when trust and goodwill were broken down (Box 6.3).

Nonetheless, for rural and agricultural communities to function and develop, efforts to rebuild trust and social cohesion are essential. This responsibility, although not traditionally expected of agricultural extension service providers, might be a new and central element of their programming in post-conflict Sri Lanka and other countries emerging from war and social upheaval.

Resource mobilization

Continued dependence on donor funding is risky, particularly if there is no long-term

Box 6.3. Respondent in focus group discussion, January 2014.

One night in June 1990, all 90 families in our village, Samalankulam, fled because two families were killed and their houses were burnt. We left our belongings, particularly our livestock. Most of us had cattle and goats. Our cattle were branded for identity. We returned two years later. We are extremely unhappy that we can recognize our cattle, which are now owned and tended by our adjacent villagers. We feel helpless that we can't claim our assets back.

vision for self-development and reducing dependency. There is a serious need for the government to recognize the importance of more sustainable funding and to commit more funds to revitalizing and strengthening agricultural extension. Any attempt to restore livelihoods and lift people out of poverty will require serious commitment.

The beneficiaries targeted in war-affected areas are different from those in other regions of the country. Their concerns will be distinct, given their experiences of living with the tension of conflict for more than 25 years. There are over 20,000 war widows in the region, as well as a great many disabled people and families with no male breadwinners. These demographic considerations will require more attention and state support to the region, in addition to extension programs tailored to their specific needs.

Market-driven extension

All agricultural and extension efforts need to create an enabling environment that supports market development and improves rural livelihoods. Farmers need market information on inputs and buyers for their produce, as well as knowledge about technology. The infrastructure currently being developed must extend to remote rural areas where farmers operate and harvests are processed. Efficient storage and transport are essential services that agricultural extension will need to coordinate if other aspects of agricultural improvement are to bear fruit.

Professional competencies

All extension professionals, trainers and program planners must develop a balance in their social as well as their technical competencies. This balance must be extended to their clients, who may be crop or livestock farmers, women farmers, young farmers or field extension officers. Women have been particularly affected by the war, and they remain a vulnerable group. When addressing their livelihood needs, a gender-and conflict-sensitive approach should be adopted. People may need special training in skills to work with women in post-conflict settings.

For example, agricultural extension personnel serving in post-conflict regions, both governmental and non-governmental, need to have strong CTPB skills to address the unique needs of farming communities in these areas. For this reason, any training program for extension officers must incorporate CTPB training. Extension programs must be designed to reflect conflict-sensitive and 'do no harm' principles. Indeed, all development interventions should be conflict-sensitive and underpinned by an awareness of how the efforts impact the different groups of beneficiaries, as well as of the benefits and costs of interventions.

Also, extension personnel must have competencies to support resettlement. Those who emigrated from the country as refugees will continue to return over the next few years. The return and resettlement process is often stressful, but can be used to build strong social relationships based on compassion, forgiveness and inclusion, all vital in CTPB, if extension personnel develop the needed competencies. In a social context where nearly 40% of the population in the north are returnee displaced persons, the emphasis on addressing their needs in a conflict-sensitive manner is paramount. Agricultural extension and other service delivery agencies in the north need capacity development support to provide them with the competencies required to address the needs of resettled persons (Buthpitiya, 2013).

Targeted programming

Agricultural extension must be properly targeted to serve the diverse groups that require services in post-conflict Sri Lanka. Smallholders are one extremely important audience. Given the size of their production units, smallholder farmers have specialized needs for enterprise promotion and market integration that must be incorporated into agricultural extension efforts.

Group approaches, collectivization and establishing federations to strengthen smallholder farmers, especially women farmers, should be actively pursued as part of any social competencies and/or development strategy in a post-war setting.

There are also several groups of people who are direct victims of the war, and organizations and agencies—particularly state agencies—need to have mechanisms and processes to deliver services to a clientele quite different from those of the pre-war period. Widows, the disabled and mobilized youth have special needs that must be recognized, and extension interventions should be crafted to respond to their needs.

For example, individuals who were directly involved in LTTE as combatants are commonly ostracized and socially marginalized. Several thousand LTTE ex-combatants have been rehabilitated and released, but most are shunned by other members of society. Many of these men and women are from poor agricultural families and will need special approaches to develop social acceptance and become mainstream members of community life once again.

However, the reintegration of ex-combatants through socio-technical agricultural enterprises can be very rewarding. Young people can be trained to develop technical skills with relative ease, but should also be encouraged to develop the social skills of communication, negotiation, bargaining, assertiveness and management. These social competencies will be highly beneficial in the reintegration and mainstreaming processes. Agricultural extension officers will have to facilitate integration and acceptance as part of their new approach to extension.

References

Abeysekera, C. (ed.) (1985) *Capital and Peasant Production: Studies in the Continuity and Discontinuity of Agrarian Structures in Sri Lanka.* Social Scientists' Association of Sri Lanka, Colombo, Sri Lanka.

Alles, A.C. (1990) *The JVP 1969–1989.* Lake House Publishers, Colombo, Sri Lanka.

Bandarage, A. (2009) *The Separatist Conflict in Sri Lanka: Terrorism, Ethnicity, Political Economy.* Routledge, London.

Buthpitiya, V. (2013) *Reconciling Rights, Responsibilities and Disjunctures: An Assessment of Sri Lanka's Post-war Development Drive.* Law and Society Trust, Colombo, Sri Lanka.

Central Bank of Sri Lanka (2014) Economic and Social Statistics of Sri Lanka 2013. Central Bank, Colombo. Available at: www.cbsl.gov.lk (accessed 19 January 2017).

Department of Census and Statistics (2011) *Enumeration of Vital Events 2011, Northern Province, Sri Lanka.* Department of Census and Statistics, Colombo, Sri Lanka.

Department of Census and Statistics (n.d.) Census and Statistics Sri Lanka. Department of Census and Statistics, Colombo, Sri Lanka. Available at: www.statistics.gov.lk (accessed 19 January 2017).

de Silva, S.B.D. (1985a) Plantations and underdevelopment. In: Abeysekera, C. (ed.) *Capital and Peasant Production: Studies in the Continuity and Discontinuity of Agrarian Structures in Sri Lanka.* Social Scientists' Association of Sri Lanka, Colombo, Sri Lanka, pp. 21–32.

de Silva, S.B.D. (1985b) Plantation investment in Sri Lanka. In: Abeysekera, C. (ed.) *Capital and Peasant Production: Studies in the Continuity and Discontinuity of Agrarian Structures in Sri Lanka.* Social Scientists' Association of Sri Lanka, Colombo, Sri Lanka, pp. 33–62.

de Silva, K.M. (2005) *A History of Sri Lanka.* Penguin Books, New Delhi, India.

de Silva, K.M. (2012) *Sri Lanka and the Defeat of the LTTE.* Penguin, New Delhi, India.

de Silva, C.R., Perera, A.R., Hangawatte, K., Chanmugam, C., Palihakkara, H.M.G.S., Ramanathan, M., Paranagama, M.P. and Bafiq, M.T.M. (2011) Report of the Commission of Inquiry on Lessons Learnt and Reconciliation. Lessons Learnt and Reconciliation Commission, Colombo, Sri Lanka.

FAO (2009) *Recovery and Rehabilitation Programme for the Agriculture Sector in Conflict Affected Areas of the Northern Sri Lanka.* Food and Agriculture Organization, Colombo, Sri Lanka.

Fernando, M.M.T. (2014) Perceived value of agriculture extension services among onion and paddy farmers in Dambulla. Unpublished report. University of Peradeniya, Sri Lanka.
Galkanda, N. (2007) *Api Nodanna Lankikayo.* Suratha Publishers, Mt Lavinia, Sri Lanka.
Gomez, M. (2002) *National Human Rights Commission and Internally Displaced Persons: Illustrated by the Sri Lankan Experience.* Brookings Institution, Washington, DC.
Gunaratne, R. (1990) *Sri Lanka, A Lost Revolution? The Inside Story of the JVP.* Institute of Fundamental Studies, Kandy, Sri Lanka.
Gunasinghe, N. (1985) Peasant agrarian systems and structural transformation. In: Abeysekera, C. (ed.) *Capital and Peasant Production: Studies in the Continuity and Discontinuity of Agrarian Structures in Sri Lanka.* Social Scientists' Association of Sri Lanka, Colombo, Sri Lanka, pp. 191–230.
Hettiarachchi, T. (1982) *The Sinhala Peasant in a Changing Society.* Lake House Investments Ltd Book Publishers, Colombo, Sri Lanka.
IFAD (2014) *Democratic Socialist Republic of Sri Lanka: Dry Zone Livelihood Support and Partnership Programme Impact Evaluation.* International Fund for Agricultural Development, Rome, Italy.
LLRC (Lessons Learnt and Reconciliation Commission) (2011) Report of the Commission of Inquiry on Lessons Learnt and Reconciliation. Government of Sri Lanka, Colombo, Sri Lanka.
Mahaweli Authority of Sri Lanka (n.d.) Mahaweli Authority. Ministry of Mahaweli Development and Environment, Colombo, Sri Lanka. Available at: http://mahaweli.gov.lk (accessed 19 January 2017).
Marikar, S.M.F. (2012) *CIDA: Midterm Assessment of Rebuilding Agriculture Livelihoods Project (RALP).* Canadian International Development Agency, Colombo, Sri Lanka.
Ministry of National Languages and Social Integration (2014) *National Policy Frame for Social Integration.* Ministry of National Languages and Social Integration, Government of Sri Lanka, Colombo, Sri Lanka.
MoA (n.d.) Ministry of Agriculture. Colombo, Sri Lanka. Available at: www.agrimin.gov.lk (accessed 19 January 2017).
Northern Provincial Council (2010) *Master Plan for Agriculture Development Sector, Northern Province.* Northern Provincial Council, Jaffna, Sri Lanka.
NPC (2001) *Cost of War: Economic, Social, and Human Costs of the War in Sri Lanka.* National Peace Council, Colombo, Sri Lanka.
OECD (2005) The Paris Declaration on Aid Effectiveness. Organisation for Economic Co-operation and Development, Paris, France. Available at: www.oecd.org/dac/effectiveness/34428351.pdf (accessed 19 January 2017).
Panabokke, C.R. (2002) Institutional infrastructure needs in agricultural development. In: Somaratne, W.G. (ed.) Sri Lanka Agriculture for the Next Decade: Challenges and Opportunities, Proceedings of a Workshop, April, Hector Kobbekaduwa Agrarian Research and Training Institute, Colombo, Sri Lanka, pp. 106–108.
Perera, S. (1999) *The Ethnic Conflict in Sri Lanka: A Historical and Socio-political Outline.* Department of Sociology, University of Colombo, Sri Lanka.
Sandaratne, N. (2004) Agriculture development: controversial issues. In: Kelegama, S. (ed.) *Economic Policy in Sri Lanka: Issues and Debates.* Sage Publications, New Delhi, India, pp. 195–212.
Saravanathan, M. (2003) Post-war development in Nepal and Sri Lanka: a comparison. *Groundviews* 14, September.
Seneviratna, A. (1987) *The Hydraulic Civilization of Ancient Rajarata in Sri Lanka.* International Irrigation Management Institute, Colombo, Sri Lanka.
Silva, K.T., de Silva, S.B.D., Kodituwaku, S., Razaak, M.G.M., Anada, J., Gunawardena, I.M. and Lux, D. (1999) *No Future in Farming? The Potential Impact of Commercialization of Non-Plantation Agriculture on Rural Poverty in Sri Lanka.* Centre for Inter-Sectoral Community Health Studies, Kandy, Sri Lanka.
Somaratne, W.D. (ed.) (2002) Sri Lankan Agriculture for the Next Decade: Challenges and Opportunities. Proceedings of a Workshop, Hector Kobbekaduwa Agrarian Research and Training Institute, Colombo, Sri Lanka.
Sri Lanka Agricultural Extension Association (1992) *Profiles of Agricultural Extension Organizations in Sri Lanka.* Vol. 1. Sri Lanka Agricultural Extension Association, Peradeniya, Sri Lanka.
UN (2010) Report of the Secretary General's Panel of Experts on Accountability in Sri Lanka. United Nations, New York.

UN (2014) Country profile: Sri Lanka. Available at: http://data.un.org/CountryProfile.aspx?crName=SRI%20LANKA (accessed 19 January 2017).

UNICEF (n.d.) *Protecting the Rights of Children and Women Affected by Armed Conflict in Sri Lanka: Programme Description and Funding Appeal 2001–2003.* UNICEF, Colombo, Sri Lanka.

Uyangoda, J. and Bastian, S. (2008) *State Responsiveness to Public Security Needs: The Politics of Security Decision-making: Sri Lanka Country Study.* CSDG Papers Number 15. Kings College, University of London, London.

World Bank (2003) *Sri Lanka: Country Assistance Strategy of the World Bank Group 2003–2006.* World Bank Group, Colombo, Sri Lanka.

7 Agricultural Extension in Post-Conflict Myanmar (Burma): Context and Lessons Learned

Joshua Ringer*
Oklahoma State University, Stillwater, Oklahoma, USA

Background on Myanmar (Burma)

Any discussion of agricultural extension requires an understanding of the long history of struggles and armed conflict in Myanmar (previously named Burma). A country on mainland South-east Asia, Myanmar has struggled to develop as a nation state. This struggle has helped define its approach to nation-building and development.

Myanmar has often been at the crossroads of several powerful empires—the Mongols, Chinese, Indian and Thai—and has been greatly affected by these countries politically, culturally and spiritually. Throughout Myanmar's development, the Burmese Kingdom was threatened and occasionally subjugated by the many groups surrounding it. During the Mongol empire, China threatened the Burmese Kingdom. Later, the influence of Indian kingdoms transformed the belief systems of Burma through the introduction of Hinduism and Buddhism (Myint-U, 2011). Many wars were fought with the Thai kings and the Mon peoples nearby until an uneasy truce was achieved (Hall, 1981).

For a time, despite the British expansion into South-east Asia, Burma was free of control by the British Empire even as the British Raj was growing in neighboring India. Eventually, though, the British East India Company saw the advantage of subduing the Burmese Kingdom and making it part of British colonial holdings. After a series of wars, this process was completed. To rule Burma, the British undertook several actions that changed the complexion of the country. Indians were brought in to provide labor, implement police and military forces and help conduct civil affairs (Hall, 1981). This action brought in many from outside who had not traditionally been part of the Burmese Kingdom's ethnic make-up. In addition, the British ruled the mountainous buffer-zone areas by empowering local ethnic leaders and bringing them into the armed forces (Ekeh and Smith, 2007). This action reinforced the perception among the mountain ethnic groups that they were different and capable of self-rule.

World War II brought about the swift destruction of the British control of Burma. Several nationalist groups emerged which attempted to develop Burmese capabilities to rule themselves, under the protection of the Japanese. Burma was on the frontline of the fighting and suffered much infrastructure damage during this time. The British and

* Corresponding author e-mail: joshua.ringer@okstate.edu

Allied forces eventually drove out the Japanese. The nationalist groups welcomed the removal of the Japanese and expected the British to grant Burma freedom to rule itself (Maung, 1971).

After the war, the many ethnic groups were well armed because they had fought alongside the Allied forces against the Japanese. These groups were not eager to disarm and place themselves under the rule of a nationalist Burmese government dominated by ethnic Burmese. With the assassination in 1947 of Aung San, the best-known nationalist leader, along with most of his cabinet, the country descended into a period of intermittent internal armed conflict. The result of this conflict was that, over time, politically key persons in the military gained more and more influence over the decision-making bodies of the state.

The period of military rule lengthened as the country first turned to the 'Burmese way to socialism' (Thuzar, 2013) and then transitioned to military rule under the State Law and Order Restoration Council (SLORC) in 1988 (Lintner and Black, 2009). Battles continued with over 20 different armed insurgent groups. These insurgent groups were divided mainly along ethnic lines. Because of the excesses of the military leaders, the USA placed a strict trade embargo on Burma. These factors led to an emphasis by the military government on national development and an effort to pacify all the rebel areas where the armed ethnic groups were based (Steinberg, 2010; Clapp, 2015). The main strategy employed by the government was the program of the 'four cuts', which was originally designed by the military leader Ne Win as a counterinsurgency strategy (Joliffe, 2015). The 'four cuts' strategy consisted of removing insurgents from their source of food, funds, intelligence and recruits (Smith, 1999). This often was done through forced relocation (Lintner and Black, 2009). This program was brutally effective and over time served to cause many factions to develop within the insurgent militia groups. Many of the militia groups were splintered along political and religious lines. Some groups began fighting alongside the government forces against the insurgent militia groups that still opposed the government. The Burmese government maintained a system of truces so that it could fight whichever insurgent group it wished when it was advantageous to do so (Lintner and Black, 2009).

In 1989, the Burmese military government changed the name of the country to Myanmar (Steinberg, 2010). Around 2005, Myanmar slowly began opening up to market interactions. After several setbacks, agricultural production areas and markets that used to be affected by armed conflict were opened up to international interaction, and restrictions on open society have been eased (Asia Society, 2010; Thuzar, 2013). The increase in demand for primary commodities in the region has meant that Myanmar has ready markets for exporting rice, maize and legumes to China and India (Byerlee *et al.*, 2014).

Yet, even now, there appears to be a balancing of the Western countries against the strong influence of China and India on Myanmar's political leaders (Myint-U, 2011). Myanmar's recent openness to opportunities with the Association of Southeast Asian Nations (ASEAN) indicates these relationships. Henley (2014) noted that Myanmar is different from other countries in South-east Asia, largely due to the long-standing economic sanctions.

Also, unlike other countries in the region such as Thailand, Lao PDR and Vietnam, the political opposition parties have strong international links and strong domestic support, and there is pressure for a federalist style of government because of the relative strength and proportion of the ethnic minority groups (Henley, 2014). Great conflicts raged between these different groups from the early 1950s up to the present day. At the time of writing, an uneasy truce has settled on the geopolitical administrative states of Myanmar as the government maintains up to 16 different peace agreements (Pick and Thein, 2010). Through this divide-and-conquer strategy the country has been able to exert more and more control over its main cities and agriculturally productive areas.

Despite increased control by the Burmese government, sporadic conflict is still common in the country. Myanmar continues to experience conflict in many borderland areas of the country under the control of armed insurgent groups (Joliffe, 2015). The Myanmar government controls the areas around the main administrative towns and along the road networks, while insurgent militias exhibit loose control over areas further from the main towns and the paved all-weather road system. In northern Shan State, for example, ongoing conflict remains highly variable. The areas of influence of the militia groups has shrunk slightly since 2007 (Joliffe, 2015). The country's status as 'post-conflict' is somewhat debatable. The patchwork of influence by the Myanmar government and the persistence of violent events also mean that the context for post-conflict agricultural development and extension in Myanmar is very complex with very different variables (Kalyvas, 2006; Eck, 2015; Joliffe, 2015).

Background on Burmese Agriculture

Myanmar's (Burma's) agricultural history is equally influential on modern extension. Agriculture during the Burmese Kingdom period was based on lowland rice paddy systems that required large inputs of labor. This need for labor, and for products from the forests in the hills and mountainous regions, meant that the Burmese Kingdom relied on the peoples of the mountain fringes surrounding the Irrawaddy Delta and lowlands (Scott, 2009). These mountain regions were the buffer boundary between the Burma, Thai and Assam lowland rice kingdoms. They were populated by many ethnic groups, the largest being the Shan, Mon, Kayin, Kachin, Chin, Wa and Rakhine peoples, who lived in the mountainous buffer zones between the lowland paddy rice-based cultures (Scott, 2009). At times these groups were subjects of the Burmese kings and forced to provide labor for maintaining irrigation and paddy works. At other times these mountain peoples, reliant on shifting agriculture, were free from close control by the lowland Burmese Kingdom. Scott (2009) suggests that continual raids by the Burmese kingdoms to capture labor resulted in mountain peoples developing conflict-resistant, shifting agriculture to avoid conflict and capture. These experiences left the Burmese with uneasy relationships with their neighbors, both near and far. This history also left the peoples of the buffer areas wary of coming under Burmese rule.

During the British colonial period, Burma became the rice bowl that fed the British Raj (Maung, 1971). Rice production, along with other agricultural and livestock production, created important ties with India, especially eastern India. British rule emphasized the extractive nature of the many different British companies that sprang up to benefit from Burmese natural resources. In addition to production of rice, staples and livestock, a major emphasis was on harvesting teak and other high-quality woods from Burmese forests (Walinsky, 1962).

During the post-war period to 1960, the USA worked closely with Burma to help it develop as a country (Maung, 1971). Agriculture was affected as battles took place close to the capital. Agricultural production began to recover from the destruction that had occurred during World War II.

As Burma converted to socialism, agricultural production suffered both from lack of investment and from the US trade embargo. Agriculture followed the socialist model of central control. Investment in agricultural technology for rice production and processing was minimal, and most technology was a carry-over from pre-socialist times.

Impacts of Conflict

Conflict has affected agriculture in Myanmar (Burma) in a range of ways.

Conflict effects on agricultural lands

At the national level, paddy production accounts for approximately half of the total production area (Kudo *et al.*, 2013), although national statistics on agricultural production during the conflict period are extremely unreliable (Haggblade *et al.*, 2014). However, defining the areas of conflict is perhaps more important for understanding which types of agricultural production have been affected most by armed conflict. During the 1950s and 1960s, armed conflict occurred in many of the regions near the capital city of Yangon. This is the central delta and dryland zone of the country. Later, in the 1970s, as the government asserted control over many areas near the capital, most of the conflict in the delta and the dry zone areas of Burma subsided. The significant conflicts have instead taken place in the upland and mountainous areas, which were traditionally controlled by the ethnic minorities.

The proportion of productive agricultural land affected by conflict is approximately 40% of the arable land in the country. This includes Shan State, Kachin State and parts of Kayah State. The most economically viable agricultural land is the delta land used for rice production, which is not in the conflict-affected areas. The higher value-added cropping areas—coffee, tea and fruit production—are in conflict zones.

Conflict effects on farmers

Myanmar is still predominantly an agrarian nation, with 63% of the population involved in agriculture in some manner (Cho, 2013). However, most farmers in Myanmar are not directly affected by conflict. The farmers in the delta areas and the central dry zone, who make up about 70% of the farmers in Myanmar and produce the majority of its rice and major export crops, did not face the same conditions as the farmers living in conflict-affected areas (Wilson and Mwee, 2013). Numbers are difficult to obtain, but an estimated 30% of the farming population, largely farmers from the ethnic minorities, were affected by conflict.

The farmers in Shan State and Kachin State comprise both paddy rice cultivators in the valley areas and shifting cultivators in the mountainside and upland areas. Data from research concerning value chains and post-conflict recovery (Ringer, 2014) reveal that there are innovators and early adopters of hybrid maize production within the ethnic minority farmers sub-group. While these farmers are not currently a large part of the country's highest earning agricultural producers, they could have high earning potential once the upland cropping areas are developed, and significant production potential due to the size of Shan State and the other mountainous regions. However, using the innovation definitions of Rogers (2003), these farmers would be considered laggards compared with farmers countrywide. Hybrid maize yields are quite low (e.g. 3 tonnes per hectare [t/ha]) compared with those of resource-rich farmers (e.g. 7–8 t/ha) who were not affected by conflict (Ringer, 2014).

There are many reasons for these differences. The following is a description of some of the main barriers these farmers face due to armed conflict.

Loss of animals and planting material

One factor affecting the post-conflict environment is that Burmese farmers have greatly diminished livestock and seeds with which to rebuild their production capacity. Without their traditional seeds and livestock, marginalized farmers have lost an important resource that they have preserved for hundreds of years. This phenomenon is common in post-conflict settings, and development practitioners have recognized that farmers recovering from armed conflict often need seed for replanting (Christoplos *et al.*, 2006). Any development approach needs to take into consideration the supply of livestock and seeds, but this assistance needs to be carefully considered as the wrong seed may be provided and at the wrong time.

Loss of markets

Markets in northern Shan State are not stable because of the ebb and flow of conflict. The potential for warfare to erupt at any moment causes farmers and farmers' groups to pause when considering how to re-engage markets after conflict has subsided in their area. As a result, most farmers complain of the lack of sustained and reliable markets for their goods (Ringer, 2014). Conflict-affected farmers in northern Shan State commonly have limited knowledge of current market rates. They are also hampered by limited buyers and have to accept low prices because of the trader's risk in coming into armed conflict zones. The traders then reduce prices to compensate themselves for the high chance of losing both transport vehicles and purchased products. These agricultural traders could be described as 'conflict entrepreneurs' (Eide, 1997).

Land and employment controlled by agribusiness, government military or insurgent militias

The actions of influential individuals within the foreign and local business community, insurgent militias and the Burmese military can also be seen as the actions of conflict entrepreneurs (Eide, 1997). Conflict entrepreneurs take advantage of armed conflicts to enrich themselves through force or influential relationships with powerful individuals.

The Myanmar military used a strategy of pacifying areas by removing villagers from remote regions controlled by armed insurgent groups and resettling them in villages near the main roads (Ferguson, 2014; Henley, 2014), thus reducing support for the insurgent groups. However, these pacification programs also impacted the ability of resettled villagers to conduct agricultural activities and produce food and export crops, and gave their lands over to military control.

Another military strategy for the pacification of conflict areas was through encouraging government soldiers to marry local women. Soldiers who complied would receive a parcel of land, and whole villages were formed to control territory with loyal soldiers (Pick and Thein, 2010). While not completely successful in all areas, the government has gained control of strategic lands including the main export route to China and to Thailand, which is vital for continued agricultural development.

Land confiscation is still a leading cause of conflict and protest countrywide (ALTSEAN, 2014). Several key laws make land confiscation a legal method of appropriating land for military, business and government use (ALTSEAN, 2014). Because farmers have been dispossessed due to forced migration and village removal from insurgent areas, they have lost access to their traditional lands. These lands provided the resources needed to sustain their families and provide sufficient food security (Than, 2005). Henley (2014) noted that there have been attempts to start large-scale land retitling, but the government did not have the capacity to implement the program effectively.

Conflict effects on food security and food aid

Food security through increased food production is vitally important for Myanmar's expanding population. Cho (2013) suggests that it will require 20 million tonnes of rice a year to feed Myanmar's people. To meet this demand plus targets for rice exports will require a major expansion of production and the rice-growing area (Cho, 2013). Although Myanmar has achieved national self-sufficiency in rice, there is still roughly 65% of the population that has inadequate food intake for 2 months of the year (Haggblade *et al.*, 2014).

Food security has been a concern of the government, and the situation was made worse by low food production in conflict areas. Due to decreases in the country's productive capacity as a result of conflict, food aid continues to be important in Myanmar, especially in conflict zones. Food aid is needed mainly for internally displaced persons (IDPs) due to armed conflict (UNCTAD, 2011; USAID, 2015a). As an example, USAID provided significant food aid during 2015 to alleviate food scarcity among IDPs in Shan State and Kachin State (USAID, 2015a). However, this is becoming less necessary as the national agricultural system begins to become more food-resilient (Byerlee *et al.*, 2014).

Conflict effects on agricultural extension, research and education

Agricultural extension

Agricultural extension has a long history in Myanmar (Burma), stretching back to the early 1920s (Cho, 2013). Extension has long been underfunded, particularly in regions far from the capital, where most of the armed conflict occurred (Robertson and Olson, 2012). The World Bank's Training and Visit (T&V) system was proposed and implemented, but never developed beyond the initial pilot projects (Cho, 2013). During the transition to a socialist political system, the extension system de-emphasized interactions with farmers and exerted most effort in collecting agricultural data for reports to the Ministry for central planning (Cho, 2013).

Agricultural extension suffered in many ways due to conflict. One direct effect of conflict was the dual reporting nature of district extension officers. These officers had to report to their superiors within the Ministry of Agriculture and to the District Government Military Commander (Cho, 2013). Effectively this meant that the Extension Director for the district had to convince the military commander of the value of extension activities in order to receive work approvals. Military commanders were rotated out of the district every 3 years, so the value of extension at the district level was undermined (Cho, 2013).

Agricultural research

A number of research institutes have been developed by the Ministry of Agriculture. These research sites provided information on the main crops produced but were generally underfunded (Cho, 2013). Increasing research capacity is a major focus of current government efforts to develop Myanmar's agriculture (Win, 2013; Byerlee *et al.*, 2014). These efforts include research sites in conflict-affected areas in Shan State (Cho, 2013).

Agricultural universities and institutes

The military government often clashed with university students (Hedlund and Daw, 2008), and in the 1980s saw the universities as sources of resistance and protest against the government. As a result, many academic years of the university system were suspended. Higher education institutions were closed during periods of civil unrest, including 2 years after student protests in 1988, 3 years after student strikes in 1996 and 1998, and in Yangon from 1988 to 2000 for a period of 10 years (Lall, 2008: 132). This was a heavy blow to the training of qualified agricultural extension staff. This university backlog and the lack of skilled agriculturalists have meant that the development of more efficient, market-oriented agriculture has been hampered by a dearth of trained personnel (Cho, 2013). A clear result of the armed conflict throughout the country, this situation is now slowly being rectified.

Conflict-Related Displacement of Peoples

Movement to safe areas and neighboring countries

The trauma of armed conflict often means that farmers lose the desire to rebuild in their previous location. Canavati *et al.* (2011) estimated that up to 2 million migrants moved to Thailand from Myanmar to escape fighting. Migration to neighboring countries is fraught with difficulties for young men and women, but despite the danger and exploitation that occurs it is seen as bearable compared with remaining in northern Shan State, where the future is not bright for young adults. The loss of the youth and young adults (aged 16–30) for the community, and the risks involved in moving, are described by the corrosive community reaction theory of Freudenberg and Jones (1991), where social capital is lost after a shock to the community. Out-migration increases financial and personal risk for farm families that stay in northern Shan State (Bryceson, 2000).

This out-migration is not promising for further development because the younger generation are a key group for helping to develop agriculture. While the agricultural opportunities in neighboring countries such as Thailand, China and Malaysia offer positive on-the-job training opportunities for these young people, the benefit will be realized in Myanmar only if they return home. Research indicates that the situation is not yet positive enough for young people to want to return to the post-conflict areas of Myanmar (Bacchin, 2012; Ringer, 2014).

Conflict and Community Resilience

Innovation and resilience

Because of poor extension services and the lack of trust in government information sources, local input dealers became the de facto opinion leaders concerning agricultural innovation in many conflict areas of Myanmar (Rogers, 2003; Borgatti, 2006). At the same time, marginalized farmers became more cosmopolitan and actively sought out agricultural information, leading to increased opportunities for extension actors to work with them as they became more open to new ideas (Rogers, 2003). There are also indications that opinion leaders within conflict-affected communities are becoming more innovative and resilient as armed conflict becomes more common in their areas (Ringer, 2014). For example, some farmers are trying to grow new crops to respond to market demand from China.

Reliance on familial networks

Farm families rely on family members to mitigate the effects of conflict in ways that farmers in non-conflict areas do not. This can be described as reliance upon familial networks to provide early warning if violence is likely to occur and/or to negotiate for their village and fields being bypassed by warring groups. This use of family members in key places within society ties in with key player theory (Borgatti, 2006). These key family members act as important nodes of information and play the role of opinion leaders in Rogers' diffusion of innovations theory (2003). Conflict management by family members using smoothing and compromising approaches (Hamad, 2005) is commonly seen among ethnic minority farmers affected by conflict (Ringer, 2014). Smoothing and compromising approaches are used by less powerful groups with more powerful groups that emphasize commonalities and minimize or suppress group differences (Hamad, 2005; Mohammed and Karami, 2007). Although conflict management theory describes several approaches, including forceful ones, the only approaches seen among these marginalized farmers in post-conflict zones were smoothing and compromising. This is due to their weak position compared with other, more powerful members of society.

Post-Conflict Contextual Factors

Positives

Transition from acute- to chronic-focused agricultural development

The Ministry of Agriculture has been tasked with alleviating food insecurity throughout the country and is also closely coordinating the activities of international and local non-governmental organizations (NGOs), which are predominantly focusing on improving food security within the areas that have experienced armed conflict (Joliffe, 2014; Ringer, 2014).

Indeed, progress has been made. As Myanmar has developed its agricultural production and more conflict areas remain peaceful, there has been less need for acute food aid. Domestic and international NGOs (INGOs) helping farmers affected by conflict are shifting from acute emergency relief to working to solve chronic agricultural development issues. The transition from acute to chronic aid assistance has been demonstrated through the World Food Programme's transition from providing food aid to helping rebuild productive capacity. Similarly, the Myanmar extension service under the Department of Agriculture (DOA) has encouraged a quick transition to farming livelihood assistance. One exception to this was Cyclone Nargis, which had a major impact on the Irrawaddy Delta rice-producing areas in 2008 (World Concern, 2013). Acute food aid was required while food supplies and production were re-established.

In this shift it is important to offer farmers 'opportunity ladders' (Christoplos and Farrington, 2004: 29). These are proactive attempts by agricultural development workers and agricultural extension staff to develop methods for farmers to engage in their own development quickly so they are less dependent upon foreign assistance.

Shift from agricultural deficiencies to surplus

Myanmar has focused on the effort to return once again to being a major rice exporter. The gains made in recent years are important and are now evident in the changes that have been made to reposition the agricultural extension service to better meet the rice production goals in the delta region (Cho, 2013). There is strong potential for growth in agricultural products. It is Myanmar government policy to see Myanmar once again as the rice bowl of Asia, and this has been the focus of many development efforts. Several of the larger agribusiness interests in the region see Myanmar, and Shan State in particular, as one of the last large areas of arable, fertile agricultural land available for development in Asia (Bert Hicks, Winrock International, Yangon, 2013, personal communication). However, rice production and agricultural production are still low compared with that of neighboring countries (Haggblade *et al.*, 2014).

Kudo *et al.* (2013) noted that the Myanmar government has prioritized agricultural growth. Currently agricultural production is 36% of gross domestic product (GDP), employs a majority of the labor force and as of 2010 generated 30% of total exports. On 19 June 2012 President U Thein Sein stated that Myanmar was entering the second phase of reform strategy. This effort by the government is part of four economic policies, the first of which is 'sustaining agricultural development towards industrialization and all-around development' (Kudo *et al.*, 2013: 1). Myanmar has 12,441,000 ha of agricultural land (of a total area of 67,659,000 ha), and a population of 48 million people, so there is a major need to increase land productivity and improve labor efficiency. The current average paddy rice yield is 4.12 t/ha, which could be improved. There are 8 million ha of rice paddy, in both conflict zones and non-conflict zones, although yields in the conflict zones need the most improvement (Kudo *et al.*, 2013).

Large trading partners in China and India

Myanmar has natural geographic advantages for expanding agricultural production and trade. The country is well placed as the western outlet in China's strategic plan to gain access to a reliable port on the India Ocean (Asia Society, 2010; Thuzar, 2013). Recent events have shown relentless growth in this area. The overland routes to India are not as developed as for China (Myint-U, 2011), but Myanmar's exports of agricultural products to India are large and growing.

Other areas of growth are in the expansion of other Asian multinational agribusinesses into Myanmar as a market to develop a production base for exporting into Asia. Some examples include the Thai multinational Charoen Pokphand Group (CP) and other, smaller Thai companies. Malaysian and Vietnamese companies also have growing interests in developing Myanmar's vast agricultural potential. For example, there has been a surge in maize growing due to demand for animal feed and regional markets (Haggblade *et al.*, 2014).

External demands and price increases

A positive for Myanmar farmers is the continued and growing demand for agricultural exports to China. This pull of the Chinese markets for agricultural products is felt not only in the post-conflict regions of Myanmar, near the border with China, but also in the dry zone, where demand for goat meat, melons, coffee, maize and soybeans means there is always a ready buyer. This does not guarantee high prices but it does provide farmers a consistent buyer of some kind.

The ongoing demand from Chinese markets, coupled with the development of transport and information thoroughfares, will make Myanmar's agricultural products more accessible for the Chinese markets, with lower transportation costs. This will eventually increase prices received by farmers. However, there are indications that marginalized population segments are being bypassed in order for this development to occur, and efforts to ensure more equitable benefits must be considered (Kabeer, 2002; Longley *et al.*, 2006).

Negatives

Persistence of conflicts

In the post-conflict areas of Myanmar there is a continued threat of renewed conflict (Ferguson, 2014). The Rohingya issue in Myanmar continues to grow as a conflict between ethnic Burmese, people in Rakhine and the Rohingya. These tensions pose a problem for stability in Rakhine State. Policies by the Myanmar government intended to stem the tide of perceived loss of ethnic Burmese/Rakhine land to the Rohingya have exacerbated and prolonged the conflict (Leavitt, 2007). These issues and instances of violence continue to cause farmers to flee their villages into IDP camps (USAID, 2015a) and set back agricultural development (Xinhua News Agency, 2013).

Large areas and population segments still under insurgent control

The patchwork nature of the post-conflict environment in Myanmar means that many areas and population segments are under insurgent militia control (AsiaPacific University, 2005; Joliffe, 2014). This may be so for the foreseeable future. There is a real possibility that armed conflict could break out again as the Myanmar government attempts to assert control over the militia areas (Bacchin, 2012). This lack of control means that many areas and population segments do not receive any agricultural assistance from the government. Instead, some NGOs are allowed access to provide information about food aid but little else. Otherwise, the only agricultural technical assistance they receive is from fertilizer or seed dealers who have reached agreements with the insurgent militias' ruling groups (Ringer, 2014). The impact of these services is not likely to provide much support to conflict-affected farmers, however.

Internally displaced persons

Many of the issues with IDPs from the larger conflict period persist today. There are still many IDPs in Myanmar at the edges of the rebel-controlled areas. USAID estimates that there are at least 230,000 IDPs currently in camps in Rakhine, Shan and Kachin States (USAID, 2015a, 2015b), and these IDP camps are filled with marginalized ethnic minority farmers. For IDPs, the decision to return is often filled with anxiety (Kedir, 2011; Bacchin, 2012). The alternative choice for farmers is to move to the regional towns that are controlled by the Myanmar government (Ibiblio, 2010). One interesting finding is that IDPs were more likely to return to their village and resume farming if they had participated in the early stages of an agricultural extension intervention—such as a Farmer Field School—either before or during their presence in the IDP camp (Ringer, 2014).

Citizen protests against land displacement by agribusiness companies

Land appropriation has occurred over decades (Henley, 2014: 4): (i) by the Myanmar military government for military purposes; and (ii) by the government for investment purposes. The main groups that benefit from this are Burmese agribusiness firms and foreign companies (TNI, 2013; Henley, 2014). Marginalized farmers in northern Shan State have a growing sense that they have been pushed away from agricultural land that they believe belongs to them (Henley, 2014; Ringer, 2014). This issue will likely grow in importance as the Myanmar government encourages foreign agricultural investment. Landlessness is estimated at 25–30% in the rural countryside, with most of the landless poor employed as agricultural laborers (Haggblade *et al.*, 2014; Henley, 2014). As marginalized farmers take part in off-farm labor, it will become more important for the agricultural extension service to provide training and skills development in worker safety and equitable wages.

Post-Conflict Extension Strategies and Actions

The slowing of conflict and the opening up of the country has meant there is a large emphasis upon developing agriculture. This has led to a range of extension activities and approaches in post-conflict Myanmar.

Public extension

Myanmar agricultural extension system efforts

The DOA is the government agency responsible for providing agricultural extension services to Burmese farmers (Cho, 2013). There are two objectives of an effective, comprehensive agricultural extension system: (i) maintaining national food security through technology transfer (mainly for staple crops) and training farmers to use sustainable natural resource management practices; and (ii) improving rural livelihoods through teaching farmers to diversify farming systems to increase incomes and organizing farmers into producer and community groups (Swanson and Rajalahti, 2010; Swanson, 2011). This is an appropriate lens through which to view the Myanmar Agricultural Extension System.

The Myanmar Agricultural Extension System has focused closely on maintaining and improving national food security. The increase in domestic production of rice, maize, and legumes such as chickpeas indicates that the technology transfer component, at least in non-conflict zones, has been a success. However, most sources indicate that natural resource management more generally has suffered in both conflict-affected areas and non-conflict zones (Robertson and Olson, 2012). This gap is evident not only in agricultural production but also in forestry management (Ringer, 2014).

Improving rural livelihoods has also been a priority. The efforts by the Myanmar Ministry of Agriculture to diversify farming production and add value along agricultural value chains have been slow to progress but are showing signs of increasing (Cho, 2013). Conflict narrows the options for smallholder farmers because only low-risk crops can be grown and many other barriers to profitable production exist (Ringer, 2014). For example, marginalized farmers' transition to growing hybrid maize provides the quickest route for cash income, but farmers in post-conflict areas struggle with access to these inputs and the costs associated with increased purchase and application of agrochemicals for optimum growth. Maize production has been growing throughout the conflict areas, but production is still below regional production levels (Byerlee *et al.*, 2014).

Only as the conflict-affected zones stabilize will diversification of agricultural production be able to improve farm livelihoods. The organization of farmers into producer and community groups has been taking place steadily over the past 10 years. The Farmer Field School methodology has helped encourage the growth of these groups and provided a form of grassroots organization that was palatable to the previous military-led government (Morris, 2009; Cho, 2013; Mai, 2014). Progress has been slow, but over the past 3 years there have been promising signs in rural livelihood growth, especially with the involvement of NGOs.

Reassignment of extensionists from upland areas to rice-producing areas

Recently there has been a reassignment (termed retrenchment) of agricultural extension staff members in order to focus on rice production in the delta and the dry zone (Cho, 2013; Ringer, 2014). This retrenchment meant that fewer extension personnel were available to assist marginalized farmers in the conflict-affected areas.

Retrenchment was part of the Selective Concentrative Strategy promoted by the government to focus development on critical areas in agriculture (Cho and Boland, 2003; Cho, 2013). This strategy of retrenchment and refocusing extension field staff has been effective in enabling Myanmar to develop rice production and become a significant exporter of rice, even though it also removed extension officers from conflict-affected areas in Shan State and Kachin State. It has also been successful in developing key areas along the agricultural production value chain.

However, the strategy has also been criticized for failing to develop other, more conflict-affected regions in Myanmar. There remains a crucial need for agricultural capacity to be rebuilt in the conflict areas in order to continue national agricultural development, although there are signs that this is beginning to happen (Cho, 2013).

Government efforts to engage agribusinesses in conflict areas with profit potential

The government's larger objective to increase agricultural production includes the expansion of agricultural opportunities for larger companies (Thuzar, 2013). To benefit from the agricultural potential of conflict areas, the Myanmar government is encouraging domestic and international agribusinesses to conduct business in conflict areas where there is moderate to intermittent stability. The Ministry of Agriculture and/or local agricultural/government officials provide opportunities for these companies to purchase land and import/export permits, and to conduct business, while the companies take on the risk of conducting business in a conflict area. This also saves government extension staff from needing to provide assistance to either the contract farmers or the businesses. If armed insurgent groups attack one of these businesses, then the only loss is to the business itself (Ringer, 2014).

This strategy of encouraging foreign investment (Cho, 2013) has been successful from the government's perspective because private businesses have encouraged the development of infrastructure and provided opportunities for employment. The influx of businesses willing to take the risk of armed conflict and bear the taxes imposed by militia groups also has slowly changed the reality on the ground for the government. If it can maintain enough stability through truces and ceasefires with the armed insurgent groups, then agribusiness activities can continue with little cost to the government.

Drug eradication and opium-substitution programs

During the conflict, and especially over the past 3 years, there has been a return to opium production (Thuzar, 2013). The growth of high-value illegal crops is a common resilience strategy in conflict and post-conflict countries.

Crop-substitution efforts have been in place for years, with mixed results (UNODC, 2011). Some of the most important agricultural extension activities in conflict and post-conflict areas occurred under the many crop-substitution programs funded by the USA and implemented by the Burmese military. For example, during the period of military rule, the USA partnered with the Burmese military government to eradicate opium production in the mountainous areas of Shan State and Kachin State, despite the ongoing US trade embargo. The Myanmar Agricultural Extension Service has also focused on crop substitution in recent years.

However, it is not known yet what impacts the government's agricultural extension service can have on this segment of the farming population over the coming years. Fuller and Moe (2015) reported that opium production has continued among many smallholder farmers in conflict areas controlled by the militias. Ringer (2014) noted that many smallholder farmers will begin producing crops other than opium only if the economic return remains high and conflict is avoided.

NGO extension

NGO and donor community involvement

The Myanmar government has also followed a strategy of allowing NGOs and INGOs (such as the UNDP, World Food Programme and World Vision) to engage farmer groups in relatively stable areas. This allows the government and the Ministry of Agriculture to have an influence over programming in areas where their extension staff are not welcome. Several local NGOs are based in the main business city of Yangon and are affiliated with one specific ethnic group. These local NGOs have inroads as 'third party mediators/providers' of extension services. This serves the purposes both of the armed insurgent groups that allow them to operate and of the Myanmar government, which does not have to risk its staff or expend resources (Ghosh, 2009).

The issue of 'donor shopping', where government officials have limited capacity to absorb donor funding, is a concern (Rieffel and Fox, 2013). There are many NGOs rushing into Myanmar, as in some other post-conflict countries such as South Sudan. This probably will result in many disjointed efforts, which will not lead to building a strong agricultural extension system (Goodhand, 2006). Many different NGOs working without coordination is a concern of both the INGO community and the Myanmar government (Saha, 2011).

To avoid potential issues, donors should slow down and conduct more joint operations. Efforts have been made to work in partnerships and to conduct joint monitoring and evaluation efforts. An example of this is the Food Security Working Group, a voluntary organization of local NGOs and INGOs that has helped provide joint learning and capacity building for NGO staff members. This organization and others like it have helped with coordination of activities, specifically in the dry zone. The potential for the Myanmar Agricultural Extension Service to step into the coordinating role is strong.

Working groups and impact area coordination

In Myanmar there are strong interactions between the Ministry of Agriculture and the INGOs providing Overseas Direct Assistance (ODA) in agricultural development. There are two main ways in which the Ministry of Agriculture has worked with INGOs and local NGOs to coordinate NGO activity:

1. Having working groups centered around important themes of the government's national agricultural development plan. The benefit of these groups is that they allow NGOs to jointly develop training material and programs, share experiences and best practices and interact with the Ministry of Agriculture in a coherent manner.
2. Coordinating impact area approaches that are in line with the Ministry of Agriculture's plan for a grouping of five to ten districts. One NGO is designated the lead NGO which reports to the division's (similar to a province) Ministry of Agriculture director. This means that programs are closely coordinated and meet Ministry of Agriculture goals (Cho, 2013). This impact area coordination is likely a carry-over from the time of the military leadership during which all NGOs were under close government scrutiny. This is now beginning to change since a new law liberalizing NGO activity in the country was passed in 2014. At the same time, NGOs do see the benefit of the impact area coordination effort and largely comply.

Accountability of NGOs

A concern often expressed in conflict areas, and in agricultural development in general, is the lack of NGO accountability to the national government, and the government extension service in particular. This is seen as undercutting the efforts of the government extension service. One major difference in Myanmar compared with other countries newly emerging from conflict is that the central government has retained control over the national agricultural structure in spite of major political turmoil. It was also a centrally planned economy in which control over every aspect of agricultural production was seen as a government prerogative. INGO approvals to work in agricultural development and extension were tedious and slow (Local Resource Centre, 2012). The government carefully controlled NGOs' development activities, and INGOs were accountable to the government; if they were not, they would lose the right to work in the country.

This is due in part to Myanmar's history as a socialist, centrally planned country without space for a civil society free from centrally planned government control. However, local NGOs have played an important part in Myanmar by delivering agricultural advisory services to minority ethnic communities that the government was unable or unwilling to reach (MacGinty and Richmond, 2013). This area will grow as the restrictions on local NGO activity are removed following the recent opening up and elections, which is the preference of staff members in the NGOs and the farmers they serve.

A growing and vibrant civil society is important for peace-building, and therefore the growth of civil society in Myanmar is an encouraging sign that a durable peace can be obtained (Christoplos, 2010; Clapp, 2015). In a country like Myanmar, expanding civil society with multiple actors leads to more voices being heard in agriculture, including those traditionally underserved by governmental agriculture services.

Technical expertise of NGOs

One concern of agricultural extension theorists and practitioners is whether NGOs and INGOs that provide agricultural advisory services through their activities have the technical agricultural expertise relevant to the Burmese context that is necessary to provide quality services (Fowler, 2000). If so, there is the potential for NGOs lacking technical capacity to crowd out actors who might be more effective. This is a legitimate concern, but the indications are that INGOs often access technical capacity through hiring retired Burmese agricultural professors, who are some of the most experienced technical agricultural staff in the country, expanding the available technical agricultural capacity for extension stakeholders. The second way in which technical expertise is provided is by hiring private consultants to provide technical backstopping for projects. A third way is through joint staff and personnel development via working group coordination, which often involves study tours throughout South-east Asia or in the West. Without this NGO activity, the quality of technical capacity in Myanmar would be much weaker.

Private sector extension

Private consulting groups

The private and semi-private consulting groups active in Myanmar consist mainly of professors and development experts who predominantly are retired and have spent time in the Ministry of Agriculture or have worked for INGOs. Some of these individuals built up their technical expertise while working for the Food and Agriculture Organization of the United Nations (FAO), the United Nations and the World Bank during the time of the military dictatorship. These consulting groups have provided valuable inputs to numerous NGO and INGO activities and to farmers. The tendency for these retired professors to move in and out of NGO jobs while maintaining close connections to the Ministry of Agriculture means there is considerable sharing of expertise and coordination through informal channels. Salaries for NGOs are much higher than for government agricultural extension positions, which does draw many exceptional persons from the government agricultural service into the NGO arena. At the same time, the government agricultural service remains attractive because of the permanence of the positions, unlike NGO project-based work.

Producer groups and cooperatives

Since opening up the economy to outside investment, there has also been the formation of private and public–private national producer groups. Some of these groups focus on key commodities, such as the Livestock Federation, while others are formed to promote agricultural development in general. The government still prefers to exercise some form of guidance over these groups. Overall, the groups have become more active in encouraging new investment opportunities in agriculture, providing information for producers and advocating for positive changes in policy.

Recommendations and Areas for Improvement

Engaging stakeholders

Coordinating agricultural extension activities

Considering the framework of Swanson (2011), efforts by the Myanmar government to play a coordination role in conflict areas can increase the effectiveness of agricultural extension. It is not clear whether the government and Ministry of Agriculture will embrace the role of coordinator of NGOs' agricultural activities, as have other countries (Cho, 2013). This is especially important given that the Ministry of Agriculture is focusing on the delta and dry zones, and relying on NGOs and agribusinesses to develop agriculture in conflict zones. Better coordination of these actors would allow more opportunities for conflict-affected smallholder farmers to increase production, improve livelihoods and market goods, even though they would have to pass through the conflict boundary between the insurgent militias and the government military forces. Improved coordination could allow more government agricultural development goals to be realized without agricultural extension field staff becoming the targets of insurgent militia retribution.

Decentralizing extension services

Research has shown that decentralized agricultural extension systems are important for integrating farmers from disenfranchised communities. Decentralized systems have therefore been seen as important in the development of countries with histories of conflict and with numerous ethnic groups (Rogers, 2003). Evans (2004: 31) described the importance of the 'process of deliberation vs. importing outside blue prints'. These processes will need to be managed by the agricultural extension services and closely integrated with NGOs.

In Myanmar there are several instances of how decentralization can change the perception of government extension workers. For example, livestock officers in Kachin State are often given safe passage into Kachin Independence Army (KIA) areas because of the KIA's concern for their own livestock, and in recognition that these extension workers are local rather than national representatives. This is an example of the value of decentralized systems and how agricultural extension and technical officers can be more welcome in insurgent-controlled areas than are other government officials, who have been kidnapped or prevented entry (Joliffe, 2014).

Community ownership of resources

An additional complexity is the need to respect the natural resources that each of the ethnic minority groups believe belong to them. A clear example of this is the development of Mithun cattle by an agribusiness in Chin State. This viable and legal business was shut down because the agribusiness was owned by individuals from the main ethnic group (Bhama). Influential members of the Chin ethnic group protested to government officials at the regional level and forced the closure of the business. The Chin saw the Mithun as belonging to them. In this case, a viable enterprise was shut down because of a sense of minority ownership of resources unique to their region. It is possible that economic ties and agricultural development will eventually overcome these concerns and ease the political issues at stake. Government agricultural extension staff could play a role in mediating these concerns by balancing the goal of agricultural business development and respect for disenfranchised ethnic groups (Myanmar Centre for Responsible Business, 2014).

Serving conflict-affected farmers

Identifying key players

In conflict and post-conflict areas, farmers should be approached with caution by both public and private entities. Potential farmer participants have a high level of distrust towards outsiders attempting to develop new projects through the government. Outside development agencies have needed to be cautious while interacting with marginalized farmers in order to build trust. Some farmers are also still involved in illicit opium growing (Lintner and Black, 2009; Fuller and Moe, 2015). These factors all lead to the need to invest in trust-building before attempting agricultural activities.

Coordinating with the farmers who are key players or opinion leaders is vital in stimulating farmer participation in agricultural projects. Key players and/or opinion leaders should be identified and engaged in the early stages of agricultural program development in order to develop local enthusiasm and support for program activities. There is an effort to develop a more participatory agricultural extension service, and the coordinating role may become even more important for the extension service (Cho, 2013). Many agricultural development projects have noticed this reality and have focused on developing equitable business ventures that involve the leaders among otherwise marginalized farmers (Ringer, 2014).

Targeting farmers' groups

More emphasis on farmer interest groups and training of key farmers could have allowed smallholder farmers and mid-level farmers to access the agricultural information and resources they needed to progress. A common complaint is that there is loss of seeds and livestock during times of conflict (Ringer, 2014). Having agricultural advisory centers in central safe areas, where agricultural extension staff could operate freely, would allow key farmers to travel from conflict zones, access what they need and then return. This would also allow the government agricultural advisory services to gather agricultural information from insurgent militia-controlled areas.

Engaging IDPs in camps

With IDPs continuing to exist in large numbers in the conflict-affected zones, it is important that NGOs and/or government extension staff begin agricultural rehabilitation planning and training while people are still in the camps. Research from the Philippines has shown that training in IDP camps helped with post-conflict recovery and reintegration (Ongsiapco *et al.*, 2007). Ringer (2014) gathered data from a local NGO that showed farmers were more likely to return to villages if they had engaged with agricultural skills training in the IDP camps using a Farmer Field School approach. Similar efforts with Burmese IDPs should be expanded.

Addressing gender issues

Gender issues in agriculture and agricultural extension are an important but often overlooked part of service delivery in post-conflict Myanmar. Traditional Burmese society does not encourage equality between men and women. However, this has been changing as gender awareness has been a component of most INGO and local NGO training programs (Metta Foundation, 2012; World Concern, 2013). Cho (2013) provided statistics on the government agricultural extension service, which showed that 60% of farmer participants are females. However, extension field staff are still predominantly male (Cho, 2013). Armed conflict-affected zones could be a factor in this tendency towards male extension staff, due to the dangers of female extension staff being alone in an intermittent conflict area and facing the potential for rape and bodily harm. Nonetheless, this gender imbalance creates social issues that affect the quality of services provided to vulnerable female farmers.

Dealing with continual conflict

Armed conflict will continue to be an issue that must be engaged within Myanmar's development. As Barron *et al.* (2011: 42) wrote: 'The development challenge is not of limiting conflict per se but of managing it in constructive ways'. Dealing with armed conflict in a constructive way will be a difficult role for the extension service.

Joliffe (2014) notes several areas of continued conflict in Myanmar that will serve to prohibit national cohesion and development. These include issues with insurgent militias in the border regions, ethnic divisions that have fueled much of the insurgent fighting, and religious divisions between Buddhists, Muslims and Christians. Agricultural extension staff will have to navigate these continued sources of societal friction. This will require attention to conflict-sensitive agricultural strategies and extension approaches that encourage equal access and participation by all types of farmers in Myanmar, including those affected by conflict.

Promoting peace-building and social cohesion

Extension can play a role in peace-building, even in a context like Myanmar, where insurgent militia groups see government organizations and the INGOs that work with them as agents of the government (Joliffe, 2014). Efforts to include multiple actors in any agricultural extension strategy are essential, so that disadvantaged and disenfranchised segments of the population can have a say in their own agricultural development. Although chaotic at times, the involvement of many local NGOs in accessing and providing agricultural advisory services is a good sign of a growing civil society and the inclusion of local perspectives (Saha, 2011).

Agricultural extension activities must also build in a social cohesion component that promotes group interaction, shared benefits and reconciliation strategies. When coupled with agricultural livelihood approaches that enhance entire regions, extension can have peace-building impacts. Furthermore, efforts to build sustainable value chains and effective institutions can contribute to stability and longer-term development. In these cases, instances of violence are less likely to occur.

Utilizing information and communication technologies and other extension technologies

Information and communication technologies (ICTs) are important for the development of agriculture in Myanmar. The coverage of smartphones has spread rapidly throughout rural areas since cell phone deregulation in September 2014, increasing the opportunities to use this technology in agricultural extension activities. ICTs have also been included in the USAID Value Chains for Rural Development Project, which has recently been funded and initiated in southern Shan State. This is especially important because agricultural information can be gathered and provided between safe zones and armed conflict zones without safety and security concerns. This is of course dependent on cell phone coverage, but, with recent government acceptance of liberalized cell phone access, ICTs have become a vital tool for the government extension service.

There is also an important role for government agricultural extension to provide simple, low-tech and appropriate technologies that fill the gap between hand tools and full mechanization. Examples include treadle pumps and simple weeding tools for upland rice and paddy rice cultivation. These types of technology could be serviced through a centralized town where the government can maintain stability, and also made available at a low cost to smallholder farmers.

Developing appropriate agriculture and extension-related policy

The worldwide trend towards a more pluralistic, participatory and demand-driven agricultural extension service is a positive development for marginalized farmers (Swanson and Rajalahti, 2010), including those in post-conflict areas of Myanmar. However, these trends must be accompanied by agricultural policy that is designed to develop a conflict-resilient agricultural extension system. For example, Bryceson (2000) recommended policies that develop the human capital of post-conflict farmers through programs that: (i) improve productivity using appropriate and inexpensive means; (ii) include participatory market assessments; and (iii) 'provide services that help marginalized farmers capitalize on their comparative market advantages' (p. 12).

It is recommended that Myanmar's agricultural extension policy makers seriously consider the focus on agricultural innovation systems as a paradigm for developing agriculture in Myanmar, both in general and specifically in post-conflict areas. Agricultural innovation systems are 'networks of organizations, enterprises and individuals focused on bringing new products, new processes and new forms of organization into economic use, together with the institutions and policies that affect their behavior and performance' (Davis and Heemskerk, 2012; World Bank, 2012: 2). These systems provide more flexibility in order to meet the needs of industrial agriculture and agribusiness, along with providing appropriate networks for marginalized resource-poor farmers who have experienced conflict (Requier-Desjardins, 2013). Table 7.1 shows agricultural innovation systems policies applied to post-conflict agricultural extension.

The Myanmar Agricultural Extension Service will continue to have to deal with armed conflict situations. It is vital that policy makers and administrators recognize the impact of appropriate policy in empowering the extension service to help farmers move beyond production constraints within their farming systems.

Conclusions and Lessons Learned

Although Myanmar is a unique post-conflict country in many ways, important lessons that apply to other situations can be found, especially for countries suffering from the realities of intermittent conflict.

Agricultural capacity needs to be rebuilt in a manner that focuses on the main areas of social friction that caused the conflict in the first place. This speaks to the important role of coordination and equity in agricultural extension. To develop the capacity for

Table 7.1. Agricultural innovation systems policies applied to post-conflict agricultural extension. Adapted from Davis and Heemskerk (2012): 188.

Policy issues	Application to post-conflict agricultural extension
Proceed with extension reform without relying on a single grand model	The Selective Concentration Strategy may not fit for intermittent conflict areas, so design several models that are appropriate to each state or province
Move towards pluralism	Allow multiple actors in the advisory services, including organizations that can bridge divides between the government and militia groups
Increase downward accountability	Include key farmers from farmer interest groups from villages in the conflict zones or insurgent militia zones to take part in participatory monitoring and evaluating activities and extension advisory councils
Create an effective and efficient market for service providers	Develop centralized locations for information-sharing in safe areas
Face the need for human capacity development	Increase use of Farmer Field Schools and other less costly extension training methods
Move away from projects to programs	Explore feasibility in post-conflict situations because of the influx of aid, and improve coordination so that international NGOs contribute to the overall government agricultural extension effort
Focus on institutions	Identify viable institutions with a focus on plurality, and encourage government extension staff to interact at a distance because of political considerations
Move from standard packages to tailored services	Tailor services by working more with farmers' groups and farmer group leaders, and support this by training field extension staff
Address equity issues	Find ways to include different ethnic groups in the development of local extension strategies to avoid government extension staff being excluded from playing a role in intermittent conflict areas

a coordination role, it is important that field-level agricultural extension units are set up as learning organizations. Each conflict area requires a different approach, so it is vital that agricultural extension program leaders encourage an atmosphere of identifying road-blocks and areas of weakness, and adapt to changing realities on the ground. Most government agencies lack the ability to adapt to fluid, intermittent conflict situations.

Equity is also important. Agricultural extension services must be implemented that serve all segments of populations, including different ethnic groups, women, youth and other vulnerable peoples. These activities need to be carefully developed to avoid causing marginalized resource-poor farmers to become more vulnerable and impoverished. Perceptions of bias can also exacerbate conflict and lead to further divisions. Myanmar has many examples of this occurring.

Another approach could be to loosen control of agricultural advisory activity to allow for a diversity of practice and structure within an overall structure and theme of progress and development. This would create a pluralistic system where different extension providers filled different roles. In Myanmar, NGOs and agribusinesses can work in conflict areas where government extension workers are unable to function. Similar strategies might be applicable in other post-conflict settings.

Overall, though, more research is needed on how to most effectively design appropriate responses to assist in agricultural recovery after conflict. Extension will consistently need to learn from experiences in a diverse range of countries to identify effective approaches and strategies. The Myanmar experience can be one case towards this goal.

References

ALTSEAN (2014) Land confiscation in Burma: a threat to local communities and responsible investment. Alternative Asean Network on Burma, Bangkok. Available at: www.altsean.org/Docs/PDF%20Format/Thematic%20Briefers/Land%20Confiscation%20in%20Burma_5%20May%202014.pdf (accessed 23 January 2017).

AsiaPacific University (2005) Shan State. CartoGIS maps online—ANU. 05-057b_Burma_Shan_state.jpg. Available at: https://www.google.com/search?q=Mapofnorthernshanstate (accessed 23 January 2017).

Asia Society (2010) Current realities and future possibilities in Burma/Myanmar: perspectives from Asia. Asia Society Policy Institute, New York. Available at: http://asiasociety.org/policy-institute/current-realities-and-future-possibilities-burmamyanmar (accessed 23 January 2017).

Bacchin, V. (2012) Internal displacement in Kachin and Northern Shan states, Myanmar: a protection assessment. Metta Foundation, Yangon, Myanmar. Available at: www.metta-myanmar.org/publications/Internal%20Displacement%20in%20Kachin%20&%20Northern%20Shan.pdf (accessed 23 January 2017).

Barron, P., Diprose, R. and Woolcock, M. (2011) *Contesting Development: Participatory Projects and Local Conflict Dynamics in Indonesia.* Yale University Press, New Haven, Connecticut.

Borgatti, S.P. (2006) Identifying sets of key players in a social network. *Computational Math and Organizing Theory* 12, 21–34.

Bryceson, D. (2000) *Rural Africa at the Crossroads: Livelihood Practices and Policies.* Natural Resource Perspectives 52. Overseas Development Institute, London.

Byerlee, D., Kyaw, D., Thein, U.S. and Kham, L.S. (2014) Agribusiness models for inclusive growth in Myanmar: diagnosis and ways forward. MSU International Development Working Paper 133, May 2014. USAID/Michigan State University/MDRI, East Lansing, Michigan. Available at: http://fsg.afre.msu.edu/papers/idwp133.pdf (accessed 23 January 2017).

Canavati, S., Plugge, E., Suwanjatuporn, S., Sombatrungjaroen, S. and Nosten, F. (2011) Barriers to immunization among children of migrant workers from Myanmar living in Tak Province, Thailand. Lessons from the field. *Bulletin of the World Health Organization* 89, 528–531.

Cho, K.M. (2013) Current situation and future opportunities in agricultural education, research, and extension in Myanmar. Background Paper No. 5: Strategic Agricultural Sector and Food Security Diagnostic for Myanmar. USAID/Michigan State University/MDRI, Yangon, Myanmar.

Cho, K.M. and Boland, H. (2003) Participatory learning for agricultural extension and future development in Myanmar. *Proceedings of the AIAEE Annual Conference on Tropical and Subtropical*

Agricultural and Natural Resource Management. University Press Kassel, Germany, p. 280. Available at: http://www.tropentag.de/2003/abstracts/full/233.pdf (accessed 23 January 2017).

Christoplos, I. (2010) *Mobilizing the Potential of Rural and Agricultural Extension.* Office of Knowledge Exchange, Research, and Extension, Food and Agriculture Organization of the United Nations, Rome.

Christoplos, I. and Farrington, J. (eds) (2004) *Poverty, Vulnerability, and Agricultural Extension: Policy Reforms in a Globalizing World.* Oxford University Press, New Delhi, India.

Christoplos, I., Longley, C. and Slaymaker, T. (2006) *The Changing Roles of Agricultural Rehabilitation: Linking Relief, Development, and Support to Rural Livelihoods.* Report of the Humanitarian Policy Group. Overseas Development Institute, London.

Clapp, P. (2015) Myanmar: anatomy of a political transition. USIP Special Report 369. US Institute of Peace, Washington, DC. Available at: www.usip.org/sites/default/files/SR369-Myanmar-Anatomy-of-a-Political-Transition.pdf (accessed 23 January 2017).

Davis, K.E. and Heemskerk, W. (2012) Investment in extension and advisory services as part of agricultural innovation systems in the World Bank. In: *Agricultural Innovation Systems: An Investment Sourcebook.* World Bank, Washington, DC, pp. 179–193.

Eck, K. (2015) Repression by proxy: how military purges and insurgency impact the delegation of coercion. *Journal of Conflict Resolution* 59(5), 1–23. DOI: 10.1177/0022002715576746

Eide, E.B. (1997) Conflict entrepreneurship: on the 'art' of waging civil war. In: McDermott, A. (ed.) *Humanitarian Force.* PRIO Report 4/97. International Peace Research Institute, Oslo, pp. 41–70.

Ekeh, C. and Smith, M. (2007) *Minorities in Burma.* Minority Rights Group International Briefing. Minority Rights Group International, London.

Evans, P. (2004) Development as institutional change: the pitfalls of mono-cropping and the potentials of deliberation. *Studies in Comparative International Development* 38(4), 30–52.

Ferguson, J.M. (2014) The scramble for the waste lands: tracking counterinsurgency and international investment laws in Burma/Myanmar. *Singapore Journal of Tropical Geography* 35(3), 295–311.

Fowler, A. (2000) *The Virtuous Spiral: A Guide to Sustainability for NGOs in International Development.* Earthscan, London, UK.

Freudenberg, W.R. and Jones, T.R. (1991) Attitudes and stress in the presence of a technological risk: a test of the supreme court hypothesis. *Social Forces* 69, 1143–1168.

Fuller, T. and Moe, W. (2015) Myanmar returns to what sells: heroin. *The New York Times*, 3 January. Available at: www.nytimes.com/2015/01/04/world/myanmar-returns-to-what-sells-heroin.html?action=click&contentCollection=Asia%20Pacific&region=Footer&module=MoreInSection&pgtype=article (accessed 23 January 2017).

Ghosh, S. (2009) NGOs as political institutions. *Journal of Asian and African Studies* 44(5), 475–495.

Goodhand, J. (2006) *Aiding Peace? The Role of NGOs in Armed Conflict.* Lynne Rienner, Boulder, Colorado.

Haggblade, S., Boughton, D., Cho, K.M., Denning, G., Kloeppinger-Todd, R., Oo, Z., Sandar, T.M., Than, T.M., Wai, N.E.M.A., Wilson, S., Win, N.W. and Wong, L.C.Y. (2014) Strategic choices shaping agricultural performance and food security in Myanmar. *Journal of International Affairs* 67(2), 55–71.

Hall, D.G. (1981) *A History of South-East Asia*, 4th edn. Palgrave Macmillan, London.

Hamad, A.A. (2005) The conceptualization of conflict management. *Peace, Conflict, and Development* 7, 1–30.

Hedlund, K. and Daw, M.S. (2008) Support to local initiatives in the Nargis response: a fringe versus mainstream approach. Report of the Humanitarian Policy Group, Overseas Development Institute, London. Available at: http://odihpn.org/magazine/support-to-local-initiatives-in-the-nargis-response-a-fringe-versus-mainstream-approach/ (accessed 23 January 2017).

Henley, G. (2014) *Case Study on Land in Burma. Evidence on Demand.* Overseas Development Institute, London.

Ibiblio (2010) Displacement and causes of displacement in eastern and northern Shan State. Available at: www.ibiblio.org/obl/docs4/Shan-docs.pdf (accessed 23 January 2017).

Joliffe, K. (2014) Ethnic conflict and social services in Myanmar's contested regions. Asia Foundation, San Francisco, California. Available at: http://asiafoundation.org/resources/pdfs/MMEthnicConflictandSocialServices.pdf (23 January 2017).

Joliffe, K. (2015) Ethnic armed conflict and territorial administration in Myanmar. Asia Foundation, San Francisco, California. Available at: https://asiafoundation.org/resources/pdfs/ConflictTerritorialAdministrationfullreportENG.pdf (accessed 23 January 2017).

Kabeer, N. (2002) Safety nets and opportunity ladders: addressing vulnerability and enhancing productivity in South Asia. *Development Policy Review* 20(5), 569–588.

Kalyvas, S.N. (2006) *The Logic of Violence in Civil War.* Cambridge University Press, New York.

Kedir, A.M. (2011) Donor coordination in a fragile states of Africa: capacity building for peace and poverty reduction. *World Journal of Entrepreneurship, Management and Sustainable Development* 7(2), 307–356.

Kudo, T., Kumagai, S. and Ishido, H. (2013) Agriculture plus plus: growth strategy for Myanmar agriculture. IDE Discussion Paper 421. Institute of Developing Economies, Chiba, Japan. Available at: www.ide.go.jp/English/Publish/Download/Dp/pdf/421.pdf (accessed 23 January 2017).

Lall, M. (2008) Evolving education in Myanmar: the interplay of state, business, and the community. In: Skidmorem, M. and Wilson, T. (eds) *Dictatorship, Disorder, and Decline in Myanmar.* ANU E Press, Canberra, Australia, pp. 127–150. Available at: http://oapen.org (accessed 23 January 2017).

Leavitt, S.R. (2007) Persuasion, coercion, and neglect: understanding state policy and the mobilization of Muslim minorities in Asia. PhD thesis. Georgetown University, Washington, DC.

Lintner, B. and Black, M. (2009) *Merchants of Madness: The Methamphetamine Explosion in the Golden Triangle.* Silkworm Books, Chiang Mai, Thailand.

Local Resource Centre (2012) Bridging the gap between the donor community and local organizations in Myanmar. Local Resource Centre, Yangon, Myanmar. Available at: www.lrcmyanmar.org/en/system/files/book/bridging_the_gap_between_donor_community_and_local_organizations_in_myanmar_eng.pdf (accessed 23 January 2017).

Longley, M., Christoplos, I. and Slaymaker, T. (2006) Agricultural rehabilitation: mapping the linkages between humanitarian relief, social protection, and development. Humanitarian Policy Group, Overseas Development Institute, London. Available at: www.odi.org.uk/sites/odi.org.uk/files/odi-assets/publications-opinion-files/271.pdf (accessed 23 January 2017).

MacGinty, R. and Richmond, O.P. (2013) The local turn in peace building: a critical agenda for peace. *Third World Quarterly* 34(5), 763–783.

Mai, L. (2014) Farmer Field Schools: farmer to farmer experience sharing, mutual learning and dissemination of good practices for an improved agricultural production. Report from Food Security and Livelihoods Program (2011–2014), ACF Myanmar Mission, Kayah State. Available at: http://www.themimu.info/sites/themimu.info/files/assessment_file_attachments/140725_FFS_capitalization_Final_2.pdf (accessed 23 January 2017).

Maung, M. (1971) *Burma and Pakistan: A Comparative Study of Development.* Praeger Special Studies in International Economics and Development. Praeger, New York.

Metta Foundation (2012) Annual Report: April 1, 2011–March 31, 2012. Metta Development Foundation, Regeneration Initiative, Yangon, Myanmar.

Mohammed, M. and Karami, E. (2007) Sustainable agriculture: towards a conflict management based agricultural extension. *Journal of Applied Sciences* 24, 3880–3890.

Morris, H. (2009) Agricultural extension experiences through farmer field schools in Myanmar. Presentation at ECHO Asia Symposium, June 2009. Available at: https://c.ymcdn.com/sites/echocommunity.site-ym.com/resource/collection/7B4F6F18-749A-46C0-A127-904FAA07B7F8/Agriculture_Extension_Experiences_in_Myanmar_Through_Farmer_Field_Schools-Heather_Morris.pdf (accessed 19 January 2017).

Myanmar Centre for Responsible Business (2014) Civil society organizations and the extractive industries in Myanmar—a brief overview. Myanmar Centre for Responsible Business, Yangon, Myanmar. Available at: http://myanmar-responsiblebusiness.org/pdf/2014-10-11-Briefing-Paper-CSOs-and-Extractives-Industries-Myanmar.pdf (accessed 20 February 2017).

Myint-U, T. (2011) *Where China Meets India: Burma and the New Crossroads of Asia.* Farrar, Straus and Giroux, New York.

Ongsiapco, L.M., Santos, C., Muncy, S., Frederico, M.M., Picton, C.J. and Hernandez, V. (2007) Building community following displacement due to armed conflict: a case study. *International Social Work* 50(2), 171–184.

Pick, D. and Thein, H.H. (2010) Development failure and the resource curse: the case of Myanmar. *International Journal of Sociology and Social Policy* 30(5/6), 267–279.

Requier-Desjardins, D. (2013) Innovation and social inclusion: how to reduce the vulnerability of rurals? In: Coudel, E., Devautour, H., Soulard, C.T., Faure, G. and Hubert, B. (eds) *Renewing Innovation Systems in Agriculture and Food: How to Go Towards More Sustainability?* Wageningen Academic Publishers, Wageningen, the Netherlands, pp. 109–130.

Rieffel, L. and Fox, J. (2013) Are aid donors repeating mistakes in Myanmar? *Global Post* 12 March. Available at: www.globalpost.com/dispatches/globalpost-blogs/commentary/Myanmar-aid-donors-Civil-War-conflict-zones-emerging-markets (accessed 23 January 2017).

Ringer, J. (2014) A phenomenological study of individual strategies for rebuilding food production ability in resource poor post-conflict farming communities in Myanmar. PhD thesis. Oklahoma State University, Stillwater, Oklahoma.

Robertson, A. and Olson, S. (2012) Adapting agricultural extension to peacebuilding. Report of a workshop by the National Academy of Engineering and United States Institute of Peace Roundtable on Technology, Science, and Peacebuilding. US Institute of Peace, Washington, DC. Available at: www.usip.org/events/adapting-agricultural-extension-peacebuilding (accessed 23 January 2017).

Rogers, E.M. (2003) *Diffusion of Innovations*. 5th edn. Free Press, New York.

Saha, S.R. (2011) Working through ambiguity: international NGOs in Myanmar. Hauser Center for Non-Profit Organizations at Harvard University, Cambridge, MA. Available at: www.themimu.info/sites/themimu.info/files/documents/Report_Working_Through_Ambiguity_-_INGOs_in_Myanmar_Harvard_2011.pdf (accessed 23 January 2017).

Scott, J.C. (2009) *The Art of Not Being Governed: An Anarchist History of Upland Southeast Asia*. Yale University Press, New Haven, Connecticut.

Smith, M. (1999) *Burma: Insurgency and the Politics of Ethnicity*. Politics in contemporary Asia. Zed Books, New York.

Steinberg, D.I. (2010) *Burma/Myanmar: What Everyone Needs to Know*. Oxford University Press, New York.

Swanson, B. (2011) Developing innovative extension systems within a rapidly changing global economy during the 21st century. MEAS–USAID presentation. Available at: www.youtube.com/watch?v=8fgRtg0fANI (accessed 23 January 2017).

Swanson, B. and Rajalahti, R. (2010) Strengthening agricultural extension and advisory systems: procedures for assessing, transforming, and evaluating extension systems. Agriculture and Rural Development Discussion Paper 45. World Bank, Washington, DC.

Than, T.M.M. (2005) Dreams and nightmares: state building and ethnic conflict in Myanmar (Burma). In: Thompson, W.S. (ed.) *Ethnic Conflicts in Southeast Asia*. Institute of Southeast Asian Studies, Singapore, pp. 65–108.

Thuzar, M. (2013) Myanmar: facing the future. *Regional Outlook* 2012/13, 115–118.

TNI (2013) Access denied: land rights and ethnic conflict in Burma. Burma Policy Briefing 11. Transnational Institute, Amsterdam. Available at: https://www.tni.org/files/download/accesdenied-briefing11.pdf (accessed 23 January 2017).

UNCTAD (2011) The Least Developed Countries Report 2011. United Nations Conference on Trade and Development, Geneva, Switzerland.

UNODC (2011) *South-East Asia Opium Survey 2011*. United Nations Office on Drugs and Crime, Vienna.

USAID (2015a) Burma—complex emergency. Fact Sheet 3. USAID, Washington, DC. Available at: www.usaid.gov/sites/default/files/documents/1866/burma_ce_fs03_06-19-2015.pdf (accessed 23 January 2017).

USAID (2015b) Ongoing USG humanitarian assistance to Burma. USAID, Washington, DC. Available at: www.usaid.gov/sites/default/files/documents/1866/burma_map_06-19-2015.pdf (accessed 23 January 2017).

Walinsky, L.J. (1962) *Economic Development in Burma: 1951–1960*. Twentieth Century Fund, New York.

Wilson, S. and Mwee, N.E. (2013) Food and nutrition security in Myanmar. Michigan State University, Myanmar Development Resource Institute Centre for Economic and Social Development and USAID Bureau of Food Security. Available at: http://fsg.afre.msu.edu/Myanmar/burma_background_paper_4_food_security_rev.pdf (accessed 23 January 2017).

Win, U.K. (2013) Situations and issues of Myanmar's agriculture: focusing on critical areas of cooperation. Symposium at meeting of Myanmar ODA Investment Group, Yangon, Myanmar. Available at: http://www.krei.re.kr/web/eng/oda;jsessionid=37D2D61F8065C75EB0C377CA30A320D1? (accessed March 2017).

World Bank (2012) *Agricultural Innovation Systems: An Investment Sourcebook*. World Bank, Washington, DC. DOI: 10.1596/978-8213-8684-2

World Concern (2013) Myanmar (Burma) country profile. Available at: www.worldconcern.org (accessed 20 February 2017).

Xinhua News Agency (2013) 1 killed, 4 injured in riot in Myanmar, Northern Shan State. *Xinhua News Agency*, 29 May.

8 Agricultural Extension in Iraq, 2003–2012: Perspectives of US Partnerships during Conflict

Edwin C. Price,[1]* Joseph King[2] and Kate Whitney[1]
[1]Texas A&M University, Texas, USA; [2]Conflict and Development Foundation, Castroville, Texas, USA

Background to Conflict

Agriculture first began in the eastern branch of the Fertile Crescent along the Tigris and Euphrates rivers 7000–8000 years ago, giving nurture to the Babylonian civilization. A feudal system of agriculture evolved and fed the region for centuries; even into the late 1950s 'Iraq was self-sufficient in agricultural production' (Tharp, 2009). In 1958, however, the Iraqi monarchy was overthrown by Abdul Karim Qasim, who undertook far-reaching land reforms. In 1963, fearing communist influence, Western allies supported the Ba'athist political party in overthrowing Qasim. A progressive leader and strongman in the Ba'athist government, Saddam Hussein—who eventually consolidated his presidency in 1979—fostered sound investments in agricultural infrastructure, research and extension. The country remained 67% self-sufficient in food production until the 1970s (Tharp, 2009), at which time defense investments began to take priority. By 2009 Iraq imported 80% of its food.

Iraqi Agricultural Extension and Conflict, 1980–2002

Investment in agriculture declined sharply with the start of the Iran–Iraq War in 1980. The conflict ultimately developed into the 20th century's longest interstate war, resulting in over 1 million civilian and military deaths (Iran–Iraq War, n.d.). The increasingly brutal oppression of segments of Iraqi society brought trade sanctions in 1994, to which the regime responded by placing most sectors of the agricultural economy under government control. During this time, the government provided free or heavily subsidized agricultural inputs, purchased most of the country's agricultural products and even distributed a monthly basket of food to every child and adult. The role of agricultural extension was less that of farmer education and information dissemination, and more that of distributing farm inputs and managing a heavily government-dependent agricultural sector. Extension agents were tasked with organizing farming communities and implementing

* Corresponding author e-mail: price.condev@gmail.com

Ministry of Agriculture (MOA) strategic priorities throughout the country.

With restricted policies and markets for petroleum, coupled with declining agricultural production, deprivation and hunger became severe both in Iraqi cities and in the countryside. In response, the United Nations mounted an oil-for-food program that opened import channels for food and food production inputs. While well intended, the program eventually became corrupt and further complicated an already weakened and distorted agricultural sector, with fraud permeating agricultural input supply as well as output management and utilization.

In the months leading up to the intervention of multinational forces in 2003, the Iraqi extension service, as administered by the MOA and State Board of Extension Cooperation, was an agricultural sector management organization that helped to provide agricultural inputs and services to farmers. The extension service linked farmers to state-owned seed companies, fertilizer import and manufacturing, grain warehouses and mills, cotton ginning and oilseed processing, cold storage facilities, custom tillage, transport, slaughterhouses, veterinary services, diagnostic laboratories and other state-run services. The extension service was associated with a structure of farmer cooperatives that served as a state-run apparatus for input supply and production management. The term agricultural 'cooperative' was later deleted from the lexicon of some international agencies that were attempting to assist in rebuilding Iraqi agriculture, because of the political and socio-economic connotation of the term.

A number of international security concerns led a multinational force—including the US military—to overthrow the Ba'athist regime in a military campaign that began on 19 March 2003. The Coalition Provisional Authority (CPA) undertook all centralized government functions and administered agriculture through military civil affairs units, advised by a joint expatriate civilian–military agricultural council, which was convened monthly in Baghdad (US DoD, 2007). The MOA was returned to Iraqi hands on 5 May 2004 (Downer, 2004). Multinational military units continued their assistance to Iraqi agricultural authorities at all levels, particularly at provincial and local levels, and were joined by international organizations, including the US Department of Agriculture (USDA) and the US Agency for International Development (USAID). The CPA and US agencies sought to return the Iraqi economy to a market-oriented approach in the hope of improving production efficiencies, creating opportunities for international trade and reducing centralized control. As part of this process, extension agencies and personnel no longer held managerial roles in the agricultural sector and were faced with learning new service roles in technology delivery and information dissemination.

Iraqi Extension in the Conflict and Post-Conflict Environment, 2003–2014

Overview of international cooperation

Iraqi extension services were peripherally involved in many of the agricultural assistance programs that emerged from multinational donors and agencies. A series of interrelated agricultural development programs, funded by the US government between 2003 and 2014, sought increased cooperation with Iraqi agricultural agencies, colleges and universities. Key USAID programs included: Agricultural Reconstruction and Development for Iraq (ARDI, 2003–2008), the Iraq Inma project (2008–2013) and the Harmonized Support for Agricultural Development (HSAD) project (2013–2014). The US Department of Defense (DoD) cooperated with USDA to place agriculturalists in Provincial Reconstruction Teams (PRTs). These, coupled with agricultural components of the DoD Task Force for Business and Stability Operations (TFBSO), deployed to military forward operating bases (FOBs) in combat zones throughout the country. The capstone effort by the US government in rebuilding agricultural extension came with the launch of the USDA Iraq Agricultural Extension Revitalization (IAER) project, which was led by the Texas A&M

University System's Texas A&M AgriLife Extension, in collaboration with extension programs at New Mexico State University, the University of California at Davis, Utah State University and Washington State University.

Three phases of extension rehabilitation in Iraq took place in the succession of US government-supported agricultural programs implemented from 2003 to 2014: (i) technology verification; (ii) needs assessment and planning; and (iii) capacity building and information dissemination. The USAID/ARDI program, which began in 2003, focused on applied research and extension. It made inroads in assessing infrastructure, building Western relationships with Iraqi government and community organizations, and testing and introducing improved agricultural technologies (new crop varieties, intercropping practices, plastic tunnel horticulture, etc.). The program was created as a technological and institutional foundation on which to rebuild extension, and while a variety of objectives was achieved, ARDI can be categorized as a more limited 'technology verification platform'. ARDI was constrained by insecurity in the countryside and therefore had limited access to farmers and rural areas in the north, central and southern regions of Iraq (USAID, 2006). Much of its work was carried out in Kurdistan, the autonomous region in the north of the country, especially after 2005.

The subsequent TFBSO/Agricultural Team—consisting of personnel from DoD, USDA and the Texas A&M University System's Borlaug Institute (also referred to as Team Borlaug)—was supported by multinational military assets and was able to reach farmers, communities and regional offices throughout the country. While this collaboration introduced a number of advanced technologies (improved fish stock, drip irrigation, a regional farmers' and wholesale market, agricultural curriculum and others), its main contribution was 'needs assessment and planning': working with provincial councils to formulate provincial development plans for agricultural infrastructure, institutions, technology and extension capabilities, based on more than 10,000 interpersonal contacts across 13 provinces (excluding Kurdistan). The senior author of this chapter was the principal investigator for this program.

The USAID Inma project continued to support technology improvement initiatives throughout this period. It focused on agribusiness, with a particular emphasis on agricultural enterprises without government involvement. It was not formally associated with the rehabilitation of government extension programs.

The USDA IAER project began in 2007 and focused on rebuilding extension systems and capabilities, in partnership with Iraqi agricultural universities and the MOA. Training programs for extension workers were conducted mostly outside of Iraq due to the logistics associated with persistent security concerns. This phase was dedicated to 'capacity building and information dissemination' (USDA, 2012).

After 2003, violent conflict continually erupted and abated in various Iraqi localities, but never fully disappeared. Domestic agricultural programs, as well as international assistance in agriculture, were tempered by the special constraints imposed by political insecurity. Iraq's current system of agricultural extension, and the Iraqi agricultural enterprise in general, owes much to the resilient farmers and entrepreneurs, a system of universities with many colleges of agriculture and the ministries of agriculture (MoAs) of both Iraq and Kurdistan, which asserted their own identity, authority and responsibility despite the challenges of conflict. Kurdistan is a Federal Region under the Iraqi constitution. The central government of Iraq based in Baghdad is responsible for international affairs of Iraq and for agricultural affairs outside of Kurdistan, while the regional government of Kurdistan administers agricultural affairs of Kurdistan.

In the following sections, the Iraqi institutional environment and the constraints on agricultural extension posed by conflict are discussed in more detail.

Public administration in Iraqi agriculture reflects the greater divisions that prevail throughout the government and society. At the top, the MoA of the Iraqi central government in Baghdad regards itself, and is regarded externally, as the country's legitimate national agricultural authority. Even so, the predominantly Kurdish provinces of Irbil, Dohuk and Sulaimaniya come under the rule of the Kurdish Regional Government (KRG). As a 'federal entity' under the Iraqi constitution, with a capital in Irbil, the KRG has authority over certain internal responsibilities, including agriculture. As such, the KRG has its own Ministry of Agriculture and Water Resources (MAWR), which includes an extension service. The respective authority of the Baghdad-based and Irbil-based governments was and is continually debated, but as a practical matter, the KRG operates independently of Baghdad.

Separately from this regional political distinction, the agricultural extension service itself is divided between federal and provincial extension employees. To the farmer constituencies they are indistinguishable, but salaries and reporting structures are administered separately, at federal and provincial levels. Farmers' associations have long been formally affiliated with government involvement in agriculture, but associations formerly dealt primarily with input supply and product management.

Each of the 18 Iraqi provinces has a provincial agricultural office and a lead agricultural officer, and each office maintains a roster of farms. These rosters, in turn, guide agricultural input distribution and farm produce purchases. Prior to 2003, input distribution would have been the primary function of the provincial offices. Provincial agricultural extension workers still report to the provincial office and, more recently, have been developing capacity for farmer technical advisory services, in addition to fulfilling their former responsibilities of input distribution.

Outside the KRG, some Iraqi provinces—including Anbar, Basra and Najaf—are also prominent in their own governance and exhibit some independence from federal and regional government influence. In Basra, a major petroleum production and shipping hub, there is a close alliance between private agriculture, business and the clergy. This alliance has been a greater source of stability and agricultural leadership than public institutions, and it is said that government officials occasionally take instructions from the clergy and private sector.

In Najaf province, the main city mosque provides training in business and agriculture. It is important to note that Najaf houses the Holy Shrine of Imam Ali, which is the holiest place of Shia Islam. The Imam of the shrine, Sheikh Dhia Zain Aldeen, is highly knowledgeable about agriculture and associates with the Dean of Agriculture of the University of Najaf.

The Governor of Anbar is an agricultural entrepreneur who is highly influential in the sector. For example, he is able to import and introduce improved crop varieties, occasionally bypassing the 'mandatory' 4-year national testing and approval process for all genetic material. Anbar agricultural extension workers balance their allegiances between federal and provincial authorities.

Political divisions

An additional complication in internal Iraqi agricultural administration was that the leaders of Iraq's State Board of Extension Cooperation—one of the administrators of the Iraqi extension service—were also members of the opposition Sadrist Party. The party had come to occupy the extension office following the assassination in 2005 of the previous Director General of Extension, Dr Awad Issa, outside his office. Dr Issa's assassination was part of a campaign against intellectuals. Of unattributed origin, by June 2006 the campaign had claimed more than 300 lives (Stone, 2006). At the first organizational meeting of the USDA-supported IAER project, held in Amman, Jordan, in March 2007, the Sadrist-led contingent of the Baghdad MoA extension staff exited the meeting on the first day of the event, protesting the order of speakers. They were persuaded to stay only after apologies were made and accepted. The leader of the walkout, the Director General of Extension, gave a magnanimous address at the end of the workshop.

He was later murdered at the Sadrist Party offices in late 2007 or early 2008 (K. Whitney, personal communication, 24 July 2015). Sectarian issues also influenced selection of trainees for IAER training programs.

Tribal leaders

Most Iraqi farmers claim to have had little contact with or assistance from extension personnel in recent years. Many lament the decline in government distribution of inputs—chiefly seed, fertilizer and custom tillage services. While the CPA attempted to elevate the relevance and authority of governmental agencies, tribal leadership remained strong throughout the presence of multinational forces. In rural areas, tribal leadership remains the dominant influence and authority in the daily lives and affairs of farmers. Sheikhs—tribal leaders—are highly active in soliciting technical assistance for farmers in their communities, often investing their own assets in public goods and introducing new technologies to their communities. Among matters arranged by tribal leaders is land access.

Iraqi government land institutions are in great disarray. Overlapping and contradictory systems of land rights and access date back to the Ottoman Empire, overlaid by successive regulations promulgated by the monarchy, the revolutionary government of 1958, the regime of former President Saddam Hussein and ultimately by the adjudication and edicts of tribal Sheikhs. In a private conversation with E.C. Price in 2008, former Prime Minister (now Vice President) Ayad Allawi, himself a major landowner, asserted that the most serious problem facing Iraqi agriculture was insecure rights to land. Insecure land rights led to deficiencies in the maintenance of farm resources, as illustrated by widespread soil salinization, caused by the poor upkeep of drainage systems. When farmers are not assured of long-term access to land, they are less willing to invest in its improvement. Incidentally, the second most serious impediment, according to Vice President Allawi, was the lack of improved crop genetics.

Farming families

About two-thirds of Iraqis derive their livelihood from agriculture. When farmers were asked where they learned about agricultural practices, they invariably answered that they learned from their parents. During visits to farmers in their homes, they were often surrounded by their children. Farmers generally expect their children to be farmers, and these children often confirm this career path—both young women and young men. Children expressed faith in the technologies handed down to them and often provided expansive views of how they would increase the size of their farms or flocks. Iraqi farming families in the Central Region appear to be a modern exception to the pattern of dispersal of youth away from the family farm, based on a decision to leave agriculture. Loyalty to the land, to family and to community is strong in rural Iraq. This does not mean that there is no rural-to-urban migration. While farms have been repeatedly subdivided to accommodate successive generations, families are large and some family members move away.

Colleges of agriculture

Iraqi provinces typically have at least one public university, including a minimum of one faculty (college) of agriculture. Hitherto, universities had no official role in agricultural extension, as they are administered through the Ministry of Higher Education rather than the MoA. Formal degree programs in extension education are offered at the University of Baghdad at BSc, MSc and PhD levels. Briers *et al.* (2011), however, observe that the degree programs offer insufficient practical application and state that there is little trust between the MoA and the university. Universities occasionally provide training for extension workers under agreements with the MoA, but operationally they are not well connected. Still, international agencies have occasionally tried to broker closer working relationships between universities and Ministries, and these efforts have led to some useful informal professional relationships.

Extension Conditioned by Conflict

Conflict has placed a number of constraints on agricultural extension in Iraq. Some of these affect extension services indirectly because they impact the scope of technologies that are most needed, and that can be viably extended and supported. Other constraints directly affect the conduct of extension personnel and their operations. In this chapter we examine official government agricultural extension, as well as extension provided by international aid projects, non-governmental organizations (NGOs) and the private sector. We commonly distinguish between extension with an upper case 'E', and extension with a lower case 'e'. Extension with an upper case 'E' includes formal government extension services provided by line agencies, such as the federal MoA and State Board on Agricultural Extension, along with provincial agricultural offices. Extension with a lower case 'e' consists of agricultural training and technical assistance provided by Iraqi NGOs and international agencies. For example, the Washington, DC-based United States Peace Institute (USIP) mounted a poultry project in Baghdad province (on-site interview of USIP staff by E.C. Price, 2008).

Resource Constraints

Direct constraints on Iraqi government extension services during conflict and early post-conflict include budgetary restrictions on salaries and operations, less obvious communications and security concerns, and high-level political differences regarding agricultural policies. For example, salary payments to government extension staff were irregular during conflict, and workers complained of long periods without pay. Travel and program support funds were also scarce. This was all further complicated by uncertainty concerning the purpose of agricultural extension. At the highest level of the Iraqi MoA, there were those who held out for continued programs of input subsidy and delivery, and others willing to undertake the shift to technical advisory services. Such discord made programming government budgets for extension difficult. In short, it was impossible to actualize any proposal for national campaigns to revitalize Iraqi agriculture.

Cost of doing business

Conflict imposes additional costs on most public and private transactions and also has an impact on agricultural extension. Simply put, everything costs more, including all transactions in agriculture. The additional costs come in many forms. The decline of law and order gives rise to illegal charges, such as bribes and protection fees paid by businesses. Insecurity requires increased vigilance, so that more people are needed to carry out routine tasks. More secure routes must be followed, which in turn increases travel times and restricts periods of travel. Costly and multiple face-to-face meetings are required, rather than simpler telephone or written communications. Personal security concerns at production sites, in transit and at markets reduce productive activity and therefore the supply of goods and services, increasing their prices as a result (US DoD, 2007).

Not only does the supply of services decline, but demand may also increase dramatically, as international agencies with thousands of personnel and ample funding move into conflict and post-conflict economies to render assistance. Requirements for food, housing, transport, fuel, energy and labor for new expatriate communities in urban areas initially drive price increases in the cities, and then in rural markets as well. In short, the cost of doing one's job increases during conflict and carries over into the post-conflict environment. It reduces what extension workers can accomplish with their limited incomes and resources.

Content of extension programming in conflict

Cost increases tend to shift the types of enterprises and technologies that farmers can profitably use, requiring revised content in extension programs. For example, insecurity in Iraq contributed to a shift away from extensive production systems towards intensive production systems.

Greenhouses and integrated poultry production systems flourished. One extension worker in the Central Region of Iraq introduced intensive cattle feeding in 2007, and later cattle and sheep feedlots spread into Kurdistan. The guideline expressed by several farmers was that they sought enterprises entirely independent of the government, so that they could control as much of the value chain as possible. Government involvement in a supply chain could make businesses vulnerable to the extraction of rents by public officials. Also, the impairment of public services during conflict necessitates greater self-reliance.

Another feature of intensive agriculture that is advantageous in conflict regimes is the greater possibility of concealing or protecting an integrated value chain over a relatively small space. It was observed that horticultural and poultry production met these requirements and flourished.

A less positive dynamic was observed in the broad collapse of public services in many rural communities, with the result that no single intervention was viable. The case of a village in Wasit province is instructive. In 2008, schools had been closed since the war began in 2003, and the future appeared dim for a generation of children. Water treatment had ceased, and drinking water was scarce and impure. Crop yields were exceedingly low at 0.5 mt/ha; a crust of salt covered much of the land. Finding feed for a few thin dairy cows was clearly a source of concern for their owners. People were hungry, because they had sold their food allotments to pay medical bills, and many women were said to be ill. It was clear that any approach to improving agriculture and livelihoods would have to be comprehensive, touching every facet of rural life. Interrelated weaknesses caused by conflict can implode rural economies. In such cases, integrated rural community development may offer the best opportunity for improvement.

Suppressed communications

Communications, the lifeblood of extension, were constrained by an induced culture of reticence and insecurity and a weak communications infrastructure. Iraqi dissent was suppressed during the 40-year Ba'athist rule. Civil organizations not directly managed by the government were discouraged, including professional organizations, as these were regarded as a possible threat to the regime. Interpersonal and intra-organizational communications were suppressed. Agriculturalists could not easily share knowledge or build professional networks. Scientists, teachers, researchers and extension personnel worked in relative isolation. When Western development organizations first became active in the economy and society after 2003, it was difficult for them to identify key players, convene groups or distribute information.

When the TFBSO/Ag Team members visited communities, offices and institutions, it was difficult for them to uncover institutional linkages. Over several weeks spent in various provinces in 2008, the team often inquired about the existence and whereabouts of extension workers, but none of these workers was ever acknowledged or introduced. At the University of Babil, 3 months into their rural reconnaissance, during a meeting with faculty and administrators, team members noticed a back row of five individuals who had said nothing throughout. Upon inquiry, it was discovered that they were provincial extension staff. Once they began talking, it became clear that extension workers were well dispersed in the countryside but simply not linked into a communications network with one another or to universities. Instead, their communications typically consisted of one-way contact, receiving information solely from their regional directors or Baghdad. Thereafter, by careful and focused inquiry, the US agriculture teams began to learn about and see more of the work of rural extension workers. On the part of extension workers, there was likely some reluctance to meet with outsiders because the mission of their organization and their own responsibilities had become unclear and they were at a loss to explain their purpose.

Security and communications

Among Iraqis in general, and especially among extension workers, security concerns remained about meeting with Westerners or being seen in places away from their home localities. In order to travel to meetings convened by Westerners, they traversed various waypoints to disguise their ultimate destination. When the call went out for applicants to a USAID-supported scholarship program to study in the USA, it took months for applicants to dare show their faces, and even then, they also took circuitous routes to reach the venues for interviews.

Translators who travelled with Western agriculturalists often had to take precautionary measures as well, especially those who sought to contact farmers and extension workers in local communities; some wore masks so they would not be recognized. Even so, there were many cases of Iraqis who paid with their lives when insurgents discovered they had been working with Westerners.

While internet connectivity grew slowly, communications benefited from the increased availability of cell phones. Still too costly for most farmers and extension workers, sufficient phones were distributed among leaders for contacts to be made for scheduling meetings. For security reasons, care was taken not to use place names, personal names, times or routings when communicating by phone or in writing. Among both Western workers and Iraqi personnel there was a heavy reliance on discreet word of mouth to set up meetings and appointments.

Communications across cultures

The challenge to foreigners attempting to elicit and understand the 'felt needs' of a rural community, in which trust has been eroded by violent conflict or oppression, is to achieve confidence that what one hears and articulates truly reflects the feelings of a client farmer or community. This is a complicated issue because extension services (and other development assistance), when rendered by international donors, are not linked to recipients through markets as is the case, for example, in the USA. There, recipients pay for extension services through taxes and signify their satisfaction (or lack of it) through their votes. When a provincial agricultural needs assessment was completed, TFBSO/Ag Teams first delivered their reports in the local language to communities before submitting them to regional, national or international bodies. One team appeared to hit the mark when, upon one such delivery, a local leader gave a tearful thank you and said of the report: 'It sounds as if we are speaking'. The teams listened to thousands of farmers, members of rural communities and local leaders and strove to relay their message in words that local community members called their own.

Foreign agricultural extension communications in Iraq attempted to bridge ethnic divides as well as the gap between civilian and military identities. International extension programs were often carried out in Iraq by multinational military civil affairs units and by other foreign workers accompanied by military or military-style escorts. It is sometimes suggested that military-style protective gear, uniforms and weapons can present a barrier to communicating with farmers and other members of the rural community. There have been a few cases where high-level officials refused discussions in the presence of weapons and protective gear. However, rural community members showed little aversion to military clothing and assets, and this issue was quickly overcome. Curious children relished the opportunity to associate with visitors, and adults soon followed suit. Contact and communications with farmers were unimpeded by military armaments and uniforms.

United States Government Efforts to Rebuild Extension

Four key agricultural initiatives were undertaken by the US Government between 2003 and 2012. Three of these included a focus on extension:

1. USAID ARDI program, led by Development Alternatives Incorporated (DAI) with Texas A&M University AgriLife Research as a partner, 2003–2008.
2. US Department of Defense Task Force for Business and Stability Operations agricultural component (TFBSO/Ag), led by Texas A&M AgriLife Research, funded by the DoD through USDA and known in the field as Team Borlaug (2007–2009).
3. USDA IAER project, led by Texas A&M AgriLife Extension (2007–2012).

A further USAID effort, implemented through the International Center for Agricultural Research in the Dry Areas (ICARDA), was the Inma project, which continued until 2014 (HSAD Project, 2014; ICARDA, n.d.). This fourth project was largely directed towards private agribusiness, with minimal emphasis on formal capacity building in extension or other government agencies.

Agricultural reconstruction and development for Iraq (ARDI)

In 2003, the ARDI project began to support rebuilding the agriculture sector in Iraq. It focused on many aspects of agriculture, including field testing, the formation of NGOs and farmers' organizations, strengthening the private sector and building the capacity of the extension service. It also introduced improved crop technologies, animal production and health, cadastral surveys and water management systems. Initially, all project activities were led from Baghdad. However, in December 2005 the center of activity was moved to Irbil in Kurdistan due to increased violence. Thereafter, much of the field activity took place in the provinces of Dohuk, Hawler (Irbil) and Sulaimaniya.

The principal theme in ARDI's approach to strengthening extension, termed 'participatory extension', consisted of 'taking extension to farmers rather than bringing farmers to extension' (USAID, 2006: 249). Visioning workshops were held at national and provincial levels, with the MoA and State Board of Extension involved, as well as universities and related government staff. By the end of 2005, farmer workshops were held to identify priorities and key needs. The history of strong government involvement in agriculture emerged clearly through observations made by farmers and extension staff, who commented that the government should solve farmers' problems and that the MoA should return to subsidizing tillage services, equipment, fertilizer, seed, pesticides and other inputs (USAID, 2006). State Board of Extension leaders and Ministry officials in Irbil, Sulaimaniya and Dohuk agreed that people experienced in participatory methodologies—not university faculty members—should train extension staff in participatory methods. Old institutional rivalries subsequently surfaced, with MoA officials asserting that universities should not play a role in extension.

Training-of-trainers in participatory methods was held in Egypt and subsequently in-country. Field extension workers were trained and the methods applied in several technical areas: sheep breeding, crop production, horticulture, honey production and integrated plant protection. At first, progress was slow because, although leaders had agreed to the new approach, the MoA was slow to approve the movement of personnel and resources to implement the program. An essential point of participatory extension was that demonstrations should take place on farmers' fields, not at Ministry facilities. This was unprecedented. Finally, in September 2006 the MoA issued instructions to the governorates to implement the participatory extension system. In the space of 3 years, an official transformation of extension methods had begun. By the end of the project, ARDI had trained 191 extension workers in participatory methods in all governorates of the country (USAID, 2006).

Despite these successes, old habits and policies proved hard to change, and subsequent programs implemented by others had difficulty in finding evidence of these earlier in-field programs. It is likely that the new methods took hold only among a few progressive personnel. In the follow-on USAID agricultural program Inma, USAID eschewed contact with the ministry, opting instead for an entirely private sector-oriented program.

ARDI laid the groundwork for a new phase in farmer-oriented extension programs in Iraq. In cooperative crop trials with universities in Mosul, Irbil, Sulaimaniya and Baghdad, a number of verification trials provided new tools for extension in the ARDI project as well as for subsequent programs. Some of the advances were in greenhouse horticulture crops, field production of tomato under plastic, improved potato varieties, barley–vetch mixes that proved advantageous for livestock production and saline-tolerant wheat production. These were not all entirely new technologies, but war had isolated scientists from each other, from extension services and from farmers. ARDI provided the opportunity for these technologies to be further field tested and disseminated to farmers.

DoD/TFBSO agricultural teams

From among the multinational forces operating at about 170 forward operating bases (FOBs) in Iraq, selected combat units were designated for 'civil affairs'. Their job was to 'win hearts and minds', which in its most productive interpretation was taken to mean economic development. The problems and opportunities seen by civil affairs units assigned to rural communities were often related to agriculture, but the units rarely had personnel who were trained in agriculture. This dilemma was compounded by the fact that the civilian agriculturalists working with USAID in the ARDI project were not accessible to the multinational forces, nor did they have the assets to operate in less secure areas. The State Department and the US military sought to remedy the issue in 2007 by hiring agriculturists through USDA to serve on military–civilian hybrid PRTs, but the flow of personnel was slow.

Also in 2006 and 2007, a unit within the DoD—the TFBSO—was working to broadly revive Iraqi business and industry, but it too found agriculture to be especially problematic due to a lack of capable experts on hand to meet the population's needs. In March 2007 the TFBSO supported a team of agriculturists from land grant universities and the USDA to assess the agricultural situation at selected US military FOBs around Iraq (US DoD, 2007). Their report to the DoD demonstrated the utility of having agriculturalists stationed at FOBs.

In early 2008, the US military command in central Iraq, through TFBSO, invited Texas A&M AgriLife Research to field an agricultural team for a 6-month assessment of agriculture, to assist civil affairs units in framing agricultural development plans and to implement selected programs in technology dissemination. The northern, western and southern commands soon followed with similar requests. Some of the team members had already worked with the ARDI project and were beginning to work with the IAER project as well. The difference with this TFBSO/Ag initiative lay in the fact that when embedded with the military, agriculturalists had access to the most remote and least secure parts of the country. The system enabled the TFBSO/Ag Teams ('Team Borlaug') to conduct what was probably the most comprehensive field-level external assessment of agriculture ever conducted in modern Iraq. The work of the TFBSO/Ag Teams provided insights concerning the impacts of conflict on extension, as presented earlier in this chapter.

The framework for conducting needs assessments and technology dissemination varied between different FOBs. At some FOBs, PRTs had been established. PRTs typically incorporated some or all military civil affairs units at a base, but also occasionally included personnel from other parts of the US government, including the Department of State, the Department of Justice, the Army

Corps of Engineers, USDA and USAID. Even so, it was rare that all these government units were represented. The PRTs, for instance, were responsible to the US Department of State, while the military civil affairs units and TFBSO/Ag reported to the military command. The PRTs were designed to plan and execute integrated, comprehensive development strategies. If a FOB did not have a PRT, TFBSO/Ag worked solely with the military personnel. Regardless of the organizational structure, the TFBSO/Ag Teams mobilized with combat units about 5 days a week, going into communities for discussions, training and project implementation.

Training was delivered on poultry management, animal health and nutrition, tillage practices, oilseed processing, mechanization, extension methods, aquaculture, water management, produce handling and grading, marketing and drip irrigation. Four key projects implemented were:

- Construction and management training for the Central Euphrates Farmers Market and Training Center;
- Fish breed improvement for pond aquaculture;
- Drip irrigation production and demonstration;
- Curriculum improvement for agricultural education.

Students were recruited and sent for graduate studies in the USA, with many drawn from the ranks of extension staff in the MoA (Baghdad) and the MAWR (Irbil). The aquaculture project was especially noteworthy, because it drew on substantial resources and cooperation from the US DoD, USDA and USAID.

TFBSO/Ag had excellent access to the MoA in Baghdad and served to inform the Ministry of opportunities and needs in regions and subject matter areas to which it had limited access. The top ten priorities in agriculture that TFBSO/Ag summarized in its final report were:

1. Crop seed quality and varieties;
2. Animal breeds—cattle and fish;
3. Diagnostic laboratory capability;
4. Farmer knowledge:
   (i) tillage
   (ii) crop varietal management
   (iii) efficient water use;
5. Youth programs;
6. Land rights;
7. Livestock forage and feed;
8. Oilseed production and processing;
9. Farmer credit;
10. Management of saline soils.

TFBSO/Ag fielded 40 professionals across many disciplines of agriculture and made contact with more than 10,000 farmers, business persons and leaders. Many contacts had been out of touch with agricultural professionals throughout the period of conflict. One contribution seemed clear: the simple act of recognizing farmers, community leaders and families who had been isolated for years, often in fear, gave hope and stimulated new determination to improve agricultural techniques and rural lives.

Iraq Agriculture Extension Revitalization (IAER)

In 2006, the US Department of State funded the USDA, National Institute for Food and Agriculture (NIFA) and Foreign Agriculture Service (FAS) to work with US universities in strengthening the Iraqi extension system. The USDA entered into an agreement with a consortium of land grant universities led by Texas A&M AgriLife Extension Service (TAMU) to '(1) Increase agricultural capacity through science based extension training to Iraqi extension personnel; and (2) Foster effective interagency cooperation and collaboration among Iraqi government agencies to benefit Extension' (USDA, 2012: 3). Other university partners in the USA included New Mexico State University (NMSU), the University of California–Davis (UCD), Utah State University (USU) and Washington State University (WSU). NMSU also worked with Diné College, a Native American 1994 land grant school in New Mexico.

Adopting a train-the-trainer approach and operating largely outside of Iraq, the consortium trained 720 Iraqi extension and university professionals, who in turn each reported training an average of 40 additional colleagues. Training focused on five main areas: dryland cropping (WSU), water management (USU), horticulture (UCD), extension methodology (NMSU) and livestock (TAMU). For security reasons, the training was conducted in Egypt, Jordan, Lebanon and Syria. Later, training was conducted in the KRG provinces of Irbil, Dohuk and Sulaimaniya, and also in the USA (USDA, 2012).

An effort was made throughout the project to include representatives of Iraqi universities and Ministries in training and planning exercises, in order to build professional relationships and increase cooperation. Over the course of the project, 45 Iraqi university faculty members participated as trainers or trainees, along with government extension professionals. IAER's most significant achievement was to foster cooperation between the Ministries of Higher Education (Baghdad), Agriculture (Baghdad), Water Resources (Baghdad) and Agriculture and Water Resources (KRG). The project also succeeded in establishing a 1-year diploma program at Baghdad University and at Salahaddin University in Irbil (USDA, 2012).

Extensive materials on technology and extension methods were made available to the various Ministries in both Arabic and English. An innovative element to help ensure their use was a competitive mini-grant program sponsored by IAER during the early phase of training in Egypt and Jordan. Trainees were offered encouragement and assistance in writing proposals for demonstrations and other farmer training events, when they returned to their posts. The awardees, in turn, used their grants to reach more than 900 farmers. The program built trust and credibility for the extension workers. However, attempts to persuade the Ministries to extend and enlarge the initiative were unsuccessful (USDA, 2012).

Evaluations conducted throughout the program (Abi-Ghanem *et al.*, 2013) documented the following achievements (USDA, 2012):

- Increased capacity of higher education institutions to support agricultural extension.
- Increased knowledge and skills of extension agents.
- Improved extension services and materials targeting women, youth and other specific audiences.
- Improved quality of extension services, with most agents saying they used the information they had gained from training to help their clients.
- Increased understanding of technologies among farmers and other clients.
- Increased use of improved agricultural practices by farmers.

IAER collaborating institutions made a number of recommendations for continued improvement of agricultural extension in Iraq (USDA, 2012), including:

- Continued strengthening of university relationships with extension.
- Improved internet connectivity for international learning.
- Reorienting Ministry programming to be driven by farmer needs and extension agent feedback, rather than by Ministry priorities alone.
- Improved programming for women and increased numbers of female extension professionals.
- Introducing programming for youth, previously regarded as unimportant.
- Exposing mid-level extension managers to policy and management environments in the USA and other countries.
- Making resources available to extension workers for field demonstrations.
- Continued training in farm management, marketing and agribusiness in order to replace dependence upon government with free-market principles.
- Recognizing the greater responsiveness in KRG towards modern extension methods.
- Continuing to train, elevate and increase programming support to extension within the MAWR.

Opportunities and Limitations

The discussion above identifies a number of areas in which various interventions achieved success. For example, the IAER project evaluations identified six areas of achievement and made nine recommendations for improvement. The following paragraphs outline the key concepts related to agriculture and extension that emerged during Iraq's period of conflict between 2003 and 2012.

Transition from material inputs to educational services

Extension staff in the Iraqi MoA based in Baghdad, and those of the KRG MAWR, continued to serve the agricultural sector through the period of warfare that began in 2003, which is an achievement in itself. Encouraged by international partners, these personnel transitioned from distributing farm inputs to guiding farmers towards improved farming practices. Along with that shift, notable changes in production technologies were also introduced—albeit often with external assistance—including horticultural production under plastic, improved aquaculture and beef feedlot production. Nevertheless, on-farm extension services to farmers, youth and women remain weak and need to expand and improve. Better cooperation between government extension agencies and university faculties of agriculture could be a resource for improved agricultural extension in Iraq.

On-farm extension services

A key achievement of the DAI-led ARDI project was to plant the idea at the highest levels of the MoA that extension services needed to shift towards providing on-farm education to farmers through participatory methods. The concept began to take hold and continued to be emphasized through subsequent international partnerships. This important approach is limited in Iraq, however, by the low level of funding for field operations for extension workers.

Extension on the ground during conflict

The DoD/TFBSO Agricultural Teams fostered agricultural planning at the provincial level, based on intensive farmer contacts. The teams demonstrated that extension programs can work in areas of active conflict, with the cooperation of military personnel and civilians at the community level and in farmers' fields. The assessment and planning teams each worked for a maximum of 6 months in the central, northern, western and southern regions of Iraq. Nevertheless, it would have been useful if the farm, community and provincial presence of the teams could have continued through the implementation of the agricultural plans that had been developed cooperatively by the TFBSO teams, community members and provincial authorities.

Competitive grants for extension services

A key innovation undertaken by the IAER project was the provision of small competitive grants to Iraqi extension workers to carry out new initiatives in their respective communities. After receiving the grants, the workers were guided by project staff. This program energized extension staff and achieved rapid impact in communities. As a result, participating staff expressed a desire for the program to be expanded. This is a model that merits being resumed, but the MOA declined to advance the program, ending it after only one round of grants.

Challenges to extension posed by conflict

Conflict regimes impose special challenges requiring methods and technologies that differ from those appropriate in a more peaceful context. Increased costs imposed by conflict shift the conditions that underlie supply and demand, which in turn may require changes in products and production methods. This means that extension workers need to retool for new conflict-responsive agricultural enterprises. Conflict also increases the cost of conducting extension activities on farms and in communities, straining local extension budgets.

Conflict is a public cost that cannot be recovered by firms in an economy with open borders and may require foreign assistance or other injections of capital to assist in the transition to more conflict-viable enterprises. Conflict-resistant products and methods (such as intensive poultry production) need to be identified and introduced. Mistrust and insecurity alter how extension workers communicate with one another and with clientele.

International agricultural partnerships during conflict

During conflict, parents want to feed their children, farmers want to produce and entrepreneurs want to thrive. It is incumbent on the international community to partner with agencies, communities, families and individuals in conflict regimes, and to use the most direct and effective methods possible to aid in their survival. This may involve the use and further development of communication technologies appropriate for areas in conflict. It can also require the formation of, and reliance on, policies and programs that support and protect those in the international community who are willing to work in communities during conflict. In the USA, it means there should be more effective cooperation between civilian and military agencies for the purpose of assisting communities on the ground in conflict regimes.

References

Abi-Ghanem, R., Carpenter-Boggs, L., Koenig, R.T., Ullman, J.L., Murphy, K.M. and Pannkuk, C. (2013) Access to agricultural inputs, technology and information, communicating with farmers, and the role of women in agriculture: perceptions of Iraqi extension agents. *Journal of International Agricultural Extension Education* 20(1), 6–18. DOI: 10.5191/jiaee.2013.20101

Briers, G.E., Balschweid, M. and Al-Ajeeli, S.A.Y. (2011) Review of the Department of Agricultural Extension, College of Agriculture, University of Baghdad. Iraq Agricultural Extension Revitalization Project. Foreign Agricultural Service, United States Department of Agriculture, Washington, DC.

Downer, A. (2004) Iraq Ministry of Agriculture back in Iraqi hands. Report of Australian Government, 5 May. Available at: http://reliefweb.int/report/iraq/iraq-ministry-agriculture-back-iraqi-hands (accessed 23 January 2017).

HSAD Project (2014) Final stakeholder's meeting: achieving long-term sustainability. USAID/ICARDA HSAD Iraq Project: Harmonized Support for Agriculture Development. Available at: http://hsad.icarda.org/blog/final-stakeholders-meeting-achieving-long-term-sustainability (accessed 23 January 2017).

ICARDA (n.d.) Strengthening Iraq–ICARDA partnerships—new Baghdad office and hub for HSAD Project. International Center for Agricultural Research in the Dry Areas. Available at: https://www.icarda.org/blog/strengthening-iraq-icarda-partnerships-%E2%80%93-new-baghdad-office-and-hub-hsad-project (accessed 23 January 2017).

Iran–Iraq War (n.d.) History 351, Iran-Iraq War. Available at: www.saylor.org/sitemap/

Stone, R. (2006) Targeted for murder, Iraqi scientists named on a hit list. *Science* 312, 30 June, 1857–1858. Available at: http://science.sciencemag.org/content/312/5782/1857.1.short (accessed 23 January 2017).

Tharp, M. (2009) Once world's bread basket, Iraq now a farming basket case. McClatchy Newspapers, Sacramento, California, 17 July. Available at: www.mcclatchydc.com/news/nation-world/article24546748.html (accessed 10 April 2017).

USAID (2006) Final report: agriculture reconstruction and development program for Iraq ARDI. United States Agency for International Development, Washington, DC. Available at: http://pdf.usaid.gov/pdf_docs/Pdacj285.pdf (accessed 23 January 2017).

USDA (2012) Iraq agricultural extension revitalization. Final report to the Department of State and USDA. Foreign Agricultural Service, United States Department of Agriculture, Washington, DC. Available at: https://nifa.usda.gov/sites/default/files/resources/IAER%20Final%20Report%20(2).pdf (accessed 23 January 2017).

US DoD (2007) Iraq business transformation assessment, agriculture team report. Task force for business and stability operations, Office of the Deputy Under Secretary of Defense for Business Transformation, United States Department of Defense, Washington, DC.

9 Post-Conflict Rebuilding of Afghanistan's Agricultural Extension System

Nicholaus M. Madden[1], Christopher D. Pannkuk[2,*] and Linda R. Klein[3]
[1]University of California at Davis, Davis, California, USA; [2]International Agriculture Research and Development Consultant, Moscow, Idaho, USA; [3]LRK Consulting, Pullman, Washington, USA

> The success of the aid system should be measured by its contribution to building the state rather than substituting for it.... As there is an emerging consensus that technical assistance is a very expensive, and yet ineffective instrument for addressing the problem of governance, the issue of capacity-building must be examined afresh.
>
> Ghani *et al.* (n.d.)

Agriculture and Extension before the Conflict, Pre-1978

Dating back to 350 BC, Afghanistan has endured frequent foreign invasions, civil wars and conflicts (GFRAS, n.d.). Although it is one of the least developed countries in the world, Afghanistan's people and rural farming systems have met these hardships with robustness, resilience and adaptability. Agriculture (including crop/cereal production, horticulture and livestock) remains the largest and most important sector of the national economy. Of the employed population, approximately 40% are engaged in agricultural work and nearly 80% of all households depend on agricultural income. However, agricultural productivity remains low due to Afghanistan's limited water resources, infertile soils and harsh climate (CSO, 2014).

Although public agricultural extension activities began in the 1920s, Afghanistan's agricultural extension system was first formally organized under a directorate within the Ministry of Agriculture in 1958. In 1959, this position was elevated to the level of General Directorate. In 1963, the name of the section was changed to the Production and Extension Department (PED); and in 1966 it became the Research and Extension Department (RED). In 1970, RED was divided into the Department of Agricultural Extension and Production and the Department of Soil Sciences. In 1972, in response to a severe drought, agricultural extension was briefly relocated to the Rural Development Department, under the direct control of the prime minister, but was subsequently returned to the Ministry of Agriculture (Wesa, 2009). At this time, the country's agricultural research and extension system included 24 research stations, more than 400 extension offices and more than 1000 research and extension professionals, distributed across all provinces and provincial districts (World Bank, 2005).

By the 1970s, the Afghan government was providing agricultural research, extension and advisory services that allowed the country to meet national demand for cereals

* Corresponding author e-mail: chrispannkuk@gmail.com

and to sustain an export market in horticultural products (Emadi *et al.*, 2009). From 1968 to 1978, the production of wheat increased by about 13%, from 2,354,000 to 2,652,000 tonnes (t). In 1978, an estimated 85% of the 15 million people lived in rural villages, and most of the other 15% were connected with rural enterprises in some way (UNDP, 1993). Cereals were grown on approximately 87% of all cultivated land, with wheat (57%) as the principal crop. At 180 kg/year, the per capita consumption of wheat in Afghanistan was one of the highest in the world (UNDP, 1993).

By the mid-1970s, the country was self-sufficient in food grains (Christensen, 1995). Agricultural products were major exports and totaled nearly 65% of pre-war export earnings, 30% of which came from dried fruits, principally raisins (UNDP, 1993). From 1974 to 1978, agricultural exports increased in value from US$158 million to about US$232 million (FAO, n.d.).

Background to the Conflict, 1978–2002

Chronology of conflict

A timeline of key events during Afghanistan's most recent conflict years is presented in Fig. 9.1.

Agricultural extension in conflict, 1978–2002

Soviet influence

After the Soviet invasion in 1979, Afghanistan's agricultural sector experienced a multitude of development challenges:

1. Damage to human resources. In addition to the huge loss of life, many Afghans were displaced or forced to leave their homeland. Technical professionals either fled or were killed, and a whole generation grew up without access to proper education.
2. The conflict limited the active participation of women—major stakeholders in Afghan agriculture—in agricultural employment.
3. The damage to infrastructure and land destroyed ancient underground irrigation systems, as well as roads and bridges that once carried agricultural goods to market and fertilizer to farmers. Arable lands were indiscriminately bombed, trees and orchards were burned, livestock once used to till the land vanished and landmines turned productive land into dead zones. Moreover, a decline in surface area of arable land, coupled with high inflation, high unemployment and the central government's inability to curb cultivation for illicit purposes, prompted farmers to grow large quantities of opium poppy in place of staple crops (Wesa, 2009).

In 1987, the direct effects of war caused an estimated decline in agricultural production to 45–53% of 1978 levels. Availability of family labor fell by 20% during the same period and numbers of draft oxen fell by about 40%. The percentage of farmers using fertilizer declined from 76% to 53% in the case of urea, and from 57% to 33% for diammonium phosphate (DAP). The total area of cultivated land fell by 18%, and cereal imports increased by a factor of 52. Small stock numbers fell by 70% in the case of karakul flocks, and by 67% for sheep and goats. Opium poppy cultivation increased from an estimated 6000 hectares (ha) in 1978 to about 57,000 ha in 1992 (UNDP, 1993).

The Afghan agricultural research and extension system also experienced many challenges under the Soviet-sponsored Afghan government, the People's Democratic Party of Afghanistan (PDPA). The PDPA's agrarian land reform policies (e.g. Decree No. 8) were viewed by many farmers as anti-Muslim. For example, in 1979, agricultural lands were expropriated from any Afghan who owned more than 6 ha and redistributed to farmers who had priority for redistributed land. However, Afghan Muslims are forbidden to own others' property without his or her permission (Rubin, 2002). Also during the Soviet occupation, the number of agricultural research stations fell from 24 to 12. Only 80–90 extension units were functional—most of them without a full complement of staff—and professional staff strength decreased from more than 1000 to an estimated 500. Furthermore,

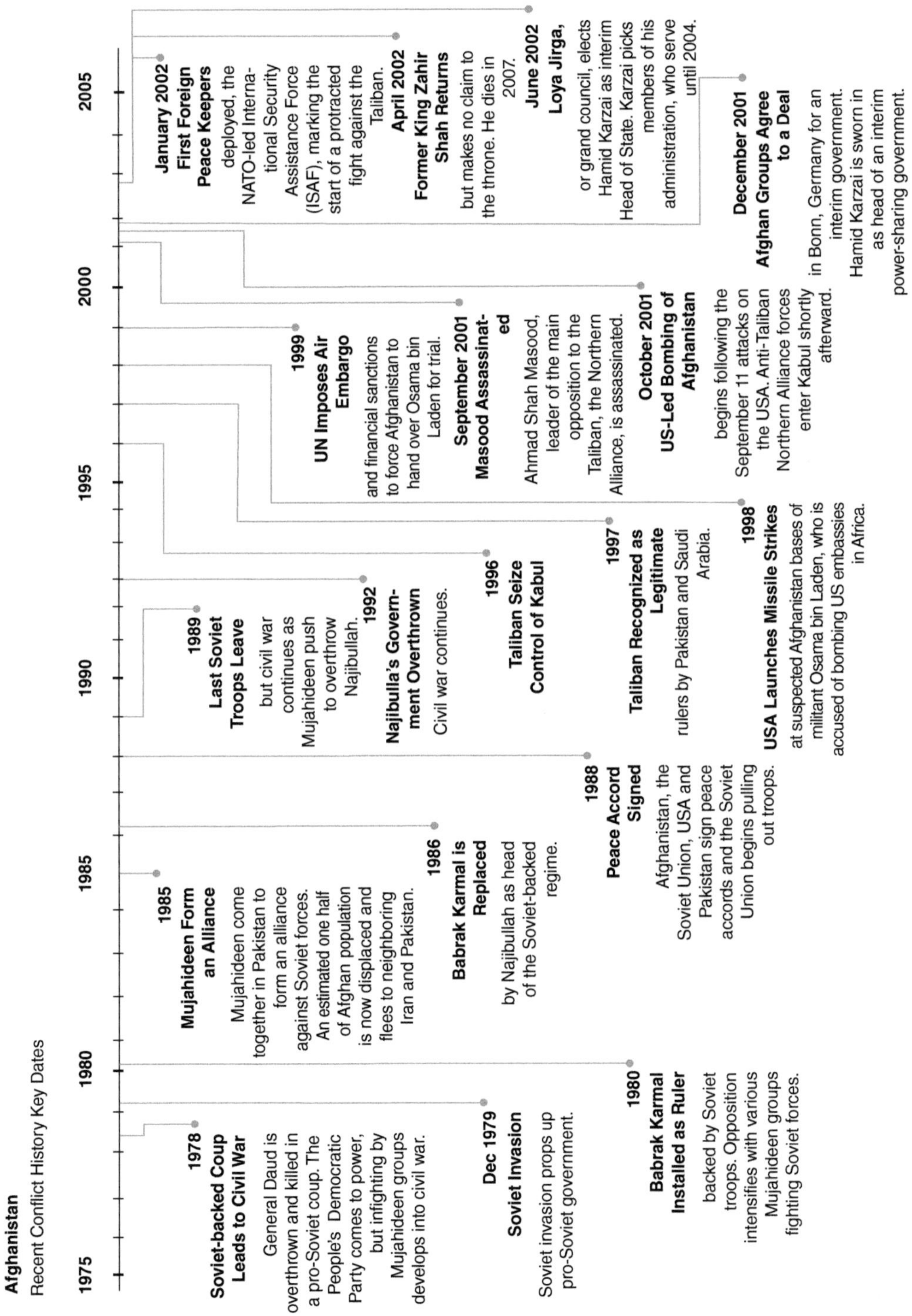

Fig. 9.1. Timeline of key events during Afghanistan's most recent conflict years, 1978–2002.

the quality of trained staff was depleted. During the 1970s the Department of Agricultural Extension had over 40 postgraduates, whereas during the occupation postgraduate numbers dropped to 20 (Wesa, 2009).

Civil war influence

After the withdrawal of the Soviet Union from Afghanistan in 1989, the PDPA was left to fend for itself against the Mujahideen. The years of factional infighting that followed during the Afghan civil war destroyed critical research and extension infrastructure and continued to drain qualified professionals from the service. Applied research efforts ceased, research and extension stations were looted, salaries were withheld from staff and the exodus of research and extension professionals accelerated.

Taliban influence

After 1996, under the Taliban government, the quality and coverage of the extension service continued to decline. Crop and animal production fell by 50% and the number of extension workers dropped below 650 countrywide (Emadi *et al.*, 2009). Most of the extension workers were Mullahs, with limited agricultural technical knowledge or skills. The teaching of agricultural subjects was undermined in favor of religious learning. No provision of inputs, such as seed or fertilizer, was provided to farmers from the skeleton extension service, nor were markets or private organizations regulated. Farmers were left in a weak, uncompetitive position with private businessmen (i.e. middlemen) and were unable to contribute to the national economy. The agricultural economy in general, and smallholder farms in particular, declined to subsistence level production (Emadi *et al.*, 2009).

The Post-Conflict Extension System, 2002–2014

Afghanistan is experiencing a lasting food deficit and is primarily dependent on humanitarian food aid. The country's rain-fed cereal productivity—on which more than 80% of smallholder farmers depend—is as low as 0.6 t per hectare (t/ha) and has resulted in a large group of vulnerable rural poor. In fact, because of the persistent decline in overall agricultural sector productivity, 35–45% of the population were unable to grow enough food for a healthy and productive life in 2007–2008 (Emadi *et al.*, 2009).

In the following sections we describe the post-conflict state of the national research and extension services, as well as several initial efforts to restore Afghanistan's agricultural capacity, infrastructure and self-sufficiency.

Afghanistan's public agricultural research and extension system

Currently, Afghanistan's research and extension system is administered under the Ministry of Agriculture, Irrigation, and Livestock (MAIL). MAIL has offices in all 34 provinces and in 364 of the 398 provincial districts. However, only 136 of these district offices have functioning extension units, and less than 5% of the entire workforce have specialized training. The national research system has 18 research stations and substations, located in or near the provincial capitals. Unfortunately, staff at these stations have limited formal education, experience and/or knowledge of modern applied research and extension methodology. The same conflict-related issues that have weakened overall Afghanistan public institutions over the past 30 years have also severely diminished the rigor of the country's university education. As a result, the few professionals with graduate degrees from the national universities make up the bulk of professional technical staff within the research and extension system, but are ill-equipped to undertake effective applied research and extension activities (Emadi *et al.*, 2009).

During the 1970s, when the country reportedly attained self-sufficiency in wheat and had a vibrant agricultural export market, more than five extension workers served each district. Currently, some provinces are without district-level extension workers; in other provinces only one or two extension workers serve each district. The exception is the Kabul extension service—the best resourced extension service in Afghanistan—which has 76 extension workers covering the province. However, despite the large staff, the Kabul-based service faces the same challenges (lack of transportation and other critical operational resources) that other provincial services endure. According to the 2007 Afghan National Risk and Vulnerability Assessment (NRVA) survey, 'only 1.3% of Afghan farmers had contact with agricultural extension services, including both government and non-governmental organizations (NGOs). Of this 1.3%, 60% received advice only once' (WFP, 2009).

The various directorates in MAIL's Kabul headquarters decide the technical programs. MAIL's research and extension programs cover agronomy, horticulture, livestock, forage and forestry, plant protection, soils and agricultural mechanization. Some programs include international and private sector cooperation and collaboration. For example, maize, wheat, rice and legume genotypes are provided by collaborating international research centers, such as the International Center for Agricultural Research in the Dry Areas (ICARDA) and the International Maize and Wheat Improvement Center (CIMMYT). The Food and Agriculture Organization of the United Nations (FAO), in collaboration with provincial Directorates of Agriculture, Irrigation, and Livestock (DAILs) and private sector seed systems, obtains and multiplies the best released seed materials to produce high-quality foundation and certified seed, which farmers can then purchase (WFP, 2009). However, the beneficiaries of this highly subsidized operation are largely medium- and large-scale producers involved in mechanized farming—not the smallholder producers who make up the majority of Afghan farmers and who need the most assistance.

MAIL's Agronomy Program conducts research on a limited number of thematic areas; the benefits of outputs to local communities have been insignificant. The Agronomy Program needs to integrate more effectively with other programs in key areas, such as that for cereal and horticultural products, and to emphasize on-farm adaptive research. The scope of plant protection research is narrowly focused on disease evaluations in the wheat nurseries.

The Soils Program is at an early stage of rebuilding and developing national plant and soil laboratories. The existing laboratories lack both skilled laboratory technicians and properly functioning analytical instrumentation for providing effective analytical services. Thus, at present, plant and soil analysis capabilities are limited in MAIL and DAILs.

MAIL's Horticulture Program is responsible for research on trees and vines (almonds, apples, stone fruit and grapes) and vegetables (tomatoes, onions, carrots and cucurbits, especially melons). In 2005, Afghanistan reportedly exported US$127 million worth of horticultural products, but this export value has the potential to be much greater (Emadi *et al.*, 2009). However, horticulture sector productivity is constrained by antiquated production methods, inadequate water supplies and lack of modern technologies (e.g. improved varieties, processing infrastructure and storage facilities).

Research and extension interventions in the livestock sector focus on animal health, range management and dairy production, through externally funded projects such as FAO's Integrated Dairy Development Project (2005–2008). These projects train farmers in improved dairy production and animal husbandry, vaccinate animals against known animal health threats and, in rare cases, provide drugs to treat common animal health conditions. Yet these activities are neither adequate nor strategically coordinated. The interventions, largely supported by external sources and implemented by expatriate employees on limited appointment contracts, fall far short of meeting national demand for such critical services. The national livestock research and extension system lacks human, financial and material resources to undertake meaningful extension or research interventions.

International research interventions (2002–present)

After the fall of the Taliban in 2002, the Future Harvest Consortium to Rebuild Agriculture in Afghanistan (FHCRAA) was established, with support from the United States Agency for International Development (USAID), and in collaboration with the Afghan national research and extension system. The consortium consisted of more than 12 institutes from the Consultative Group for International Agricultural Research (CGIAR), FAO and several US land-grant universities—all with the capacity to rebuild MAIL. FHCRAA reactivated a series of wheat varietal selection and seed multiplication centers and relaunched a seed distribution system (Emadi *et al.*, 2009). It also implemented a series of resource assessments for MAIL, including soil, water, livestock and rangeland. In addition, FHCRAA rebuilt five agricultural stations, located in Baghlan, Kabul, Kunduz, Nangarhar and Takhar (GFRAS, n.d.).

By 2003, experimental plots of wheat, barley, chickpeas, lentils, fava beans, tomatoes, onions and peppers were established at the rehabilitated agricultural stations. Seeds lost during the past decades of conflict were also repatriated from duplicate collections from around the world. By 2007, 41 barley landraces and 250 kg of seed from several Afghan cereal and legume landraces were returned, including 60 almond, 47 pistachio, 14 pomegranate and 16 melon. Additionally, fruit tree nurseries were established at research farms in Baghlan, Kunduz, Nangarhar and Takhar (WFP, 2009).

Non-governmental organizations, 2002–present

NGOs have played an important humanitarian and development role in Afghanistan for many years. During the 1980s, more than 200 NGOs provided assistance to refugees fleeing Afghanistan following the Soviet invasion (Waisová, 2008). NGOs were also among the only providers of aid during the Taliban regime, as many donors, including the United Nations, withdrew support from Afghanistan. After the fall of the Taliban in 2001, the international community pledged large amounts of funds and forces for the reconstruction and peace-building in Afghanistan. This effort was coupled with a renewed influx of NGOs, and by 2005 an estimated 2400 NGOs were working in Afghanistan (Olsen, 2006).

Because of the high visibility of NGOs and the public confusion between NGO staff, foreign government contractors and private security companies, NGOs were being blamed for the slow progress of reconstruction efforts and the squandering of billions of dollars. In 2005, partly as a result of lobbying from NGOs and the donor community, the Afghan government passed a new Law on Non-Governmental Organizations, which required all NGOs to officially register. This registration helped differentiate NGOs from private sector contractors and remove any 'fake' NGOs (Olsen, 2006). As of 2013, the country had 1857 registered, active national NGOs and 293 registered, active international NGOs (MoE, n.d.). Most NGOs are involved in providing emergency relief and establishing health, education and agricultural programs.

The Afghan government has also been very critical of the large amounts of off-budget donor funds (i.e. donated funds that are not managed through the Afghan treasury system) devoted to aid organizations. The government argues that these funds could instead be used by local firms or other government programs, such as the National Solidarity Program created in 2003 by the Ministry of Rural Rehabilitation and Development (MRRD) (NSP, n.d.), to provide similar services at lower costs. Additionally, the government believes that treasury management of these funds would likely prevent capable Afghans from moving away from government positions to take higher-paid positions with foreign aid agencies (Olsen, 2006).

From 2002 to 2012, the total external assistance pledged to Afghanistan was US$119 billion, of which US$70 billion was actually disbursed. The USA is by far the largest provider of assistance in Afghanistan at about US$48 billion (2002–2012); the next largest is Japan at about US$4 billion (2002–2012). However, not all external assistance goes directly to development. For example, of the US$13 billion disbursed in 2011, 68% was spent on security. Of the remaining 32% (US$4.1 billion), only 27% (US$1.1 billion) was delivered on-budget and mainly used to operate the government, while the major portion was delivered off-budget and was managed by the development groups. Off-budget aid is a major funding source for many NGOs and can complement Afghan government development projects, but can also lead to conflicting agendas because the government has little control over off-budget projects.

In 2005, to avert the potential 'Paradox of Plurality' caused by the multitude of NGOs performing similar activities without mutual coordination, the Agency Coordinating Body for Afghan Relief (ACBAR) released a set of regulations. Called the Code of Conduct, these regulations aid communication among the Afghan government, media and NGOS, and among NGOs themselves for greater transparency, accountability and coordination of NGO activities (Waisová, 2008). The ACBAR inspired other platforms and networks for Afghan and international NGOs, namely the American Council for Voluntary International Action (InterAction), the European Network of NGOs in Afghanistan (ENNA), the Afghan NGOs Coordination Bureau (ANCB), the Islamic Coordination Council (ICC), the South West Afghanistan and Baluchistan Association for Coordination (SWABAC) and the Japan Afghan NGO Network (JANN) (Waisová, 2008). Of these, only ACBAR has a formalized collaboration with the UN Assistance Mission in Afghanistan (UNAMA), while the others have established links with Afghan Transitional Authority (ATA) or local authorities.

Provincial reconstruction teams and agribusiness/agriculture development teams, 2002–2014

The US government also established the Provincial Reconstruction Teams (PRTs) and the Agribusiness (or Agriculture) Development Teams (ADTs), to provide technical support or reconstruction in agriculture.

PRTs were small teams (100–250 people) of military and civilian personnel who began working in Afghanistan's provinces in 2002 to provide security for aid workers and to help with humanitarian reconstruction tasks. The teams did not engage in combat operations, but the integration of military and humanitarian work enabled civilian experts to receive protection as they conducted activities in hostile regions. All PRTs eventually fell under the operational command of the International Security Assistance Force (ISAF), while individual nations led the teams. ISAF commanded 26 PRTs with 14 separate lead nations; the USA led 12 of these teams. PRTs had left Afghanistan by the end of 2014 (Mitchell, 2015).

Conceptually, developing a team comprising military personnel and civilian experts seemed like a good way to distil the best from both worlds. However, differences in ideology, training methods and funding sources often caused internal imbalances and friction between military and civilian personnel. Also, PRTs were perceived by Afghan officials and NGOs as bureaucratic, top-down units that were too prescriptive and donor-driven. Both groups criticized 'militarized aid' as 'fast-win' charity-like projects that were poorly executed, and they viewed this meddling as using 'aid money as a weapon system' (Stewart, 2014).

The Afghan government and the international aid community believed that development efforts should be a 'bottom-up' grassroots-style process, involving local communities in small-scale projects. The USG acknowledged these criticisms and began improving the PRT mission. But, in 2008, the US Army, in conjunction with various Army National Guard commands, also started anew by establishing specialized, grassroots US Army Agriculture (or Agribusiness) Development Teams (ADTs) (Stewart, 2014). These specialized National Guard teams comprised 12 soldier-experts who worked, when not deployed, as professionals in an agribusiness field such as veterinary science and pest management. Based on their expertise, ADTs trained and advised Afghan Ministries and universities. Additionally, ADTs operated from the bottom up in rural areas, working with local communities on small, easily replicated projects that included educational components with follow-up assessments. All ADTs received months of pre-deployment training in their home state or at training centers, for example, the Agricultural Development for Afghanistan Pre-Deployment Training (ADAPT) at California State University, Fresno, California. ADTs were also able to 'reach back' to home states that were committed to remote assistance. For example, ADT members used their experience and professional connections to access expertise from their state's agriculture industry or land-grant universities (LGUs) and cooperative extension services. In February 2008, the US Army deployed the first ADT in Afghanistan to augment the PRT located in Jalalabad. By 2014, nine states supported the ADT mission and provided a total of 49 teams that operated in 15 provinces (Stewart, 2014).

US government efforts to rebuild agricultural extension

The US government's agricultural strategy for Afghanistan is 'to support the rapid transition of Afghanistan to a more stable and productive state through the promotion of democracy, rule of law and sustainable economic and social development that is responsive to citizens' needs' (USAID, 2010). Programs designed to improve Afghan extension services supported this strategy through direct contribution to USAID's 5th strategic objective: 'a sustainable, thriving agricultural economy', linked to intermediate result (IR) 5.3, 'improved delivery of agriculture-related public services as a result of USG assistance', through sub-IR 5.3.2, 'improved government of the Islamic Republic of Afghanistan agricultural research and extension services' and sub-IR 5.3.3 'improved MAIL core functions'. Specifically, the USG's agricultural extension rebuilding efforts focused on (USAID, 2010):

- Properly training, equipping and deploying hundreds of MAIL's own extension workers and working with other extension providers.
- Rehabilitating MAIL research/extension stations and demonstration farms; improving linkages between agricultural education facilities and government institutions; and training staff to support the transfer of improved agricultural production and on-farm water management methodologies to farmers.
- Modernizing agricultural faculties in the public university system; building the capacity of faculty and staff with advanced degree educational training; transferring technology; modernizing agricultural science curricula; and mentoring faculty members from partner universities, both in the USA and in the region.
- Implementing a change-management program that will help MAIL to define its roles and priorities, and providing assistance to increase institutional capacity (e.g. in administration, project development and oversight or procurement).
- Improving the capability of DAIL directors to plan budgets and implement programs, and providing resources required to manage and coordinate program execution on the ground.

Since 2002, USAID has supported the distribution of vouchers for seed, fertilizer, tools and technology to hundreds of thousands of farmers in an effort to jumpstart production. By 2014, USAID interventions had generated more than US$306 million in sales and services for farmers and agribusinesses. USAID also trained more than 1.5 million people and connected more than 24,600 households with access to credit, while providing financing to agribusinesses generating 2913 jobs. In total, USAID investments helped to create more than 358,968 new agricultural jobs. Furthermore, USAID investment in rehabilitating irrigation infrastructure has increased water availability for approximately 106,000 ha of land (USAID, n.d.).

Early programs included the Rebuilding Agricultural Markets Program (RAMP,

2003–2006), implemented by Chemonics International, Inc.; the Dairy Industry Revitalization Project (2004–2007), implemented by Land O Lakes and Mountain Pastures Dairy Company; and the Participating Agency Services Agreement (PASA) with the US Department of Agriculture (USDA), which allowed USDA's Agricultural Research Service (USDA-ARS) to staff USAID with agricultural science experts to oversee USAID programs. Current ongoing projects include:

- Afghanistan Agricultural Extension Project II (AAEP II, 2014–2017), implemented by a consortium of US universities.
- Capacity Building and Change Management Program II (CBCMP II, 2014–2017), implemented by the Volunteers for Economic Growth Alliance (VEGA)—consisting of 23 economic development organizations.
- Commercial Horticulture and Agricultural Marketing Program (CHAMP) (2010–2016), implemented by Roots of Peace.
- Digital Integration to Amplify Agricultural Extension in Afghanistan (DIAAEA, 2014–2015), implemented by Digital Green.

To date, a total of 44 USAID-funded projects have been completed, and 11 projects are ongoing (USAID, n.d.).

Successes in rebuilding agricultural extension

Many US government and other programs have successfully addressed some of the farmers' needs in Afghanistan. These successes include identifying and implementing alternatives to poppy production, establishing water management programs to assist farmers in efficient resource use and introducing new technologies to help farmers gain better access to international markets.

Afghanistan agricultural extension project

The AAEP was a program that developed out of USAID initiatives for building the capacity of MAIL and DAIL to deliver extension services and agricultural research. USAID hired Checci and Company Consulting, Inc. to assess research and extension needs and to design capacity-building programs for the request for applications. AAEP and two other programs resulted from the assessment and design process. Of the three programs, the US$14 million AAEP was the largest and longest, spanning 3 years from 2011–2014, with the goal of building the capacity of MAIL to deliver demand-driven extension services to four target provinces (Balkh, Herat, Kabul and Nangarhar). With USAID funds, AAEP was administered by the USDA and implemented by a consortium of four universities—the University of California Davis (UC Davis), Washington State University (WSU), Purdue University and the University of Maryland. The six primary objectives of the AAEP were to:

- Strengthen technical skills of extension workers.
- Teach demand-driven extension methods.
- Elevate the role of women in agriculture.
- Rehabilitate agricultural research stations.
- Establish workgroups focused around priority commodities and livestock.
- Launch a mini-grant system to encourage agricultural innovation and technology transfer.

UC Davis served as the lead institution. In addition, the consortium contracted the Dutch Committee for Afghanistan (DCA) to implement the project's livestock programming. AAEP's contract under USDA expired on 30 September 2014.

UC Davis implemented AAEP activities first in Balkh and Kabul, before adding secondary locations in Badakhshan, Jowzjan, Kapisa, Panjshir, Parwan and Samangan. WSU was responsible for programming in Nangarhar and later added activities in Kunar and Laghman. Purdue operated out of Herat and also trained extension workers from Badghis, Farah and Ghor. Finally, Maryland was responsible for the Women in Agriculture (WIA) program, which had activities in Kabul, Kapisa, Parwan and Samangan.

Aside from working in a specific geographic area, the consortium partners each specialized in a particular subject. The aim was for each partner to provide training on that subject across the entire program. UC Davis focused on post-harvest and protected agriculture, WSU focused on Conservation Agriculture (CA), Purdue focused on Integrated Pest Management (IPM) and grain storage, and Maryland focused on WIA. Training sessions and demonstrations were provided across programming areas by permanent program staff and experts from abroad.

The AAEP developed an overall model of extension delivery that was adaptable to the Afghan context and capable of reaching a large number of farmers. The program also provided a wide variety of training initiatives in both technical skills and extension delivery for extension workers. Since few Afghan extension workers had significant experience with livestock, the program provided basic training in livestock care and created linkages among communities, extension workers and private and public veterinary services. The AAEP's work with women developed a network of community leaders who were skilled in organizing kitchen gardens that provide households with improved nutrition and supplementary incomes.

The AAEP extension-delivery model consisted of:

- Thematic workgroups to facilitate communication within and across the agricultural community and to implement mini-fund projects.
- Provincial model teaching farms (PMTFs) in each target province, where AAEP staff could demonstrate, and extension workers could practice, new techniques and technologies for farmers.
- Farmer Field Schools (FFSs) and farmer field demonstrations (FFDs), so that extension workers trained at the PMTFs could provide the same training and demonstrations for a group of district farmers, enabling them to duplicate new techniques on their own farms.

The AAEP's accomplishments are listed in Table 9.1.

The program was also able to reach women in target areas, who built sustainable community organizations. Because Afghanistan has very few female extension workers—accounting for less than 1% of the total extension workers currently on file—the AAEP WIA program used a broader approach to extension for wider impact. In addition to training female extension workers, AAEP also trained women community leaders to train and form FFSs. These training sessions were designed both for novices and for those with more experience, as the education level for women in Afghanistan is very low. 'Support Group' training was available for FFS that focused on developing skills and practices for women outside the agriculture sector, such as computer literacy and savings box programs. Savings boxes are community savings and loan programs which provide members access to capital for buying needed agricultural inputs or rebuilding livelihoods. AAEP also engaged female agricultural students to assist with research at the PMTFs.

In addition to the on-the-ground components of the AAEP, two Annual Extension Conferences were organized. The objective of these events was to showcase AAEP and MAIL/DAIL activities and to improve communication across the national extension community. The first conference emphasized the concepts behind a demand-driven extension model and workgroups. Sixty participants from Balkh, Kabul, Kandahar, Kunduz, Nangarhar and Paktia DAILs and representatives from the Afghan Agriculture Research and Extension Development Program (AGRED), DCA, USDA, the US Embassy, VEGA/CBCMP and USAID attended the 3-day conference. The second conference

Table 9.1. Outcomes of the AAEP (2011–2014).

Outcome	Number
Extension workers trained	355
Training sessions provided	344
PMTFs established	10
FFSs established	185
FFDs established	507
Workgroups operating	39
Mini-funds approved	26

FFDs, farmer field demonstrations; FFSs, farmer field schools; PMTFs, provincial model teaching farms.

attracted more than 300 people from DAIL REDs, NGO representatives and AAEP staff. The event showcased the work done by AAEP in different provinces and fostered dialogue among extension workers from around the country.

AAEP also sponsored a study tour that took place from 1 June to 21 June 2014 for 19 participants, including representatives from AAEP, DAILs and MAIL. A majority of the participants were high-level officials from the Ministry and staff from AAEP. The objective of the tour was to increase participants' understanding of the US extension system and to interact with extension experts on topics relevant to Afghanistan. The tour successfully accomplished two goals: (i) improving participants' understanding of the US extension system; and (ii) helping MAIL employees to develop extension work plans for their respective provinces. However, implementing these work plans will be difficult for extension workers, unless MAIL sets aside the resources necessary to carry them out.

Ultimately, AAEP helped to build the technical and functional capacity of individual extension workers to operate through its well-adapted model. AAEP's success in implementing these models was reflected by the decision to name it as winner of the 2014 USDA Secretary's Honor Award for Exceptional Service.

AAEP II

AAEP II is the ongoing, 3-year extension phase of AAEP. AAEP II will further build public confidence in the Ministry and demonstrate that the Afghan government is developing the capacity to provide services that make a difference to local livelihoods. AAEP II land-grant university partners will further build the extension capability of MAIL and DAIL staff and contribute additional technical resources for expanding program implementation activities. In addition, AAEP II will work with MAIL and DAIL to scale up these activities from the current 17 provinces to a total of 25, as the Ministry develops the capacity to work with the land-grant university team.

Challenges and lessons learned in rebuilding agricultural extension

Research and extension practice indicates that a single provider, whether a public extension agency, private firm or NGO, is generally unable and ill-equipped to deliver the full range of rural knowledge and information services required. Rural farm community needs are likely to be best served by a mix of private and public services. For example, public institutions could provide knowledge and information services, while the private sector teaches the technologies that are most appropriate for farmers' needs.

MAIL

MAIL's research and extension systems are centralized, with programmatic, budgetary and personnel decisions generally originating from the upper administration. The extension service, in particular, was established as a top-down, public sector agency to transfer new technologies to farmers. However, so far field-level extension services have been largely dominated by the private sector or NGOs, due to the influx of externally funded development programs and projects. The Afghan government competes with NGOs for these foreign funds, but because its capacity is so restricted, its competitiveness to secure and implement projects is also low. There is considerable rhetoric regarding cooperation and coordination, but little on-the-ground evidence exists for either.

Non-governmental organizations

Although the majority of NGOs are Afghan, the largest programmes are implemented by established international relief and development NGOs, which are overwhelmingly staffed by Afghan nationals and a handful of expatriate staff. Some of the largest NGOs employ more than 1000 people, making these organizations a significant source of employment for Afghans (Olsen, 2006).

This employment opportunity has proved detrimental to government extension services, as many qualified and professional staff have chosen to work for better pay with the NGOs (Wesa, 2009). Another factor is that, prompted by NGO programs, MAIL uses extension workers for many activities unrelated to its mandate.

Provincial reconstruction teams

The PRT approach was met with much cynicism in the NGO sector, particularly on the part of Afghans working in the same regions. Distinguishing between the intentions of uniformed soldiers and civilian aid workers was difficult for Afghan farmers. Such blurring of military and civilian efforts compromised the neutral, non-political image so carefully nurtured by the NGOs. The 2014 Aid Worker Security Report (Stoddard *et al.*, 2014) revealed that 251 incidents of major violence against aid workers occurred globally in 2013, resulting in 460 workers being killed, kidnapped or seriously wounded. Afghanistan alone was the scene of 81 such attacks, the highest number in 2013. Certain NGOs specifically attributed decreased aid worker security to the PRTs operating in Afghanistan (Mitchell, 2015).

Additionally, few PRTs understood the delicate tasks of building trust, sustaining community cooperation and following other humanitarian principles. Without a sophisticated understanding of local power dynamics, these PRTs missed vulnerable groups and strengthened local entities to the detriment of the central government. The PRTs required no community contributions, so some NGO programs were abandoned for the 'free handouts' given by the PRTs. Thus, the PRTs probably undermined some ongoing projects that had taken years to establish.

Afghanistan agricultural extension project

The AAEP was widely appreciated by trainees and by both MAIL and DAIL leadership. Anecdotal evidence strongly suggests that the program has had a positive effect on extension workers. However, in order to establish the actual impact with any certainty, AAEP (and AAEP II) must create a strong record of activities, costs and results. This documentation will be vital to the sustainability of activities and their transfer to MAIL. Improving communication across the program by further empowering in-country leadership will be crucial for transitioning activity implementation to MAIL.

Recommendations and Opportunities for Improvement

Many recommendations for improving Afghanistan's extension services can be found in the literature. Often with each new donor project, whether on- or off-budget, yet another extension model is developed and put forth. Currently, the Afghan government has been trying to institutionalize locally elected development bodies known as Community Development Councils (CDCs). CDCs were initially created under the National Solidarity Programme (NSP) and are found in about 85% of Afghan communities (361 districts in all 34 provinces) (World Bank, 2015). The CDC election process is currently organized by national and international NGOs acting as facilitating partners that have a contract with the MRRD. The specific functions of the CDCs are to identify community development priorities, apply for project funding from the national development fund, and contribute to project monitoring and evaluation.

The recently elected National Unity Government plans to make the CDCs the primary vehicle to launch a Citizens' Charter—a commitment by the government to all its citizens guaranteeing a minimum set of core services. Ideally, this core set of services would be defined for each community, including the types, access, timing and duration of the services, and the officials responsible for service delivery. The CDCs will be the government platform and entry point through which other agencies, including MAIL and agricultural extension and international NGOs, deliver local-level services. This approach will help coordinate efforts and concentrate resources from international donors, but how agricultural extension and its role as rural educator and organizer will be integrated is yet to be determined. Whatever the future brings, an effective Afghan extension system must be created that is realistic, given the country's socio-economic situation. Furthermore, capacity-building measures must be included to reduce the substitution of temporary staff funded by off-budget donor projects for the system's long-term sustainability and resilience. A few proposals that advocate for such improvements now follow.

Increase on-budget spending for MAIL

The majority of development expenditures for agriculture extension come from off-budget, bilateral donor-funded projects, leaving MAIL little discretion over their allocation. Consequently, coordination efforts become complicated when deciding which extension services should be administered through MAIL and which should be supplemented by outside donors. This situation also contributes to the transmission of confusing extension messages to farmers, as each extension group works with a different extension protocol for implementing projects. Ideally, international donors and MAIL should jointly plan the transfer of donor projects to the Ministry's operational directorates. Hopefully, this problem will be addressed as the CDCs are institutionalized within the government and serve as a focal point to prioritize and coordinate extension services.

Increase spending at the district level

Too much of the available funding is centralized and spent on administrative functions at central and provincial levels, rather than at the district level. As a result, many extension workers are paid a salary but do not have an operations budget for field interactions with farmers, and become demoralized. MAIL needs to develop a more transparent mechanism for allocating its budget to district-level departments for service delivery. One creative way might be to establish a competitive mini-fund system, similar to that operated in AAEP II. In this system, provincial workgroups of district extension workers, researchers and farmers submit proposals with a detailed budget of up to US$5000 for training and extension activities with farmers. Projects must be completed in a short time frame (less than 1 year) and must include a presentation, final report and extension materials that can be used by other extension workers. Another mechanism to obtain money for extension projects could be working with the CDCs to apply for funding from the Afghanistan Reconstruction Trust Fund via the NSP.

Build capability of extension staff

MAIL has an ambiguous national training plan to improve the skills of existing extension staff. The inadequate technical capacity of most extension workers and their limited capacity to disseminate skills severely constrain agricultural extension. Part of the problem is attributed to recruiting extension workers from Afghanistan's 25 agricultural vocational high schools (grades 10 [ages 14–15] and 12 [ages 17–18]) and two training institutes (more advanced training through grade 14 [grade 20–21]) (World Bank, 2014). These training centers are neither well equipped nor adequately maintained, and graduates lack effective extension skills. A curriculum or certified crop advisor-licensing program should be developed, guaranteeing minimum agricultural education standards for important commodities. A crop advisor-licensing program could be organized in a manner that generates income for MAIL through a fee system for initial licenses and renewals. Also, a joint work plan between the research and extension directorates should be developed to encourage the undertaking and sharing of new applied research that directly benefits farmers. The research directorate is currently delivering some training to extension staff at the Badam Bagh Research Station in

Kabul (and at its six other stations across the country). However, because very little applied research is actually being performed at these research stations, the extension training is minimal. Continued support for staff capacity building, as provided by the more successful international government, university and NGO extension programs, can help this transition to a more effective and resilient extension system.

Information and communication technologies

Afghan extension services are poorly equipped and have inadequate training to take advantage of modern mass communication technologies, even though these are now important sources of information for many farmers. Information and communication technologies (ICTs) include social media groups, e-learning, radio and video streaming and cell phones, all of which can be used to engage farmers with extension activities and information. According to a survey by GSMA Intelligence (GSMA, n.d.), 68% of the Afghan population have a cell phone—about the same level as is observed in India. MAIL should begin dialogue with farmers through these newer communication platforms.

A web-based repository of agricultural knowledge is currently being adapted to the Afghan context, using UC Davis' e-Afghan Ag site (UC Davis, n.d.) and other donor-funded websites as models. The General Directorate of Extension has established a link to the MAIL homepage (MAIL, n.d.) called the Agriculture Knowledge Bank, but the site is not maintained and content is rarely added. Concrete plans should be developed by provincial offices to collect, vet and electronically post agricultural information that could be easily accessed and used by extension workers and farmers nationwide.

Increase the number of women in extension

As noted above, there are few female extension workers, and none at all in some provinces. Without the capacity to reach women in farming communities, MAIL and DAILs will fail to reach half of the population. Attracting more women to agricultural education is essential for increasing the number of female research scientists and extension staff. Moreover, 80% of Afghan women either own or have access to a cell phone (World Bank, 2014). Use of cell phones could serve as a powerful channel through which female extension workers could reach women farmers, but female MAIL staff require training if they are to be successful in tapping into this new communication medium.

Strengthen crop improvement programs

Developing a crop improvement program will be essential to advancing new varieties and eventually to Afghan-based breeding programs, which currently do not exist in MAIL's research system. Furthermore, MAIL urgently needs the capacity to form partnerships with relevant international and regional research institutions (e.g. CIMMYT and ICARDA) so that it can procure, evaluate, select and deploy improved varieties and agronomic best practices in a sustainable manner. Priority thematic areas for extension and applied research interventions by the crop improvement program could include agronomy, varietal improvement, pest and disease management, weed control, seeding regime, nutrient management and comparative analysis of conservation agricultural practices.

Improve security

Obviously, MAIL/DAIL's capacity to deliver services would be bolstered by ensuring a secure environment. A recent survey of extension workers at the fourth annual MAIL extension conference (February 2016) suggested that security could improve if DAIL extension objectives were clear to religious leaders, shuras (councils or assemblies of those who will be affected by a decision), tribal elders and media. Too often, extension workers feel their safety is compromised when their work is mistakenly linked to political agendas. The current ambiguous nature of extension workers in the eyes of villagers could also be clarified through the help of the CDCs.

Conclusions

Key to successful agricultural development in Afghanistan is an effective extension system that provides farmers with the knowledge and technologies needed to increase production, identify markets and sustain limited natural resources. The 2009 Afghanistan National Development Strategy stated that 'agriculture will determine whether Afghanistan will succeed or fail'. This statement rings true, given that an estimated 75% of Afghanistan's 33 million people live in rural regions, where agriculture is the principal means of livelihood. Additionally, recruiting adequate numbers of female extension workers and community leaders will be imperative for serving the large percentage of female growers. Regardless of political or moral agendas, all players should ensure that all Afghans have the best possible access to sustainable agricultural resources and knowledge and may again face the future with dignity.

While much progress has been made, the Ministry must still demonstrate its ability to continue delivering services to Afghan households across many provinces and districts, with limited donor funding and intervention. In 2009, MAIL's former Minister Mohammad Asif Rahimi stated: 'Donors helped us make plenty of plans for agriculture, it's time for us to implement'. A new role for MAIL will need to be defined within the Citizens' Charter and implemented through the CDCs. Continued outreach—including conferences and demonstration and teaching farm tours—will be vital to expanding the use of an adaptable and realistic extension model. Outreach can also provide participating DAIL staff with training and extension materials, using ICTs to enhance knowledge sharing.

References

Christensen, A. (1995) *Aiding Afghanistan: The Background and Prospects for Reconstruction in a Fragmented Society*. Nordic Institute of Asian Studies, Copenhagen.

CSO (2014) *National Risk and Vulnerability Assessment 2011–12*. Afghanistan Living Condition Survey. Central Statistics Organization, Kabul, Afghanistan.

Emadi, M.H., Adlparvar, N., Soltani, S., Mossavi, M. and Godarzi, M. (2009) *Agricultural Extension in Afghanistan: Review and Recommendations*. SSPSRL Project, GRM International. Department for International Development (DFID), London.

FAO (n.d.) FAOSTAT database. Food and Agriculture Organization of the United Nations Statistics Division. Available at: http://faostat3.fao.org (accessed 23 January 2017).

GFRAS (n.d.) Worldwide extension study: Afghanistan. Global Forum for Rural Advisory Services, Lindau, Switzerland. Available at: http://g-fras.org/en/knowledge/world-wide-extension-study/92%20world-wide%20extension-study/asia/southern-asia/282-afghanistan.html (accessed 23 January 2017).

Ghani, A., Carnahan, M. and Lockhart, C. (n.d.) Stability, state-building and development assistance: an outside perspective. The Princeton Project on National Security. Institute for State Effectiveness, Washington, DC. Available at: http://effectivestates.org/wp-content/uploads/2015/01/Stability-State-building-and-Development-Assistance-The-Princeton-Project-on-National-Security.pdf (accessed 23 January 2017).

GSMA (n.d.) Measuring mobile penetration. Available at: https://gsmaintelligence.com (accessed 23 January 2017).

MAIL (n.d.) Islamic Republic of Afghanistan, Ministry of Agriculture, Irrigation and Livestock. Available at: http://mail.gov.af/en (accessed 23 January 2017).

Mitchell, D.F. (2015) Blurred lines? provincial reconstruction teams and NGO insecurity in Afghanistan, 2010–2011. *International Journal of Security and Development* 4(1), 1–18. Available at: http://dx.doi.org/10.5334/sta.ev (accessed 23 January 2017).

NSP (n.d.) National solidarity programme, Afghanistan. Available at: http://www.nspafghanistan.org/ (accessed 23 January 2017).

MoE (n.d.) Ministry of Economy, Afghanistan, NGO Department. Available at: http://moec.gov.af/en (accessed 23 January 2017).

Olsen, L. (2006) Fighting for humanitarian space: NGOs in Afghanistan. *Journal of Military and Strategic Studies* 9(1), 1–28.

Rubin, B.R. (2002) *The Fragmentation of Afghanistan: State Formation and Collapse in the International System*, 2nd edn. Yale University Press, New Haven, Connecticut.

Stewart, A.K. (2014) U.S. army agriculture development teams: a grassroots effort in Afghanistan supporting development and tackling insurgency. *Science and Diplomacy* 3(1). Available at: www.sciencediplomacy.org/article/2014/us-army-agriculture-development-teams (accessed 23 January 2017).

Stoddard, A.H., Harmer, A. and Ryou, A. (2014) Aid worker security report 2014: unsafe passage: road attack and their impact on humanitarian operations. USAID, Washington, DC.

UC Davis (n.d.) e-Afghan Ag. Available at: http://afghanag.ucdavis.edu/ (accessed 23 January 2017).

UNDP (1993) *Afghanistan Rehabilitation Strategy: Action Plan for Immediate Rehabilitation*. Vols I and IV. United Nations Development Programme, Rome.

USAID (n.d.) Afghanistan agriculture. Available at: www.usaid.gov/afghanistan/agriculture (accessed 23 January 2017).

USAID (2010) U.S. foreign assistance for Afghanistan post performance management plan: 2011–2015. United States Agency for International Development, Washington, DC. Available at: www.usaid.gov/sites/default/files/documents/1871/Approved PMP updated 10-02-2010.pdf (accessed 23 January 2017).

Waisová, Š. (2008) Post-war reconstruction in Afghanistan and the changing NGO-government relationship. *China and Eurasia Quarterly* 6(3), 65–85.

Wesa, T. (2009) *Afghan Agricultural Extension System: The Impact of Neglect and Prospects for the Future*. LAP Lambert Academic Publishing, Saarbrücken, Germany.

WFP (2009) *National Risk and Vulnerability Assessment 2007/8: A Profile of Afghanistan*. European Commission, National Surveillance System Project. World Food Programme, Rome. Available at: www.wfp.org/content/afghanistan-national-risk-and-vulnerability-assessment-2007-8-2009 (accessed 23 January 2017).

World Bank (2005) *Afghanistan—State Building, Sustaining Growth, and Reducing Poverty*. World Bank Country Study. World Bank, Washington, DC.

World Bank (2014) Islamic Republic of Afghanistan agricultural sector review: revitalizing agriculture for economic growth, job creation and food security. Report No. AUS9779. World Bank, Washington, DC.

World Bank (2015) Afghanistan country summary brief: national solidarity program I, II, III. Available at: http://siteresources.worldbank.org/EXTCDD/Resources/430160-1435154813801/Afghanistan_Country_Brief.pdf (accessed 23 January 2017).

10 Tajikistan: Creating Agricultural Extension Services in a New Post-Soviet, Post-Conflict State

Don Van Atta*
University of North Carolina at Chapel Hill, North Carolina, USA

Introduction

Tajik farmers confront multiple technical, economic and environmental challenges. Their knowledge of animal and crop husbandry is limited to 'traces and fragments' (Shtaltovna, 2016: 25). Farmers lack access to input and output markets and the country does not have enough skilled agricultural personnel. An evaluation of the knowledge of agronomists employed by the United States Agency for International Development (USAID) Farmer Advisory Services in Tajikistan (FAST) program as extension agents by a senior faculty member from the University of Illinois found very large gaps in those professionals' knowledge, even though they were competitively selected and had, as a group, considerable experience with previous technical assistance projects (Babadoost, 2014).

There was no need for any extension system of the sort familiar in market economies under Soviet collectivized agriculture. The creation from scratch of extension systems must be part of the broader reform of the whole national economy and polity to move away from a command system (e.g. Karcz, 1974) to a market-based one. The collective and state farms that were the foundation of the Soviet command economy and polity and were the key organizations in the countryside until the end of Soviet rule in 1991 have been largely dismantled. Decollectivization, however, also requires the creation, repurposing or strengthening of all other rural social and political institutions. Stronger, more effective local government would play a very important role in preventing future conflict in Tajikistan. Extension, by providing a visible and valued service to village residents, can increase the legitimacy of the state and strengthen local government.

Tajikistan desperately needs modern agricultural education and extension. But so far, despite hard and dedicated work by many people and the expenditure of considerable amounts of money, especially foreign aid, little progress has been made in building viable agricultural extension structures in Tajikistan.

Background

Soviet Tajikistan was effectively a colony. The Soviet authorities arbitrarily drew the boundaries of what is now independent Tajikistan in 1924. Only after the collapse of

* Corresponding author e-mail: davanatta@gmail.com

the USSR in 1991, however, did those lines on the map became real territorial boundaries affecting people's lives (Reeves, 2007). They enclose a heterogeneous population and territory with little historical unity bordering China, Afghanistan and two other new post-Soviet states, Uzbekistan and Kyrgyzstan.

Tajikistan is a mountainous country with abundant water but relatively little land that is suitable for agriculture. Most agricultural land requires irrigation. The Soviet authorities systematically built irrigation systems and regularly brought large new areas into cultivation. As new agricultural regions were developed, the planners shipped people to them to provide the needed labor. The southwestern part of the Khatlon province, now the country's principal agricultural region responsible for about 40% of total agricultural output, was reshaped into an irrigated agricultural area optimized for cotton monoculture in a tremendous, 60-year construction effort (Aminjanov, 2012). Whole villages from elsewhere in Tajikistan and other Soviet republics were forcibly relocated to Khatlon and other areas of Tajikistan newly opened for cultivation (Kassymbekova, 2011).

Tajikistan suffered a post-independence civil war in 1992–1997 (e.g. Atkin, 1997). As many as 60,000 people perished violently in the war (Kuvatova, 2001: 128). Much of the built environment, particularly in the southern Khatlon province, was destroyed. The educational system was particularly affected. The fighters wrecked rural schools and murdered or drove away teachers. The restored national government was crippled by the loss of skilled personnel, what amounted to national bankruptcy was caused in part by its attempt to maintain subsidies to production agriculture during the war and suspicion between the former parties to the conflict. Under the circumstances, the United Nations, the International Monetary Fund and the World Bank were unusually important in supporting and funding the post-civil war government, and they acquired an unusual amount of influence over the new state. Almost 20 years after the peace settlement, the government remains highly dependent on the international donor community.

Armed clashes between government forces and insurgents continue periodically. The most recent significant fighting—between supporters of a mutinous Deputy Minister of Defense and government forces—occurred in and near the capital, Dushanbe, in September 2015. The state and civil peace are sufficiently fragile that a return to conflict is a real concern among analysts and Tajik citizens (e.g. ICG, 2016).

Tajikistan almost entirely lacks any tradition or experience of civil society or the rule of law. Until 1991 members of the Tajik national elites, unlike those of most African countries under colonial rule, were not free even in theory to study in or learn about metropolitan powers other than Russia. Their knowledge of anything other than the Soviet way of doing things was and is very limited—and all too often based on superficial understanding of the non-Soviet world. Groups opposed to the late Soviet-era *perestroika* reforms—with their limited acceptance of liberal political values and market mechanisms—were the victors in the post-independence civil war. A quarter-century after independence, Tajikistan is still in an early stage of decolonization when former colonial and first-generation post-colonial elites are trying to maximize their own control of resources left from the colonial period, destroying many of those resources in the process.

Emomali Rahmon, first elected head of state during the civil war in late 1992, remains the country's leader after winning presidential elections in 1994, 1999, 2006 and 2013. Since his first inauguration, President Rahmon has steadily strengthened his position (Global Voices, 2016; Lemon, 2016). The US-based Freedom House characterizes Tajikistan as 'not free' (Freedom House, 2016).

The contemporary Tajik state is formally centralized and practically authoritarian, largely retaining Soviet institutional patterns. The national government is extremely suspicious of any reforms that appear to devolve real power to regional or local governments, as the rulers' nightmare is a return to the territorial fragmentation of the civil war (Matveeva, 2009). Although outwardly strong, the national Tajik state is actually quite fragile. It depends on citizens' habits of obedience and its coercive apparatus to ensure civil order and political stability, having little ideological appeal and providing little economic advantage to the vast majority of citizens. The national government largely relies on international donors to improve medical, educational and other social services, and to maintain and develop infrastructure, particularly in the countryside.

The specific political economy of Tajikistan is crucial to the country's continuing stability but also greatly hinders economic and political development. Tajikistan's patrimonial political regime relies for its stability on distributing state and international donor-provided resources to a myriad of local and regional supporters in a complex and shifting set of rewards and punishments. The government faces growing difficulties mobilizing enough resources—whether cash, government jobs, construction and development projects or control of agricultural and industrial output—to keep all elite claimants satisfied.

The Republic of Tajikistan is the poorest and has been the least stable of the states that emerged from the collapse of the USSR in 1991. It is still the poorest post-Soviet state. In 2014, 32% of the population lived below the poverty line, including 16.8% of the population who lived in extreme poverty, unable to supply themselves with enough food to meet the minimum daily energy requirement (Presidential Agency on Statistics, 2015: 21).

Agriculture

In his 2016 annual message to Parliament, President Emomali Rahmon stated that agriculture contributes on average 23% to the annual gross domestic product (Khovar, 2016). More than 73% of Tajikistan's population lives in rural areas, where there is no work but primary agriculture. Agriculture provides 45% of total employment in Tajikistan (Presidential Agency on Statistics, 2015). Unpaid rural employment is hard to disentangle from paid rural labor. Most village residents work continuously on their own household farms and also receive nominal cash wages—and the right to cotton stalks and other waste which they can burn for cooking and heating—for work done on commercial farms at peak times such as harvest.

Tajikistan has been conducting a land and agrarian reform since 1992. It is now expected to be completed in 2020. From fewer than 1000 collective and state farms in 1991, some 143,653 farms had been created in Tajikistan as of 1 January 2015 (SCLM, 2015). Farm size data for the entire country are not available, but in the districts of the Khatlon province for which information was available as of 1 January 2012, more than 60% of all commercial farms held less than 5 ha of arable land (Aliev and Van Atta, 2013). The reform's design had been largely developed in the Russian Federation, where irrigated agriculture is much less prevalent and where the political purpose of the reform was to eliminate the old communist elite's power base in the countryside (Van Atta, 1993, 1994).

The reform's design and implementation have not focused on ensuring that farm operation could be profitable for farmers. The commercial farms created by ongoing land reform from the Soviet-era collective and state farms have struggled to operate profitably, contributing substantially to the continuing weakness of the country's economy (Van Atta, 2009; Eurasianet, 2016). Before 2007, the government cited the need for job creation to justify maintaining subsidies to the cotton sector, even as it was seeking donor support to pay the snowballing debts of farms to private suppliers of inputs and cotton traders (Van Atta, 2009). Following the financial collapse of the sector in 2008 (IMF, 2008; Minder and Gorst, 2008; Ernst and Young, 2009) the government agreed to freeze cotton debts, systematically reduce its subsidies to the sector, pursue crop diversification and increase efforts to make agriculture profitable for farmers over a 5-year period in return for donor support (Alimardon, 2009; Rahmon, 2009; World Bank, 2011; GoTJ/DCC, 2012; World Bank, 2013: 26). The government largely carried out these commitments.

Household agriculture—very small-scale 'subsistence' agriculture carried out by the families of collective farm members and state farm workers—was always a very important source of family food supplies and income in the Tajik SSR, as it was throughout the Soviet Union (Wädekin, 1973). Household farms have become even more important for rural livelihoods in Tajikistan since 1991. They hold about 25% of the country's total arable land (Van Atta and Safaraliev, 2015). President Rahmon stated in 2014 that household farms produce more than 60% of the country's gross agricultural output by value. Household farms hold 90% of all livestock and, in addition to producing about half of the country's total wheat, also contribute very substantially to the production of labor-intensive horticultural crops (e.g. Muminov, 2011).

The Village Community

As a result of the civil war and the accompanying collapse of government, village communities and local notables such as former civil war commanders, businessmen or former heads of successful state and collective farms—often the same persons—became the key providers of services and support to local residents.

Village organization and culture varies considerably within the country, and everywhere there is an unstable amalgam of pre-Soviet Central Asian forms and attitudes, Soviet ones and post-Soviet forms of life (Ibañez-Tirado, 2015). However, traditional, semi-formal community organizations, particularly the *mahalla* (neighborhood, also originally the area whose residents all attended the same mosque), are the basis of community life beyond the household (Kikuta, 2016: 91–93). *Mahallas* exist in all residential areas of the country. They are legally recognized and regulated as 'community self-activity organs' (Zakon, 2008). That awkward phrase roughly approximates to 'community-based organizations'. Every *mahalla* has a leader (*raisi mahalla*) who may simply be acknowledged by general consensus or who may have been elected by the community. Higher authorities confirm the *mahalla* leaders in their positions. Most *mahallas* have a council that includes the *mahalla* leader, the former collective farm brigade leader, the schoolteacher, the mullah and others. In many locations, such as generally throughout the Zarafshon and the Rasht Valleys, the *mahallas* have become legal entities. As such, they are 'community organizations' (in Russian, '*obshchestvennye organizatsii*'), the Tajik equivalent, legally, of non-governmental organizations (NGOs) elsewhere. This status allows them to have bank accounts and handle money, including donor funds.

Ethnographic observation suggests that in most places where there is an elected or informally agreed local leadership, whether as a *mahalla* committee, a 'village development organization', a Water Users' Association (WUA) management board or other community agency, the same people will be on all such local bodies as a result of community consensus or acquiescence (Sehring, 2009; Levi-Sanchez, 2015). As John Heathershaw noted, community-based organizations 'were typically composed of pre-existing groups within the community that were invisible to international programme officers (though not their local staff) and made their decisions informally, over tea in the mosque or community center out of the sight of the donors' (Heathershaw and Megoran, 2011: 608–609).

Migration and the Role of Women

The civil war and the lack of decently paying jobs in rural areas have caused massive emigration by skilled and unskilled labor alike to seek employment in the Russian Federation. The conventional estimate is that 1 million of the 8.5 million people of Tajikistan, mostly men, are migrants working outside the country. In recent years, Tajikistan has depended more on migrants' remittances than any other country in the world. Current economic difficulties in Russia caused by international sanctions and collapsing oil prices have caused a decline in the number of people leaving Tajikistan to work in Russia in 2014 and 2015 compared to 2013 (MEDT, 2015). But the expected wave of returning migrants has not yet occurred in the numbers expected—or feared—most probably because there is no work other than subsistence agriculture for them in Tajikistan (Asia-Plus, 2016).

There is some room for doubt that men would take most of the jobs available in agriculture. Women do most of the work on household farms. They also do almost all of the manual and seasonal labor on the commercial farms, where there is a preference for hand-picked cotton and an acute absence of machinery. Most of the seasonal and permanent migrants to Russia are men, so women are even more important in agricultural labor than they would be if the migrants were home. However, in most places they do not have formal power, and stories of husbands telling their wives how to manage the farm by phone from Russia are common (Giovarelli and Undeland, 2008; Mukhamedova and Wegerich, 2014). Although it is reasonable to assume that women have become more important as decision makers, USAID's index of women's empowerment in agriculture in Tajikistan notes particularly that women have little autonomy in production decisions in agriculture (Malapit *et al.*, 2014: 29), reflecting their status as casual workers, not skilled labor or managers on the commercial farms. Observations by the FAST project indicate that women have some autonomous decision making power over household farms. Their freedom in that sphere seems to increase when they know more about how to successfully manage them.

Government Administration of Agriculture

The national government has been continually pressured to reduce its spending and its staff, yet continues to see its role in agriculture largely as directly organizing production. It has never taken the lead in developing or supporting public or private extension, although individual senior officials have at times attempted to do so.

Coordination among cabinet-level agencies on cross-cutting issues is supposed to be facilitated by multi-agency councils, including a Council on Water Policy and a Council on Food Security (GoTJ, 2011). These cabinet-level councils rarely meet. As of the end of 2015, the Food Security Council had met only once since the Law on Food Security mandated its establishment in 2010.

Several cabinet-level agencies deal with agriculture in Tajikistan. The Ministry of Economic Development and Trade develops the annual official forecast for crop-sown areas. The Ministry of Agriculture exercises regulatory and production-monitoring functions, although the Presidential Agency for Statistics reports on the fulfillment of production forecasts. The State Committee on Land Management (SCLM) classifies and regulates the use of all agricultural land, conducts the land reform and ensures that agricultural land is used 'for its purpose' by all farmers.

The Ministry of Energy and Water Resources is responsible for overall water policy, including the timing of water release from the country's reservoirs for irrigation. It, however, reports to the prime minister or his deputy concerned with industry, not to the deputy prime minister responsible for agriculture and the environment. Although agriculture consumes about 95% of Tajikistan's total water usage, the Ministry of Agriculture does not have a seat on the Cabinet Water Policy Council. The Presidential Agency for Land Reclamation and Irrigation constructs and manages all irrigation and drainage facilities except for those located directly on the territory of the former collective and state farms.

With the exception of the Ministry of Economic Development and Trade, which grew out of the Soviet-era State Planning Committee, all these cabinet-level agencies are implementing bodies. They have little capacity to develop policy, to advocate the interests of their constituents within the government or to provide public services.

The national agencies have sub-units in the country's three provinces and 59 districts. The provincial and district agricultural agencies are small, with no more than a dozen staff members each. Agricultural agency personnel have subject matter expertise but are largely trained to oversee the operation of large farms. They generally consider their primary responsibility to be making sure that farmers fulfill their production plans. Most of the 367 rural councils (*jamoats*) have a staff member responsible for statistics, another for land surveying and titling and a third for agriculture. Many of the *jamoat* agricultural 'specialists' have little or no formal training in agriculture. Because of their limited budgets and consequent inability to travel, provincial, district and *jamoat* agricultural specialists are generally restricted to telephone contact with 'their' farmers and farms. No government employees at any level have training in or formal knowledge of small-farm development or business management, although many are also farmers themselves. Even if their jobs were redefined to require them to provide farm business management advice, they would not be competent to do more than recommend individual sources of inputs and possible market opportunities.

The Ministry of Agriculture has no legal responsibility to assist or oversee household farms, and normally it does not work with them, although this situation is slowly changing. The job descriptions of the *jamoat*-level agricultural specialists include many duties, one of which is to provide information and assistance to household farms. Interviews with those officials, however, suggest that they are mainly concerned with production plan fulfillment and tax collection as opposed to providing direct services and assistance to farmers.

Local WUAs manage tertiary irrigation facilities. Although they fulfill a governmental function and their collection of water charges and membership dues are effectively taxes, they are legally considered to be non-governmental, voluntary organizations. Many WUAs do not really operate.

Agricultural Research and Education

Under Soviet rule, the Dushanbe Agricultural Institute (now the Agricultural University of Tajikistan) prepared students for agricultural careers. Curricula were very narrowly specialized, for instance, to be a 'farm bookkeeper' or 'tractor driver'. The Ministry of Agriculture specified the number of persons to be trained in each specialty and directed each new specialist to a specific large farm where the new graduate's services were needed. They were legally required to continue working where they had been sent for several years after receiving their diploma. The Institute also carried out in-service training and retraining, but had no research or extension function. The All-Union Academy of Agricultural Sciences organized all agricultural research in the USSR (Nikonov, 1995). The Academy's Central Asian Division and its network of research institutes and stations carried out agricultural research at the request of, and funded by, the Ministry of Agriculture. The Ministry also had its own network of laboratories and field stations, principally designed to test and certify new equipment, plant varieties and seed and livestock breeds—all developed by the Academy—for production use. The Ministry was responsible for introducing innovations it judged successful on all collective and state farms (Van Atta, 1990).

Formal agricultural education now largely runs on by inertia, mostly training students for jobs that no longer exist. The Agricultural University is still the major institution of higher education that trains agricultural officials, agricultural specialists and farm managers. Although it has added an 'agribusiness' department and its faculty now do some research, the Agricultural University's curriculum is largely unchanged from that of the Soviet era. Many of its graduates do not have the training either to operate farms in a market economy or to provide effective advice and assistance to commercial or subsistence farmers.

The conclusions of a recent USAID-funded study of the agricultural education and training (AET) system have serious implications for extension:

> [The] AET system in Tajikistan is not ready to absorb a multimillion dollar investment in infrastructure or curriculum development because it does not have the human or institutional capacity to absorb an investment. ... [T]he AET system is currently undervalued and under-resourced within the educational and occupational context of Tajikistan, and is at risk of falling even farther behind, to the point of becoming irrelevant. ... The loss of the capacity and potential at the university and technical college levels would greatly limit the possibilities for agricultural development and growth in Tajikistan, as these are the only institutions with curricula focused on the production side of agriculture.
>
> (Jones *et al.*, 2015: 17)

Without people trained to do the work of extension agents, no agricultural extension and advisory system, and no effective project-level interventions in agricultural production, are possible in Tajikistan.

The system of directed agricultural innovation and research that fulfilled some of the functions of an extension system before 1991 has largely collapsed. The Academy of Agricultural Sciences of Tajikistan, established in 1991 from the units of the All-Union Academy of Agricultural Sciences located in Tajikistan, has been greatly reduced from its former size. The Ministry of Agriculture, which is still supposed to act as the principal customer for agricultural research, lacks the funds to support much research work. Almost all of the Ministry's Soviet-era research, testing and training facilities have been abandoned for lack of funding. Their lands have largely been broken up and distributed to farmers as part of the land reform.

Efforts to build agricultural extension services in Tajikistan

The international donor community has driven efforts to create agricultural extension services in post-independence Tajikistan. Donor activities only began in 1996, and extension activities did not become very significant until a decade later. Kazbekov and Qureshi (2011: 11–14) and FAST (2014a: 17–21) briefly survey the situation.

Many humanitarian aid and technical assistance projects operating in Tajikistan have involved some form of localized agricultural extension. In October 2006, the European Union (EU) and the Canadian International Development Agency held a public conference to consider how to coordinate efforts to develop agricultural extension more widely (Sheltinga and Coene, 2006). This event was one of the earliest public consultations among donors to systematize efforts to build an agricultural extension system.

By 2008 the Ministry of Agriculture, with assistance from the EU but building on an earlier effort supported by the Food and Agricultural Organization of the United Nations (FAO), had established a working group to consider how to develop extension. The committee designed an extension system targeted at commercial farmers, proposing to create an Information Coordination Center within the Ministry of Agriculture that would work directly with a private extension providers' peak association. That association, in turn, would facilitate the provision of agricultural consulting services by agricultural cooperatives, independent farmers' associations, private businesses and NGOs (SENAS *et al.*, 2009; Dadabaev, 2010). A peak organization, AgroDonish ('Agricultural Knowledge'), established and supported by the international community, was then active.

In 2010, the World Bank Public Sector Reform Program supported functional reviews of several Ministries, including the Ministry of Agriculture (Van Atta *et al.*, 2010a, 2010b). The review recommended that the Ministry fundamentally restructure itself, including the creation of an extension unit along the lines of the 'information center' proposal. However, the Ministry of Agriculture rejected the reorganization proposals, responding that the Ministry was structured to administer crop production, animal husbandry and agricultural engineering, as well as to organize economic planning and production reporting. Its staff had no training or experience with client service, economic analysis and market research or most of the other functions proposed for the Ministry, including US- or European-style agricultural extension and consulting. Since people who did know about these subjects in Tajikistan did not exist or those who did were not willing to work for low official government salaries, and since the Ministry and government as a whole did not want to end the employment of the existing staff, the recommendations were rejected.

In 2011, continuing the activity of the Ministry working group, the government requested assistance from the donor community to create a national strategy to develop agricultural extension services for the commercial farms created by the ongoing land reform (van Weperen *et al.*, 2012). The German development agency GIZ (Deutsche Gesellschaft für Internationale Zusammenarbeit GmbH) took the lead role in this work from the donor side. According to persons involved in the discussions, the strategy was left incomplete when the Ministry insisted that it should direct all extension activities, whether they were publicly funded or carried out by NGOs. This reported Ministry position is probably a poorly explained restatement of the plan proposed in 2010 for the Information Coordination Center, but it is likely that the dispute also concerned who, if anyone, would evaluate the quality of extension agents and the reliability of their advice. Before the disagreement could be resolved, the donor technical assistance project that supported the work ended, and the ministry stopped active work on the document.

In 2013, the Ministry of Agriculture, with advice and encouragement from the FAO, established an agricultural extension department. A 'department' is the smallest recognized unit within a Tajik Ministry. It must be composed of no fewer than two professional employees. At present, the Ministry's Extension Department has two employees.

Ministry officials have indicated that if funding were to become available—which, because of the Ministry's limited budget in practice means if an international donor were willing to pay—the Ministry would be interested in developing greater extension capacity at national and regional levels. Senior officials of the Agency for Land Reclamation and Irrigation have also recently stated that they would like to develop the capacity to advise and direct farmers on crop production, viewing that as a profit center for the Agency.

Donors' approaches to agricultural extension in Tajikistan

Donors have supported a variety of approaches to extension. All of them have been tied to specific projects, although they intend to design and demonstrate activities that, after successfully showing results in a relatively small geographic area of the country, could be taken over by government or the private sector and scaled up to achieve national coverage. However, to date these efforts have almost entirely failed. Extension services have been limited in space to the areas covered by a project's funding, and in time to the life of donor support.

There is general agreement that household and commercial farmers welcome extension services and that commercial farmers would in principle pay something for them if they could not receive services free (Engel *et al.*, 2015). But the government's limited resources, the difficulties of the business climate, the general unprofitability of agriculture for producers, and farmers' expectations after 20 years of international assistance that donors should and will give them things free of charge, make it very difficult for donor initiatives in extension to take root. These limitations are generally understood by international and national staff in Tajikistan. As one international organization employee commented to a visiting project designer in 2014: 'You should still be able to find some of the extension NGOs that worked with that project. It just ended, so they won't have disappeared yet.'

The donor community in Tajikistan debates whether agricultural extension services should be self-supporting based on fees for service or paid for by the government. At one level, this reflects a difference between the US model of agricultural extension and the one now common in Western Europe. At another level it reflects a debate about the role of government. As a USAID employee commented once during a discussion of extension among USAID employees and contractors in 2010: 'If you say that the government is to support it, that's not sustainable!' This statement was made not because of the very considerable fiscal difficulties of the Tajik government but because the speaker did not believe in government provision of such services.

'Fee or free to farmers' is a false dichotomy. Any extension system should require some contribution from the farmers it serves, although not necessarily a monetary one and very likely not full cost-recovery.

The Tajik state is not willing or, as its Ministry of Finance asserts, able to fund an extension system, although some progress has been made in convincing some policy makers that it should because it would be a public good, serving almost everyone in Tajikistan. Farmers have not been able to fully pay the costs of extension, as the tendency of extension NGOs to become moribund as soon as donor funding ends shows. Whatever their stated approach, donors have paid for most of the costs of all the extension activities discussed in detail in the following section.

Extension efforts in Tajikistan have provided different kinds of services. Most offered training on specific agricultural topics, for instance, pruning or greenhouse construction. Some models followed those topics with a specific group through a growing season or production cycle. All these efforts seem to have been limited to a single crop, product or value chain. So far as can be determined, no extension effort in Tajikistan has sought to provide training on integrated farming systems or on designing, operating and managing a multi-activity farm business.

Other extension activities were narrower, or 'extension-like', in that they sought to reach a rural audience, but with messages that might not be traditional for extension. Their experience can inform extension efforts in Tajikistan. Sometimes donors have sought to copy mechanically their experience for agricultural extension proper, with mixed results.

Most extension efforts in Tajikistan have been aimed at commercial farmers who received their land through the land reform. Commercial farmers, especially those who attend training events, are overwhelmingly male. There does not seem to have been much attempt to differentiate target audiences based on farm size or specialization. Given the circumstances, it probably would have been unreasonable to require such a narrow focus, since the attendees at many extension efforts were and are those people who happened to find out about the event, not an audience or group of participants picked by the extension activity. When the audience was selected, the local authorities generally made the choices. For instance, the Khatlon provincial authorities instructed the heads of local NGOs to attend a training session sponsored by FAST and the US Farmer-to-Farmer program by calling each NGO's leader. Since it is hard to adequately publicize events in Tajikistan, the authorities were helping their farmers and the event by informing others about it. But, for a local resident, the authorities' request to attend the event was an offer that could not be refused.

Since women do most of the work but the bosses are overwhelmingly male, activities for commercial farmers train the bosses, not the laborers. The male head of household has considerable power over the household farm, but it is worked almost exclusively by women. So extension aimed at household farmers must be tailored to reach women. Doing so requires constant care, as male trainers may not always be welcome in women's groups. When groups include both men and women, they stand separately by gender with the women generally in the back or to the side, and trainers tend to speak to the men and ignore the women.

Extension services, in Tajikistan and elsewhere, can be distinguished by the length and type of the contact between the extension worker and the audience. One-off advice, answering specific questions when someone walks in to an extension center or purchases a product from an agricultural input store, or purchases of a product from an agricultural input store (an 'agro-shop') are the contacts with the shortest duration. Depending on the question and the answer, it may be very effective or have little impact. One-time classes, such as presentations by US Farmer-to-Farmer volunteers, may be very useful if the speaker and the audience are well matched. Short courses, such as those provided by USAID's Water Users Association Support Program, have longer contact time. Farmer Field Schools extending over a whole growing season involve still more contact and begin to allow follow-up. If the purpose of the extension activity is to transfer skills, contact that continues over more than one growing season, in order to allow interaction and to show improvement, is clearly preferable.

Many of the extension activities in Tajikistan have relied on town-based centers, which required farmers seeking extension to travel to them. Input-supply stores, similarly, require travel, although in their case the travel has more than one purpose (to buy an input, as well as to seek advice). It is reasonable to assume that extension activities that require the farmer to come to the extension center generate less interest and effect than activities performed at the farm or in the village.

Specific extension projects

This section briefly reviews specific extension projects carried out in Tajikistan. The analysis is based on a review of available literature, discussions with knowledgeable Tajiks and international specialists, and the author's own knowledge. The review suggests that despite repeated donor affirmations that extension programs are vitally needed, most 'extension' efforts in Tajikistan actually have been brief training sessions or public awareness campaigns with little continuity or follow-up.

Japan International Cooperation Agency (JICA)

The National Association of Family Farms of Tajikistan is a quasi-governmental body dedicated to representing the interests of commercial farmers to the Ministry of Agriculture and elsewhere in the government. Some of its district branches employ one or two extension agents, usually working from the district administration building. As part of a general commitment to strengthening farmers' associations, JICA joined with the Association to train those agents and prepare agricultural production training materials in 2008–2011. The Association still has much of the equipment, and the training materials are still in circulation. It is not clear how many of the district level advisors continue to function.

World Bank

From 1996 on, the World Bank, almost entirely through its soft-loan window, the International Development Association, has provided major funding for land reform, irrigation works and agricultural development (World Bank, 1996). Many of these project designs have included some extension services. In most cases, this requirement was satisfied by one-off training classes for direct beneficiaries, organized by the project management unit (PMU) either through direct hires or through subcontracted NGOs. Although participants in such training events always express their appreciation for them, and there is no doubt that the information provided is useful to participants, there is little reason to think that such one-time efforts have much long-term impact.

The World Bank's Farm Privatization Support Project (FPSP, 2000–2005) paid for the refurbishment of a building for use as the National Agricultural Training Center. The facility survives as a fee-for-service subsidiary of the Ministry of Agriculture. It now rents its meeting facilities to anyone who will pay and does not concentrate on extension training.

At the request of the government, the donor community devoted considerable efforts to reviving cotton production in Tajikistan in 2000–2007. One of the most recent activities devoted specifically to the sector was the joint Asian Development

Bank Sustainable Cotton Subsector Project (SCSP, 2007–2012) and World Bank International Development Association's Cotton Sector Recovery Project (CSRP, 2007–2013). The two efforts shared a PMU (GoTJ, 2007) and largely operated as a single unit.

According to the project's Implementation Completion Report (World Bank, 2013), eight NGOs each trained ten trainers-of-trainers 'to implement community outreach and farmer extension programs in project districts' (p. viii). Those trainers conducted seven awareness campaigns. A reported 6800 people benefited from public awareness campaigns (p. vii). The report said elsewhere that 10,120 people benefited from community outreach programs, and 5251 people benefited from farmer extension programs (pp. vii–viii). (It appears that these several numbers are various ways of counting the same contacts, not a cumulative number of persons reached.) The Implementation Completion Report further states that four Training Information Centers (TICs) (p. 23) were established and would continue:

> These NGOs (19) that have received technical assistance and training under CSRP will continue to work with the farmers in terms of outreach, training and advice. The TICs that have been established will continue their operation, and have already started to recover the operational costs by charging for services to farmers. The PMU has also taken a number of measures to ensure financial sustainability and transition to self-financing of all created TICs.
>
> (pp. 7–8)

The same Implementation Completion Report later said that the PMU worked with 32 NGOs (p. 19).

The Khatlon province authorities asked USAID's FAST to cooperate with the CSRP TICs to help them continue operations. However, FAST was unable to help because the TICs concentrated on cotton production, and, so far as FAST staff could determine in 2014, the TICs had ceased operations. It is likely that some of the NGOs mentioned also worked with the Tajik Agricultural Financing Facility (TAFF, see below), which assisted in disbursement of cotton farm financing provided under CRSP.

Asian Development Bank

The Asian Development Bank (ADB) has supported some extension as part of its investments in rehabilitating irrigation, the cotton sector and overall rural development. It appears that most ADB efforts, too, were one-off training events by contracted NGOs. An evaluation of its projects in agricultural and rural development covering the period 1998–2013 suggested that 'farm capacity development and the establishment of quality extension services needs to be given greater attention in order to maximize the impact of irrigation works' (ADB, 2014: 13). The ADB reduced its emphasis on agriculture in Tajikistan after 2008 and now appears to be principally impacting Tajik agriculture as part of climate change resilience activities.

Land projects

Several donors, particularly USAID, have supported land rights consulting centers at various times and for various periods since 2000. Following the US government's decision to concentrate its local efforts in 12 districts of Khatlon province under its Feed the Future Strategy for Tajikistan (Feed the Future, 2012), the USAID-funded Land Reform and Farm Restructuring Project (LRFRP, 2013–2016), supported consulting centers in the 12 district centers and representatives (*tashabuskors*—rural correspondents) in many *jamoat* buildings (LRFRP, 2014a, 2014b). Although these land specialists are not trained as extension agents, they distribute written materials and pass on questions. The LRFRP has supported Neksigol's publication of a bi-weekly paper for farmers that its centers and representatives distribute, as well as circulating extension literature produced by USAID FAST.

These centers and advisors are working on contract to a USAID project, so there is no real intention or expectation that they will endure as centers or professional land rights advisors beyond the life of USAID support.

Water projects: USAID's Water Users' Association Support Program and Family Farming Program

USAID supported the creation of Water Users' Associations through a stand-alone Water Users' Association Support Program (WUASP) (WUASP, 2004–2011) and then as a major component of its Family Farming Program (FFP) (FFP, 2011–2015). Commercial farms are, by law, the members of WUAs. WUASP conducted comprehensive training on WUA organization and management for commercial farms' managers and specialists (Mendez England and Associates, 2010, annex 8). It also provided agricultural training, both field days and short courses, for commercial and household farmers in areas where it was establishing WUAs, particularly in its initial years when it was a multi-country activity (WUASP, 2006a, 2006b).

Nutrition messaging

The US government's Feed the Future strategy foresees that nutrition should be combined with agricultural development, since the end purpose is to improve both health and income of people living through agriculture. USAID Tajikistan included a nutrition component in the FFP. However, the FFP design turned out to be too complex to execute fully, so the nutrition education responsibility was transferred to the international NGO Mercy Corps. In 2013–2015 Mercy Corps carried out Feed the Future's nutrition education as part of its long-term Maternal and Child Health program in Tajikistan. In 2016, USAID launched a Feed the Future Tajikistan Nutrition Activity building on Mercy Corps' experience, indicating that nutrition will continue to be a separate activity from agricultural development under the overall Feed the Future Tajikistan umbrella.

Like agricultural extension, USAID's nutrition work was aimed at rural residents. Health and nutrition training would ideally be offered to all villagers. However, given gender roles and sensitivities, in practice it is usually done for schoolchildren and village women. The Mercy Corps model relied on village organizations (the *mahallas*) to gather groups of women for 1-day instruction sessions on specific nutrition messages. The groups had no organized existence or continuation beyond being gathered for the specific event. The program judged and retained its instructors in part on their ability to gather large numbers of village women for these training sessions (author's interviews, 2014).

The instructors, called 'Community Health Educators,' were normally government employees, usually doctors or nurses in the local clinics or schoolteachers. They received additional training on the specific messages to deliver and some token of appreciation for their efforts (author's interviews, 2014–2015). Since the nutrition instruction was in addition to their regular duties, they were not violating the usual international prohibition on paying government employees to do their jobs, while the training helped them to improve their skills and do their regular jobs better.

According to the State Committee on Investment of the Republic of Tajikistan, Mercy Corps reported that in 2 years it had created 500 village development committees in the 12 Feed the Future focus districts of Khatlon province (SCI, n.d.). The claim that these committees were newly created is surprising and may be a translation error. It is clear from discussions with Mercy Corps management that these committees were in fact the legally mandated *mahalla* committees or groups called together on the authority of the *mahalla* head on an ad hoc basis to support the nutrition activity.

Mercy Corps recruited and trained 1000 Community Health Educators. They held training sessions on health and nutrition with 234,049 people (SCI, n.d.). There is far too little knowledge of basic nutrition and public health in Tajikistan. This was very good and useful work. USAID Tajikistan in recent years has repeatedly praised this system. However, it has also indicated that it should be applied to agricultural extension. It may be doubted that the Mercy Corps model really constitutes 'extension' as the term is usually understood in agricultural development. There appears to be some confusion of 'messaging' about specific, important, but invariant topics with the development of agricultural knowledge in a constantly changing natural and economic environment. The way in which Feed the Future results are largely expressed in, and judged by, quantitative targets for outputs may also contribute to this misunderstanding.

Even if applying the Community Health Educator model to extension is desirable, it is difficult to do. The Ministry of Health and the Ministry of Education have positions in every village—not always employees, as the pay is very low and the conditions are often very difficult. As a result of decollectivization, the Ministry of Agriculture and the Land Committee no longer have specialist employees in every village who, for relatively small compensation after brief instruction, would undertake specific training analogous to the nutrition and public health ones. Persons with some agricultural knowledge can be found in most villages in Tajikistan. When asked, they say that they provide occasional agricultural advice to their neighbors because neighbors do so for each other as a matter of course. But since they do not already have jobs, these local experts will act as the equivalent of Community Health Educators, or lead farmers, only if they are paid a living wage or given some other fairly substantial compensation.

SughdAgroServis/Neksigol

The International Finance Corporation (IFC) and other donors supported the establishment of the SughdAgroServis (SAS) company to provide commercial funding, input supply, production advice, processing and marketing services to cotton farms in the northern Sughd province (IFC, 2004). Donors supported its extension into Khatlon as described in the next section. SAS, renamed 'Neksigol' in late 2014, is now entirely Tajik-run.

Neksigol's affiliated firm, Neksigol Consulting, offers extension advice on a fee-for-service basis, largely by providing trainers for class and field instruction to anyone willing to pay (OO SAS Consulting, 2015). It also has a fee-based mobile telephone service reporting agricultural and weather conditions (PO Neksigol Mushovir, 2015). The senior management of Neksigol view the consulting operation as a risk because Neksigol Consulting remains dependent on donor projects for most of its income; so Neksigol split Neksigol Consulting from its core input-supply business to protect the overall firm if consulting and fee-for-service extension turns out not to be viable (author's interviews, 2015).

Extension provided by input suppliers ('agroshops')

One major form of agricultural extension is advice and assistance provided by input dealers to their customers. Almost all Tajik farmers get their inputs from the bazaars. Neksigol is the only private source of legally sourced, certified inputs in Tajikistan. With funding and training assistance from the USAID Productive Agriculture Project in Tajikistan (ProAPT, 2009–2014), Neksigol expanded its network of input stores ('agroshops') into Khatlon and now has six company-owned input stores and five affiliated stores there. Each Neksigol store has a staff agricultural advisor who has received at least 40 hours' training (ProAPT, 2015: 25–26). Neksigol paid 76% of the cost of these activities (ProAPT 2015: 26), indicating that the firm finds commercial value in this activity and intends to sustain it.

The USAID-funded Tajikistan Stability Enhancement Program (TSEP, 2009–2012) supported the opening of one agroshop in Khatlon and one in Sughd. Those stores planned to provide product use advice for customers. In addition to its direct partnership with Neksigol, ProAPT also worked to upgrade the five stores that eventually affiliated with Neksigol in Khatlon (ProAPT, 2015: 27), one of which appears to have been a store initially supported by TSEP. That store was planning to hire a staff agronomist when FAST staff visited it in 2013. According to a World Bank staff member, the World Bank Emergency Food Security and Seed Import Project (EFSSIP, 2008–2013) also ended with the establishment of a network of agroshops in Khatlon. However, most of those stores seem to have ceased operations fairly soon after donor funding ended.

No data are available on the number of clients reached by any of the agricultural input stores.

Farmer-to-Farmer

The US Farmer-to-Farmer (F2F) program has operated in Tajikistan for at least 10 years under the management of the same local staff. A partial survey of volunteers' reports indicates that the same specialists have returned repeatedly, some as many as ten times. The volunteers conduct field days for audiences gathered from announcements through the local authorities and by cooperating USAID projects. Given the lack of modern knowledge in the national agricultural extension and training system, F2F has been a very important means of upgrading the skills of Tajik agricultural specialists. However, the limits of having short-term volunteers conducting 1-day sessions are obvious. F2F is not a substitute for improved initial training and continuing refresher courses for extension agents.

Extension NGOs: EU SITAF and Support to the Establishment of a National Agricultural Advisory Service, USAID ProAPT and Family Farming Program

Several donors have supported projects designed to build the capacity of NGOs that provide fee-for-service extension consulting. In 2005–2010, the EU supported two such technical assistance projects: (i) the Support to the Setting Up of a Structure to Provide Information, Training, and Advice to Farmers and Other Rural Businesses in the Khatlon Region of Tajikistan (SITAF) project; and (ii) the Support to the Establishment of a National Agricultural Advisory Service (SENAS) project. These projects trained NGOs to conduct extension work and supported the creation of a national organization of private and NGO extension providers—AgroDonish—to create a 'platform' where public and private extension service providers could meet and exchange information.

A 2010 listing and map of agricultural extension providers prepared by AgroDonish and Oxfam Tajikistan lists 46 private sector and NGO agricultural extension providers. Fourteen of them were located in or around Khujand, the capital city of Sughd province; 13 others are in or around Dushanbe (AgroDonish and Oxfam Novib, 2010). The remainder were scattered in provincial capitals and cities throughout the country. From their locations, it appears that they all relied on having farmers come to them, rather than providing services at the farm. The former SENAS team leader surveyed the state of extension NGOs in Tajikistan, concluding that, in order to flourish, they would need much further support as well as a more benign business climate for agriculture in general (Geraedts, 2011). Several of the extension NGOs listed in 2010–2011 continue limited operations, but most have become moribund because of the end of donor financing. AgroDonish maintains an office but is otherwise inactive.

The USAID ProAPT and FFP projects had large training components for commercial farmers. ProAPT contracted NGOs. FFP used both its own staff and contracted NGOs. Both projects found that local NGOs required a great deal of organizational strengthening and oversight to carry out their contracts to a high standard and to meet international reporting requirements. Having found that the project design was too complex to execute well given the level of management skill and organization available, even with the best efforts of very good people, USAID Tajikistan eventually decided to end FFP's agricultural training component before the conclusion of the overall project.

TAFF and Sarob

Probably the most successful sustained extension effort for commercial farmers in Tajikistan has grown out of the donor response to the collapse of the cotton sector in 2007. The European Bank for Reconstruction and Development, with core funding from the EU and additional funds from the UK's

Department for International Development (DFID), and GIZ, supported TAFF (TAFF, 2007–2012). Initially:

> TAFF ... used specialized NGOs for advising farmers who received credit. The results were limited and the option for massively expanding the advised area would have been far too expensive and unfeasible from the NGO capacities point of view. In cotton, NGOs cost approximately 160 to 200 USD per hectare for advising farmers during one production period. The results in terms of yield increase were relatively low (10 to 15%) as this service was too training-oriented and insufficiently based on providing concrete advice for individual farmers.
>
> (GIZ Tajikistan, 2013: 1)

To improve the provision of agricultural advice to the farms to which it lent, TAFF began the 'Technical Advisory Group' system in 2011 (Dziurman, 2012; GIZ Tajikistan, 2013). TAFF hired qualified agronomists—often former collective farm agronomists who had worked on the same fields under Soviet rule—and crop scouts to work with the agronomists to improve cotton production on farms whose operations were financed by on-lending donor funds through TAFF and Tajik commercial banks. The agronomists advised on production practices and monitored their implementation, duties much as they had done under the collective farm system. Crop scouts observed fields and brought problems to the attention of the agronomists. Farms were required to act on the agronomists' advice as a condition of their loans.

At the conclusion of TAFF, the extension agents formed a cooperative, 'Sarob', named for the mountain resort where it was founded. With continued but declining support from GIZ and others, Sarob has sought to expand its model of extension to other crops beyond cotton, and to smaller farms. A study commissioned in 2014 found that commercial small farmers in non-cotton regions of the country were in principle willing to pay for extension advice from Sarob (Engel *et al.*, 2015). The viability of Sarob without donor subsidies is still uncertain.

USAID's Modernizing Extension and Advisory Services Field Support Activity and FAST

The USAID Tajikistan Country Office, as part of its preparation for implementing the US Government Feed the Future Country Strategy for Tajikistan (Feed the Future, 2012), adjusting other existing agricultural sector projects to fulfill that strategy and designing follow-on efforts, requested the Modernizing Extension and Advisory Services (MEAS) program to establish a short-term project in Tajikistan, the MEAS Field Support Activity (MEAS FSA, July–December 2012). The MEAS FSA was to support USAID in facilitating the Donor (later Development) Coordination Council's working group on agriculture in ensuring the coherence of donor activities in the sector and maximizing their common contribution to developing a vibrant, prosperous, farmer-driven agricultural sector. It was also to advise on the design of a pluralistic extension and advisory system in Tajikistan based on earlier MEAS scoping studies in Tajikistan. Those studies had emphasized the need for a combination of public and private extension and advisory providers and for an enduring system (Swanson *et al.*, 2011; Swanson, 2012).

The MEAS FSA was asked, within its 6-month life, to develop, design, document and prepare all training materials for a participatory, pluralistic extension and advisory system as outlined in the MEAS scoping studies cited above that could be pilot-tested in the 12 Feed the Future districts in Khatlon province. The FSA developed a design based on the MEAS core principles. Extension workers would travel to the farmers. With the approval and assistance of the *jamoat* authorities and the local community (the *mahallas*), farmer groups would be organized. The group facilitator would host a small demonstration plot that would be the site of a Farmer Field School, which would continue for at least the crop's growing season. The plot owner would receive a small amount of free inputs and the advice and aid of the project-paid agronomist who conducted the Field School. The system would be structured from the province level downward, with a project employee in each *jamoat* working in tandem with the *jamoat* agricultural specialist so that eventually the agricultural specialist, a government employee, could take over most of the burden of frontline support of the farmer groups. This design would provide a form of public–private partnership into which government employees and extension NGOs could be integrated. It would ensure both community and government commitment to the system and would help to strengthen local government by developing and providing a tangible and apolitical service to the rural communities.

In correspondence and discussion, USAID representatives expressed concern about the time and cost needed to implement the farmer demand-driven, participatory elements of the design. They also expressed the desire to have the groups endure longer than a single season so they could be used as the basis for many kinds of training and outreach, including more general farm management and nutrition education. They suggested that the group-building schema developed by Catholic Relief Services and MEAS in an earlier version of the Skills for Marketing and Rural Transformation ('SMART skills') (MEAS, 2016) should be the basis of the extension design. How to achieve quick results while also engaging in long-term group-building activities was a continuing tension in the development and testing of the EAS design.

USAID then approved a follow-on project, FAST (FAST, 2013–2015) project. FAST was to continue as a general Feed the Future support project. It was also, within 1 year of its inception, to complete and pilot test the extension system so a follow-on project could roll it out for the entire Feed the Future farm constituency in the 12 focus districts of Khatlon province. That constituency included both small commercial farms and household farms. FAST was also to demonstrate to the Government of Tajikistan that the state should support the extension system in the near future and organize such state support.

The FAST pilot extension system focused on household farms because of their importance for food security and gender issues, which meant that the same general system needed to be implemented very differently for household (women-run) and commercial (male-managed) farms (FAST, 2014a, 2014b). Since household plots are physically within a household's home compounds, the target group became all village women, especially those younger ones—each household's daughters and daughters-in-law—who work the plots and are the commercial farms' manual laborers. Given the very limited time available to develop the system and train initial staff, it was also felt that concentrating on household farms filled an obvious gap in USAID programming. USAID's ProAPT was already working with commercial farmers in the 12 focus districts and continued to do so through the original end date of FAST's extension pilot. The model anticipated that, eventually, village level extension agents, who would emerge from the group facilitators, could be sustained by allowing them to charge for input supplies for household farms; however, that part of the model was never successfully developed or piloted.

As the FAST cooperative agreement did not include a grant or sub-contracting component, and as it was judged that the best way to develop extension capacity was to begin by training project employees without the added difficulty of managing an NGO as an intervening entity, extension workers for the pilot were project employees. Agronomy is traditionally a male profession in Tajikistan. Most Tajik agronomists have a narrow, technically focused skill set, so the design quickly evolved to allow for the fact that the subject matter specialists who backstop the *jamoat*-level extension specialists are almost all men. They would therefore coordinate closely with female extension facilitators, who would work most closely with the farmers' learning groups, composed almost entirely of women.

Nine months after its beginning, FAST was asked to continue its extension component for a second year, as an expanded pilot, due to delays in designing and contracting the follow-on project. Over its brief lifetime, the FAST extension pilot organized and

Table 10.1. FAST HFLG pilot cumulative impact, 2014–2015. From Van Atta (2015).

	Households	Women	Men	Total individuals
Direct beneficiaries	3,795	3,758	140	3,898
Same-household beneficiaries	Not applicable	12,334	12,334	24,668
Secondary beneficiaries	30,360	30,064	1,120	31,184
Total beneficiaries	34,155	46,156	13,594	59,750

FAST, farmer advisory services in Tajikistan; HFLG, household farm learning groups.

worked with more than 200 household farm learning groups (HFLGs) in ten districts of Khatlon province. Table 10.1 summarizes its quantitative impact.

Household budget data are notoriously unreliable in Tajikistan, especially when the household is consuming much of the increase, as the project participants did. However, FAST staff estimated from observation that overall household income increased. Participants repeatedly and positively stated that they were able to produce, consume and sell more of their household farm output.

Following the conclusion of USAID's ProAPT Project, during its unexpected second year of extension activity, FAST expanded to work with small commercial farmers. However, time and resources did not allow the full adaptation of the group model to that clientele. Most of FAST's work with commercial farmers was limited to one-off training events, often in cooperation with Farmer-to-Farmer volunteers.

The Khatlon Province's first deputy governor, who is responsible for the province's agriculture, praised the work of FAST and said that it helped him to realize how important household farms were for the country. He repeatedly requested that it be extended to more learning groups and provided significant—if non-monetary—support for the project. Similar requests came from an array of other groups active in Khatlon and from many rural residents in the region.

At the end of the project, USAID stated that FAST had met all of its goals. However, as a result of budget cuts, belief that household farms were becoming less important in Tajikistan as a result of the success of the land reform and development of commercial farms, and the need to concentrate on introducing a very small number of innovations as widely as possible (EAT, 2014), the previously intended generalization of the FAST system was abandoned. The follow-on project returned to a focus on working with commercial farms using publicly announced open field days for any commercial farmers who chose to attend (USAID, 2015).

FAST emphasized important issues and sought to satisfy real needs and requirements both for the Tajik population and the funder. In retrospect its design and tasks were clearly overambitious given the constraints on international donor projects in Tajikistan at present. However, the issues the FAST experience raises, particularly about the need for systematic extension to support agricultural development in Tajikistan and how to involve farmers as more than just passive consumers of messages, remain.

Lessons Learned

The review of the situation in Tajikistan and efforts to develop agricultural extension there suggest a number of general lessons applicable to other post-conflict and post-transition countries. Many of the lessons are familiar to analysts and practitioners of social and economic development and apply in many low- and moderate-income countries.

Clearly define agricultural extension and its relation to overall agricultural development

The various attempts at 'agricultural extension' in Tajikistan have actually tried to do quite different things. The extension mechanism needed depends on the content to be 'extended', how it is to be done and the intended and actual audience. Clarity on whether what is intended is simply top-down one-time technology transfer; helping rural communities that rely on agriculture increase their capacity to help one another; making farmers the driving force of a modern agricultural knowledge system; or something else, is necessary for the activity to work.

Agricultural extension is part of an AET system. It is also part of an overall agricultural knowledge system, including primary and applied research, education, training and an institutional orientation towards providing service to farmers so they can prosper. An extension system that exists in isolation from the other institutions and processes needed for productive, prosperous agriculture has little value.

Agricultural extension needs to be part of an overall strategy for agricultural development and improving rural livelihoods. The donors, and the Government of Tajikistan, generally treat these as two different objectives. The separation appears logical because agricultural development is usually measured and judged by macro-economic indicators such as an increase in national agricultural output, while improved livelihoods are measured in micro-economic indicators such as an increase in household income or improved household nutrition. In Tajikistan, this separation is often further institutionalized because agricultural development projects aim at commercial farms and are usually implemented by major for-profit consulting companies, while livelihoods efforts are often seen as emergency, 'humanitarian' efforts that are led by international NGOs. Yet both approaches are needed in combination, and the role of extension in both should be clearly understood.

Much of what has been done in recent years has confused 'messaging'—seeking to convey to farmers a particular piece of new knowledge, specific training or a particular innovation that a government official, international donor or expert believes would be helpful—with extension, in which farmers interact with educators, researchers, officials and service providers as part of an agricultural knowledge system. Most extension efforts in Tajikistan have also been forced to accept the temporal and spatial limitations of specific internationally funded projects and have not seriously addressed how to create an extension system or systems that could endure. Given the length of an agricultural season and the need to continuously adapt to changing natural and economic conditions, the focus on single messages and the time limitations have particularly limited agricultural development and the institutionalization of extension.

Government leadership of agricultural reform and extension efforts

Unless Tajik policy makers understand reform and development programs, and unless they lead and effectively coordinate those efforts, the programs and the considerable sums of money invested in them can at best only keep things from getting worse. Since independence, the Government of Tajikistan has focused most of its efforts in the agricultural sphere on simply keeping the sector operating in order to maintain rural employment. The government has agreed to reform efforts when donors imposed conditions, but senior political leaders have never developed an understanding or vision of what a reformed, market-oriented agricultural and food sector in Tajikistan should be. They continue to think in terms of fulfilling production targets. Given their focus, typical of government officials everywhere, on just coping with day-to-day crises and maintaining the flow of resources on which political stability rests, this lack of vision is hardly surprising. But since the national leadership does not understand what the systemic result of market-oriented reforms could be, they generally give them only lukewarm support.

The centralized Tajik political system makes it necessary to convince one or more top leaders before any general change is possible. The success of those leaders in winning the civil war and maintaining power and relative stability since its end suggests that they are politically sophisticated enough to recognize when an initiative or a program is in their interest. Agricultural extension can flourish in Tajikistan if the leadership sees that, on balance, it benefits them without posing a substantial threat.

Extension can build support for the regime and reform

The Tajik regime is harsh to its opponents because its leadership knows, and fears, how uncertain the situation is. The national government has been very hesitant to support any institutional changes or economic development that could give more autonomous power to regional and local actors, because it continues to fear losing control of the countryside and a recurrence of the regional conflicts of the civil war. The current elite does not want to lose its position. Presented with reasonable alternatives, no Tajik citizens want to return to civil war. Yet the average citizen has little reason other than the desire for civil peace and order to support a state that has little contact with its citizenry except to demand production and to collect taxes from them. Villagers avoid interactions with government as much as possible, relying instead on informal local institutions for services. This hollowing out of the state threatens its legitimacy and offers an opening for insurgents, of whatever kind, to gain support.

Donor-supported reforms, especially the ongoing land reform, have inadvertently contributed to this weakening of the Tajik state's authority and legitimacy. The government runs land reform from the center, according to quantitative targets. Given that many families who are eligible for land have adult working-age males in Russia, lack knowledge of how to run a farm business, depend on uncertain input supplies, have limited ability to reach markets and that in most cases agriculture is unprofitable for small commercial farmers, there is little incentive for people to support the land reform. Officials of the SCLM, which is responsible for it, admit that their role is limited to surveying, demarcating boundaries and titling. But when a farm has been chosen to be broken up into individual smallholdings, its former workers get land plots and are expected to manage an independent farm by themselves. Without help, the new recipients of land for farming are unlikely to be successful, and they, and their neighbors, are unlikely to support the reform no matter how much donor money is spent in raising people's consciousness of their right to land.

Since 2007, donor—and to a great extent, government—attention has moved away from efforts to rehabilitate cotton monoculture to supporting other potentially more profitable agricultural value chains. However, these efforts remain top-down, with little attention being paid to the vast challenge of making more new farmers good at the technology and business of agriculture. A national policy of building profitable small-farm businesses, a policy in which agricultural extension would be key, can build support for reform. If the state develops a capacity to impartially and effectively serve those farmers, the rural population's support for the government should increase, strengthening the state overall and stabilizing the country.

Far from all rural residents will want to run their own farms beyond their existing household farms, but some will. If enough villagers see that they can make money from expanding their existing household operations and that they will be helped to do so by effective government services, then the demand for land, and the demand for land reform and effective regulation of land tenure, will increase but can be directed and channeled without fundamentally threatening existing political arrangements.

An effective agricultural knowledge system—combining research, extension and sensible public investment in the agricultural sector—would create a different kind of bond between the state and the villager, one based on the interests of the villagers in both peace and plenty. An effective extension system would be both a key institution and a visible symbol of this changed relationship between state and society. If carefully done, it could provide both employment for regime supporters and real benefits for everyone.

Given the very large number of rural residents and farmers and the importance of the agricultural sector in the national economy, it is reasonable that Tajikistan should consider the provision of extension services to be a public good, something that benefits everyone and should at least in part be funded by the state. The enormous potential demand for extension services, and the wisdom of having multiple sources of information, suggest that a directly government-run extension system could and should coexist with private extension providers.

Coordinate within government and among donors

The political system and government administration are fragmented into narrow administrative verticals ('stovepipes'), which concentrate on particular tasks rather than the overall picture. For example, Khatlon province's first deputy governor commented in an interview that there are nine deputy governors for the province. Each one, including the first deputy, is responsible for his own set of issues. The provincial governor and his very small staff must coordinate all of the activities of these subordinates since the deputies have no staff and only limited knowledge of what other parts of the provincial government are doing. As a result, agriculture and nutrition activities are poorly coordinated, because one falls under the first deputy governor and the other under a separate deputy governor. The international community has done little to analyze and help the Tajik government ameliorate such structural problems and, as a result, their efforts are fragmented and less effective than they could be.

The donor community follows the government's divisions. Since the donors must have institutional counterparts, and those counterparts are fragmented, individual projects have little choice; so the donor community should address the problems of fragmentation through higher-level consultations.

Develop institutional memory

Government personnel are focused on their immediate administrative and fiscal tasks. Most international staff in Tajikistan are only there for a 2- or 3-year tour of duty. Tajikistan lacks a generation of middle-level, middle-aged specialists because the civil war prevented them from being trained or they perished in it. Low government salaries, lack of opportunity in the private sector for specialists except in international projects and the impermanence of those projects all drive the best people to seek work abroad; so it is difficult to train and retain Tajik specialists who know their situation well and choose to stay in the country to work. The international community tends to bring in ready-made solutions designed elsewhere or by short-term consultants with limited familiarity with the country. Local specialists who know the situation are generally very careful about expressing their opinions to international project designers and managers. As a result, the same approaches to problems tend to be tried repeatedly, with little awareness of the lessons learned from previous applications of similar ideas.

Tajik leadership and ownership can be increased by directly improving the institutional memory and coordination of the government through such means as donor support for creating agricultural policy capacity. This is difficult but has succeeded in Eastern European transition economies and, to a lesser extent, in the Russian Federation, where the government structure and institutions were and are similar to those of Tajikistan.

Consider the specific situation in the country

Because of the Soviet institutional heritage and the specific political economy of Tajikistan, institutions and activities that at first glance appear to be similar to those in developed market economies or other developing states may in fact be quite different. As an international consultant considering agricultural privatization once commented over breakfast: 'farms aren't farms, banks aren't banks, government isn't government. Nothing is what it seems here.' Development tactics and suggested reforms need to be carefully considered in the Tajik context rather than simply replicated from elsewhere because they gave good results in that environment.

For instance, donor agencies and international staff who have experience with agricultural extension in post-colonial regions of the world, such as most of sub-Saharan Africa, often assume that the appropriate way to build extension capacity in Tajikistan is to work though NGOs. A considerable number of NGOs have been organized in Tajikistan to meet this donor demand. But private enterprises and NGOs are a new concept in Tajikistan, since they were forcibly repressed—and when found, treated as criminal conspiracies—until 1988 and did not really begin to spread until the international community sought them as partners after the civil war. The line between for-profit businesses and NGOs is hazy, with most organizations having dual legal status in order to maximize potential business profits while remaining able to pursue grants as NGOs. Few Tajik NGOs have solid management or accounting systems, as a number of donor projects, including several USAID efforts, have discovered to their cost. Tajik business and NGO legislation and standards are quite different from those in developed market economies. Few Tajiks have much experience with accounting and business management in market systems. The expectations of donors and local people about what for-profit businesses and NGOs can and should do are very different in Tajikistan.

Consider community politics

Extension is by definition a community-based activity. However, few extension activities in Tajikistan have explicitly paid attention to community politics, informal institutions and power relationships. Given the way in which the population has been forcibly moved from place to place in the past 85 years, considerations of the specifics of each community are particularly important for a successful development intervention.

Train agricultural specialists in modern agriculture and economics

The donor community tends to assume that technical agricultural knowledge of the kind extension requires is readily available in Tajikistan. Yet, when the level of knowledge of agricultural professionals is impartially assessed, it usually turns out to be low. Given the destruction of the rural primary and secondary education system during the civil war, the dispersal of remaining research facilities and farmland as part of the land reform and the limited abilities of the higher educational and research institutions in the sector, this should come as no surprise. Government officials generally are aware of their own limited knowledge. But they and indeed most Tajik specialists have no real knowledge of best world practices. At best, they may have been given a short course by a donor.

The problem of improving training and knowledge for people working in agricultural development, whether for government, donors, NGOs or donor projects, is rarely addressed systematically. The AET system must be fundamentally rebuilt—a 30-year project since new teachers will need to be trained first. But, until that reconstruction is underway, the fundamental requirements for successful agricultural extension—a cadre of people who can train extension workers and knowledgeable extension workers themselves—will be critically lacking.

Extension takes time

Tajikistan has never had an agricultural extension and advisory system. Given the variety of clients and needs, the 'system' should not be a Soviet-style 'one-size-fits-all'. The basis for such a system cannot be created within the lifetime of a single donor project, nor by a single donor. As a USAID evaluation of its rural advisory systems efforts worldwide commented in 2015: 'Institutional and human resource capacity building is a long-term activity that needs sustained funding for at least the medium-term to ensure sustainability of the extension and advisory system initiatives in Tajikistan' (Uphaus *et al.*, 2015: 28).

Specific efforts, such as Neksigol, a few of the extension NGOs, USAID's FAST project and, in cotton-growing areas, Sarob, have been well received by their clients and generally supported by the government. One clear reason for this acceptance and relative success in working with farmers is that they have not been institutionally limited by the short-term time horizon and narrow goals typical of development projects, seeking to lay the basis for institutions that could endure.

Multi-project, longer-term efforts such as the Feed the Future strategy and the EU's Rural Development Programmes in Tajikistan are much more likely to be successful than stand-alone projects. However, for such overarching schemes to have a chance of success, they need to be understood and developed as long-term efforts. Feed the Future in Tajikistan seems currently to be missing this opportunity, as institutional demands from the donor agency and that agency's shortage of personnel on the ground have caused it to return to a narrow project orientation.

Take the rhetoric of farmer-driven change seriously

Tajikistan is a post-Soviet, post-colonial and post-conflict country. Its integration into the world system as an independent country, its transformation to a market economy, the shift of its agriculture from a dual economy—large monoculture farms and very small subsistence farms to feed the large farms' labor force—to commercial production, and the alterations in the political system that will accompany those changes, are only beginning.

Best practice in agricultural extension begins from the recognition that farmers know best what they need, want and can do, and that their demands should shape extension activities. All the proponents of reform have argued that changes must be market-driven. That means giving the people of Tajikistan the space and the support to develop appropriate knowledge and skills and to make their own decisions. Life is messy, and not all Tajik citizens will make the 'right' decisions. But they will make better decisions about their lives than government officials in Dushanbe or development bureaucrats in Washington, Brussels or Beijing. The results will not be as clear-cut as monitoring systems might like, but they will be enduring, making the real changes for the better that everyone needs.

Acknowledgments

This chapter flows from activities funded by the United States Agency for International Development Associate Cooperative Agreement AID-176-LA-13-00003. It is dedicated, with sincere thanks, to all the staff of the FAST project, the USAID Tajikistan Country Office and the Modernizing Extension and Advisory Services (MEAS) project team at the University of Illinois whose hard work made FAST happen. The views expressed in this chapter are solely those of the author and should not be attributed to any other person or to any organization with which he has worked in Tajikistan or elsewhere.

References

ADB (2014) Country assistance program evaluation for Tajikistan: linked document 8: assessment of agriculture and rural development. Asian Development Bank, Manila. Available at: http://www.adb.org/sites/default/files/linked-documents/8-Agriculture-Rural-Dev.pdf (accessed 23 January 2017).

AgroDonish and Oxfam Novib (2010) *Общественные организации, оказывающие консультационные услуги в сельском хозяйстве Республики Таджикистан [Social Organizations Offering Consultation Services for Agriculture in the Republic of Tajikistan].* AgroDonish and Oxfam Novib, Dushanbe.

Aliev, J. and Van Atta, D. (2013) *Farm Size (Arable Land Holdings) in the 12 FTF/T Tajikistan Raions* [Administrative Districts]. USAID MEAS Field Support Activity, Dushanbe.

Alimardon, M. (2009) Agricultural reform process in Republic of Tajikistan. PowerPoint presentation, Dushanbe.

Aminjanov, M. (2012) *История развития орошаемого земледелия в Вахшской долине [A History of the Development of Irrigated Agriculture in the Vakhsh Valley]*. USAID MEAS Field Support Activity, Dushanbe.

Asia-Plus (2016) Labor migrants are not returning to Tajikistan in droves as some feared, 7 January. Available at: www.asiaplus.tj/en/news/labormigrantsarenotreturningtajikistandrovessomefeared (accessed 23 January 2017).

Atkin, M. (1997) Tajikistan: reform, reaction, and civil war. In: Bremmer, I. and Taras, R. (eds) *New States, New Politics: Building the Post-Soviet Nations*. Cambridge University Press, Cambridge, UK, pp. 602–603.

Babadoost, M. (2014) A report on the trip to Tajikistan: duration of the trip: 22 February–12 March 2014. Unpublished report to USAID's FAST Program Management. Urbana-Champaign, Illinois.

Dadabaev, N. (2010) *Система сельскохозяйственного консультирования [A System of Agricultural Consulting]*. Ministry of Agriculture of the Republic of Tajikistan, PowerPoint presentation, Dushanbe.

Dziurman, S. (2012) Tajik agricultural finance framework (TAFF). Presentation at the TAFF Forum, Dushanbe, Tajikistan, 26 September 2012.

EAT (2014) *AgTCA Tajikistan: Agricultural Technology Commercialization Assessment*. USAID Enabling Agricultural Trade Project, Washington, DC.

Engel, E., Emmerling, J., Niepel, T., Peter, A. and Techert, C. (2015) *How Much Would You Pay? Adapting Fee-Based Advisory Services to Mountainous Regions in Tajikistan*. SLE Postgraduate Studies on International Cooperation Publication Series S 262. Center for Rural Development (SLE), Humboldt University, Berlin.

Ernst and Young (2009) National Bank of Tajikistan: report of special audit: executive summary. Available at: http://nbt.tj/en/files/docs/annual_report/rep_eng.pdf (accessed 23 January 2017).

Eurasianet (2016) Tajikistan: banking crisis nears cracking point. Available at: http://www.eurasianet.org/node/77801 (accessed 23 January 2017).

FAST (2014a) *Farmer Advisory Services in Tajikistan (FAST) Life of Project Summary/Year One Detailed Work Plan, 1 September 2013–30 September 2014*. USAID's Farmer Advisory Services in Tajikistan (FAST) Project, Dushanbe.

FAST (2014b) *Manual for the Feed the Future Tajikistan Household Farm Extension and Advisory System*. USAID's Farmer Advisory Services in Tajikistan (FAST) Project, Dushanbe. Draft dated 30 November 2014.

Feed the Future (2012) *Feed the Future: Tajikistan: FY 2011–2015 Multi-Year Strategy*. US Agency for International Development Tajikistan Country Office, Dushanbe.

Freedom House (2016) Freedom in the world 2016: Tajikistan. Available at: https://freedomhouse.org/report/freedom-world/2016/tajikistan (accessed 23 January 2017).

Geraedts, P. (2011) *Survey of Existing Agricultural Extension Providers and Related Programs in Tajikistan*. USAID Family Farming Program, Dushanbe.

Giovarelli, R. and Undeland, A. (2008) *Understanding Household Level Barriers Constraining Women's Access to Land and Financial Resources in Tajikistan*. World Bank Tajikistan Country Office, Dushanbe.

GIZ Tajikistan (2013) *The Technical Advisory Group System*. GIZ Tajikistan Framework and Finance for Private Sector (FF PSD) Programme, Dushanbe.

Global Voices (2016) Is Tajikistan's president 'preparing himself a parachute' with constitutional changes? Available at: https://globalvoices.org/2016/02/14/is-tajikistans-president-preparing-himself-a-parachute/print/ (accessed 23 January 2017).

GoTJ (Government of the Republic of Tajikistan) (2007) *О создании Государственного учреждения "Центр управления проектами по разрешению долгов хлопкосеющих хозяйств и устойчивому развитию хлопкового сектора" [On the establishment of the State Enterprise "Management Unit for the Cotton Sector Debt Resolution and Cotton Sector Sustainable Growth Projects"]*. Resolution number 101, 3 March 2007. GoTJ, Dushanbe.

GoTJ (Government of the Republic of Tajikistan) (2011) *Об образовании Совета по продовольственной безопасности при Правительстве Республики Таджикистан [On the Establishment of the Council of Food Security of the Government of the Republic of Tajikistan]*. Resolution 359, 1 August 2011. GoTJ, Dushanbe.

GoTJ/DCC (Government of the Republic of Tajikistan in Cooperation with the Development Cooperation Council) (2012) *Report on the Results of the Agrarian Reform of the Republic of Tajikistan.* GoTJ/DCC, Dushanbe.

Heathershaw, J. and Megoran, N. (2011) Contesting danger: a new agenda for policy and scholarship on Central Asia. *International Affairs* 87(3), 589–612.

Ibañez-Tirado, D. (2015) 'How can I be post-Soviet if I was never Soviet?' Rethinking categories of time and social change—a perspective from Kulob, southern Tajikistan. *Central Asian Survey* 34(2), 190–203.

ICG (International Crisis Group) (2016) *Crisis Group Europe and Central Asia briefing: Tajikistan Early Warning: Internal Pressures, External Threats.* ICG briefing 78. International Crisis Group, Brussels, Belgium.

IFC (2004) *Summary of Project Information: SEF SugdAgroServe.* International Finance Corporation, Washington, DC.

IMF (2008) Republic of Tajikistan: staff-monitored program. IMF Staff Country Reports. Number 08/197, Washington, DC.

Jones, K., Richter, K. and Ludgate, N. (2015) *InnovATE Study: Agricultural Education and Training System in Tajikistan.* USAID Innovation for Agricultural Education and Training Project Report. Virginia Tech, Blacksburg, Virginia.

Karcz, J. (1974) An organizational model of command farming. In: Bornstein, M. (ed.) *Comparative Economic Systems: Models and Cases.* Richard D. Irwin, Homewood, Illinois, pp. 291–312.

Kassymbekova, B. (2011) Humans as territory: forced resettlement and the making of Soviet Tajikistan, 1920–38. *Central Asian Survey* 30(3–4), 349–370.

Kazbekov, J. and Qureshi, A. (2011) Agricultural extension in Central Asia: existing strategies and future needs. IWMI Working Paper 145. International Water Management Institute, Colombo, Sri Lanka.

Khovar (2016) За 15 лет в сельскохозяйственной отрасли реализовано 42 государственных инвестиционных проекта [42 state investment projects have been realized in the agricultural sector in the past 15 years]. Khovar (Tajik National Information Agency). Available at: http://khovar.tj/rus/2016/01/za-15-let-v-selskohozyajstvennoj-otrasli-realizovano-42-gosudarstvennyh-investitsionnyh-proekta/ (accessed 23 January 2017).

Kikuta, H. (2016) Remittances, rituals and reconsidering women's norms in mahallas, emigrant labour and its social effects in Ferghana Valley. *Central Asian Survey* 35(1), 91–104.

Kuvatova, A. (2001) Gender issues in Tajikistan: consequences and impact of the civil war. In: Reysoo, F. (ed.) *Hommes Armés, Femmes Aguerries. Rapports de Genre en Situation de Conflit Armé.* Graduate Institute Publications, Geneva, Switzerland, pp. 127–135.

Lemon, E. (2016) Rahmon uses crackdown and fear to shore up his rule in Tajikistan. *World Politics Review.* Available at: http://www.worldpoliticsreview.com/articles/17979/rahmon-uses-crackdown-and-fear-to-shore-up-his-rule-in-tajikistan (accessed 23 January 2017).

Levi-Sanchez, S. (2015) Social, economic, and gender dynamics in the districts of Khatlon Province, Tajikistan, where the US 'Feed the Future' program operates. Report. USAID FAST Project, Dushanbe.

LRFRP (2014a) *List of USAID LRFRP Legal Aid Centers.* USAID's Land Reform and Farm Restructuring Project, Dushanbe. http://pdf.usaid.gov/pdf_docs/PA00M1ZJ.pdf (accessed 22 March 2017).

LRFRP (2014b) *List of USAID LRFRP Tashabuskors.* USAID's Land Reform and Farm Restructuring Project, Dushanbe. http://pdf.usaid.gov/pdf_docs/PA00M1ZJ.pdf (accessed 22 March 2017).

Malapit, H., Sproule, K., Kovarik, C., Meinzen-Dick, R., Quisumbing, A., Ramzan, F., Hogue, E. and Alkire, S. (2014) Measuring progress toward empowerment: women's empowerment in agriculture index: baseline report. International Food Policy Research Institute, Washington, DC.

Matveeva, A. (2009) Tajikistan: 'Revolutionary situation' or a resilient state? EUCAM: EU-Central Asia Monitoring. EUCAM Policy Brief 12, Brussels.

MEAS (2016) SMART Skills for smallholder farmers. Available at: http://www.crs.org/our-work-overseas/program-areas/agriculture/smart-skills-smallholder-farmers

MEDT (2015) Monitoring and early warning in Tajikistan: monthly report. Ministry of Economic Development of the Republic of Tajikistan, Dushanbe.

Mendez England and Associates (2010) *Assessment of Water User Association Support Program (WUASP).* Prepared under USAID RFTOP EG119-10-002. Bethesda, Maryland.

Minder, R. and Gorst, I. (2008) Probe into Tajikistan's loan guarantees. *Financial Times.* 5 March 2008.

Mukhamedova, N. and Wegerich, K. (2014) Land reforms and feminization of agricultural labor in Sughd Province, Tajikistan. IWMI Research Report 157. International Water Management Institute, Colombo.

Muminov, M. (2011) *Assessment of the Level of Khatlon Households' Access to Inputs and Market Based Decision-Making at the Village Level.* USAID Family Farming Program, Dushanbe.

Nikonov, A. (1995) *Spiral mnogovekovoi dramy: Agrarnaia nauka i politika Rossii (xviii–xxvv).* [A Drama Arching over Centuries: Russian Agrarian Science and Politics in the 18th–20th centuries] AO Astra Sem, Moscow.

OO SAS Consulting (2015) Консультационные услуги ОО 'САС Консалтинг' [Consulting services offered by OO SAS Consulting]. SAS Consulting [Neksigol], Khujand, Tajikistan. Unpublished.

PO Neksigol Mushovir (2015) Materials from presentation of information products for agribusiness at Sheraton Dushanbe Hotel on 9 April 2015. Public Organization Neksigol Mushovir (formerly SAS Consulting), Khujand, Tajikistan. Unpublished.

Presidential Agency on Statistics (2015) *Poverty Measurement in Tajikistan: A Methodological Note—Prepared by the Agency on Statistics under the President of Tajikistan with Technical Support from the Ministry of Economic Development and Trade of the Republic of Tajikistan and the World Bank.* Presidential Agency on Statistics, Dushanbe.

ProAPT (2015) Productive agriculture project final report, September 29, 2009–September 29, 2014. USAID Productive Agricultural Project in Tajikistan (ProAPT), Dushanbe.

Rahmon, E. (2009) *Указ Президента Республики Таджикистана // О дополнительных мерах по поддержке сельскохозяйственной отрасли Республики Таджикистан* [Decree of the President of the Republic of Tajikistan: On Additional Measures to Support the Agricultural Sector of the Republic of Tajikistan]. Decree no. 663. Dushanbe.

Reeves, M. (2007) Travels in the margins of the state: everyday geography in the Ferghana valley borderlands. In: Sahadeo, J. and Zanca, R. (eds) *Everyday Life in Central Asia: Past and Present.* Indiana University Press, Bloomington, Indiana, pp. 281–300.

SCI (n.d.) Mercy Corps in Tajikistan: past and current programmes: USAID and Mercy Corps maternal and child health program (MCH). State Committee on Investment of the Republic of Tajikistan, Dushanbe. Available at: http://amcu.gki.tj/eng/index.php?option=com_profile&task=view_dp&tdpm_dpcode=43&Itemid=154&lang=E (accessed 23 January 2017).

SCLM (2015) Отчёт о наличии земель и распределении их по видам угодий, категориями и землепользователям республики на 1 января 2015 года [Report on the amount of land and its distribution by type of land, categories and land users of the Republic as of 1 January 2015]. State Committee of the Republic of Tajikistan on Land Management and Geodesy, Dushanbe.

Sehring, J. (2009) Path dependencies and institutional bricolage in post-Soviet water governance. *Water Alternatives* 2(1), 61–81. Available at: http://www.water-alternatives.org (accessed 23 January 2017).

SENAS, AgroDonish and AIN (2009) Report on the International Workshop on Extension in the Agricultural Sector held on 22 and 23 October 2009. European Union SENAS Project, Dushanbe.

Sheltinga, J. and Coene, F. (2006) *Workshop on Agricultural Extension/Advisory Service in Tajikistan.* E-mail, Dushanbe.

Shtaltovna, A. (2016) Knowledge gaps and rural development in Tajikistan: agricultural advisory services as a panacea? *Journal of Agricultural Education and Extension* 22(1), 25–41.

Swanson, B. (2012) *Assessment of the Pluralistic Extension System in Tajikistan.* MEAS Project, University of Illinois Urbana-Champaign, Illinois.

Swanson, B., Meyer, E. and van Weperen, W. (2011) *Strengthening the Pluralistic Extension and Advisory System in Tajikistan: A MEAS Rapid Scoping Mission October 3–20, 2011.* Revised version, 25 November 2011. MEAS Project, University of Illinois, Urbana-Champaign, Illinois.

Uphaus, C., McMahon, M., Nelson, F., Gapasin, D. and Negash, K. (2015) *An Evaluation of the Rural Advisory Services Program (RASP) and Related Activities Under Feed the Future.* USAID, Washington, DC.

USAID (2015) *Request for Proposals No. SOL-176-15-000005, Feed the Future Tajikistan Agriculture and Water Project.* United States Agency for International Development, Mission for the Central Asian Republics, Almaty, Kazakhstan.

Van Atta, D. (1990) Why do models succeed? Introducing organizational innovation in agricultural work. In: Wädekin, K.-E. (ed.) *Communist Agriculture: Farming in the Soviet Union and Eastern Europe.* Routledge, London, pp. 185–201.

Van Atta, D. (1993) Agrarian interests and the politics of reform in post-soviet Russia. In: Van Atta, D. (ed.) *The 'Farmer Threat': The Political Economy of Agrarian Reform in Post-Soviet Russia.* Westview, Boulder, Colorado, pp. 193–212.

Van Atta, D. (1994) Agrarian reform in post-Soviet Russia. *Post-Soviet Affairs* 10(2), 159–190.

Van Atta, D. (2009) 'White gold' or fool's gold: the political economy of cotton in Tajikistan. *Problems of Post-Communism* 56(24), 17–35.

Van Atta, D. (2015) FAST Quarterly Report 7: 1 January–31 March 2015. USAID FAST Project, May 2015, Dushanbe, p. 6. Unpublished.

Van Atta, D. and Safaraliev, G. (2015) *FAST Policy Note: Household Farm Landholdings in the Feed the Future Zone of Influence, 2015.* USAID FAST Project, Dushanbe.

Van Atta, D., Haidarov, B. and Shukurov, I. (2010a) *Republic of Tajikistan: Conducting of Vertical Functional Reviews of Seven Ministries: Interim Report, Ministry of Agriculture.* PDP Australia Pty Ltd and 'Silk Road' LLC for the World Bank Public Sector Reform Project, Dushanbe.

Van Atta, D., Haidarov, B. and Shukurov, I. (2010b) *Republic of Tajikistan: Conducting of Vertical Functional Reviews of Seven Ministries: Optimization Plan, Ministry of Agriculture.* PDP Australia Pty Ltd and 'Silk Road' LLC for the World Bank Public Sector Reform Project, Dushanbe.

van Weperen, W., Ungetheum, H., Swoboda, T., Sharipov, M. and Dekhonov M. (2012) *Tajikistan Agricultural Extension Policy: A Second Draft to Facilitate the Discussion.* Dushanbe. Unpublished.

Wädekin, K.-E. (1973) *The Private Sector in Soviet Agriculture.* University of California Press, Berkeley, California.

World Bank (1996) *Press Release: IDA Extends First Credit to Tajikistan.* Press release 96/43 ECA, 17 May 1996. World Bank, Washington, DC.

World Bank (2011) *Republic of Tajikistan: Country Economic Memorandum: Tajikistan's Quest for Growth: Stimulating Private Investment.* World Bank, Washington, DC.

World Bank (2013) Implementation completion and results report (IDA-43140) on a credit in the amount of SDR 10 Million (US $15 million equivalent) to the Republic of Tajikistan for a cotton sector recovery project. World Bank, Washington, DC.

WUASP (2006a) *Fruit Production in Backyard Gardens: From Planning and Planting to Harvesting, Marketing and Processing, A Practical Guide.* USAID Water Users' Association Support Project, Tashkent, Uzbekistan.

WUASP (2006b) *Vegetable Production in Backyard Gardens: From Planning and Planting to Harvesting, Marketing and Processing, A Practical Guide.* USAID Water Users' Association Support Project, Tashkent, Uzbekistan.

Zakon (2008) *Zakon Respubliki Tajikistan: Ob organakh obshchestvennoi samodeiatel'nosti* [Law of the Republic of Tajikistan on organs of community self-activity]. 5 January 2008, no. 347.

11 The Development of Extension Services in Post-Soviet, Post-Conflict Georgia

Anastasiya Shtaltovna*
Centre for International Studies (CÉRIUM), University of Montreal, Canada

Introduction

Georgia is one of the former Soviet republics, located on the southern slopes of the Great Caucasian Mountains. It occupies an area of 69,700 km^2 and shares borders with Armenia, Azerbaijan, Russia and Turkey; its western coastline lies on the Black Sea. The population of Georgia in 2013 was estimated at 4.47 million (World Bank, 2013). The climate in the western part of the country is subtropical, while it is more arid in the eastern region. The country has suffered both politically and economically since the fall of the Union of Soviet Socialist Republics (USSR), and to date it has been unable to reach fully its development objectives in the areas of market liberalization, livelihood and energy security.

Agriculture in Georgia

Georgia has a complex landscape and diverse mix of soil conditions, divided into 13 zones, each with a specialized form of agriculture which includes viticulture, horticulture, fodder production, livestock production, tea growing, subtropical crops and tobacco cultivation (Mosashvili, 2007). Hazelnuts, fruit (including citrus), spirits, wine and mineral water, as well as early-season vegetables and fisheries, were the main regional outputs in the Soviet period and have remained the country's main export crops (interviews, Tbilisi, 2013, 2015).

Agriculture has played an important role in the Georgian economy, for many years representing the single most important sector of the gross domestic product (GDP) and employing around half the population (USAID, 2011). Agricultural production in Georgia was severely disrupted, first by the collapse of the Soviet Union in 1991 and then by the civil war in 1992. The agriculture sector's contribution to the GDP over the past 20 years has declined significantly, from 30% in 1990–1991, to 21% in 2000 and to 7% in 2010 (USAID, 2011).

The potential of the sector is underutilized, with only one-third of the nation's 2,270,000 hectares (ha) of arable land currently in use (interview with S. Kereselidze, Head of the National Information Consulting Centers [ICC] Network, Tbilisi, 2015). As an important element of transitioning from a socialist state, Georgia has undertaken a series of land-reform measures involving the transformation of ex-Soviet collective- and

* Corresponding author e-mail: ashtaltovna@yahoo.com

 Building Agricultural Extension Capacity in Post-Conflict Settings (eds P.E. McNamara and A. Moore) state-owned large-scale farms, which were subject to centralized management and control, into private property and by leasing land to private individuals (Ebanoidze, 2002). Land qualifying for a privatization fund has been distributed free of charge to rural households. The maximum area transferred to private ownership in the lowlands was 1.25 ha per person, while up to 5 ha per farmer was distributed to eligible households in the highlands. Most state agricultural land was distributed to private households immediately after independence. At the current time, Georgia has 640,302 farmers, each with an average of three separate 1.25 ha plots spread across different locations (interviews, Tbilisi, 2013, 2015).

The process of land privatization itself has generally been regarded as a successful reform in Georgia, with some acknowledging it as an important step in building a democratic state and moving towards a market economy (Salukvadze, 1999). However, the pace and performance of this process have been far from satisfactory, for a range of factors (Gogodze *et al.*, 2007).

One crucial factor is the concept of the Georgia 'farmer'. In Georgia, the 'farmer' is an autonomous entity that appeared only during the transition period. Thus, farmers usually do not have the experience or capacity, in agriculture or management, to play the role ascribed to them by Western agricultural economists and development practitioners (Lerman, 2008; Beniwal *et al.*, 2010; Bliss, 2012; Djanibekov *et al.*, 2013; Van Assche *et al.*, 2013; Shtaltovna, 2015). As a result of land privatization, almost everyone in Georgia received land and so everyone has become a farmer, including doctors, teachers, accountants and others engaged in professions wholly unrelated to agriculture (Shtaltovna, 2013; Shtaltovna 2015). Those who make a profit from agriculture consider themselves to be farmers. Those who do not make a profit do not consider themselves farmers but rather as simply cultivating a small plot of land for their own consumption and maybe making a small income (interviews, 2013, 2015). Thus, the Western notion of a 'farmer' does not really match the situation of those who have emerged in Georgia and in most former Soviet countries after independence.

Also, the range of agricultural measures introduced by different governments in Georgia in the past 20 years (distribution of vouchers for inputs, etc.) demonstrates that agriculture and rural development were never top priorities for the nation's government and that agricultural development has seen a good deal of variety and experimentation (interviews, Tbilisi, 2013, 2015). Reforms have taken place without any guidance (interview with Oleg Shatberashvili, Georgian Federation for Information and Documentation, Tbilisi, Georgia, 2013). In addition, agriculture is heavily politicized and is a populist cause in terms of modernization; thus, many initiatives that take place are strongly linked to political campaigns. These campaigns seemingly target improving the livelihoods of rural dwellers, but are usually launched just before parliamentary or presidential elections with the main goal of collecting votes. They are not designed to contribute to the welfare of the people or to improve agriculture (Roeing, 2016).

As a result, the agricultural sector has struggled to develop. Farmers find themselves operating small-scale and fragmented farms (some 90% of farmers have less than 1 ha of land) (interviews, Tbilisi, 2015), constrained by limited knowledge and experience, insufficient access to agricultural information, poor access to markets, minimal access to financial resources, a deteriorating road network, lack of technical equipment and inadequate infrastructure (processing, storage and irrigation), all of which result in low prices, low productivity and poor sector growth (interviews, 2013, 2015; Lerman and Sedik, 2009; Pavliashvili, 2009; USAID, 2011; Van Assche *et al.*, 2016b; Roeing, 2016).

Agricultural production is far from sufficient to provide food security to the country. The agricultural sector has become synonymous with poverty, effectively serving as an 'employer of last resort' (interview with O. Shatberashvili, Director of the Georgian Federation for Information and Documentation, Tbilisi, 2013; USAID, 2011). The majority of farmers pursue a mix of different maintenance strategies, such as subsistence farming, migration, claiming pensions, engaging in other professions or barter (Trevisani, 2008). The situation has become unsustainable both for the agricultural sector and for the rural population, where farmers occupy the poorest segment of the population and are sometimes regarded as a cause of instability (USAID, 2011; Fey, 2012; interview with O. Shatberashvili, 2013). Also, there is a large outflow of young people from rural areas to the cities, as they do not want to be farmers, and people very often start to practice agriculture only when they lose their job (interview with Soso Tedeashvili, farmer, Breti village, 2013).

However, since 2013 agriculture has become one of the most important development areas for the Georgian government, for reasons linked to the development of the Association Agreement between the EU and Georgia in 2012. One of the Association *Agreement*'s priorities is to improve agriculture and rural development. The document states that its objectives include 'improving the competitiveness of agricultural production, by fostering economies of scale via market-oriented agriculture cooperatives, by developing advisory and systems to increase production and augment exports' (EU, 2014: 24–25). Based on this document, a strategy for agricultural development for 2015–2020 was developed (MoA, 2015), which elaborates in detail the development of efficient agricultural extension services, and thus ways of increasing farmers' knowledge. Georgia's recent willingness to collaborate with the EU has resulted in a tremendous shift in the development of Georgian agriculture, and of the agricultural extension and advisory services in particular.

Agricultural Extension in Georgia

Georgia has a unique history with agricultural extension and advisory services. During the Soviet period, instead of extension Georgia had a different system of knowledge production and sharing. A comprehensive network of agricultural universities, colleges, research institutes, the Ministry of Agriculture (MoA) and other actors involved in agricultural production were responsible for knowledge production and were linked directly to *kolkhozes* and *sovkhozes*—the main production units of collectivized Soviet agriculture. However, since the end of the Soviet Union this system has undergone many changes.

Formal extension is a new concept in Georgia and was introduced by the international donor community in the 1990s. Numerous international donor organizations offered ground-based relief and development programs in agriculture during this period. In addition, a number of domestic NGOs emerged, funded by donors through different development projects. As a result, extension services are currently provided in Georgia largely by international donors, and more recently by the government, NGOs, the Academy of Agricultural Sciences and commercial input providers (Goldberg *et al.*, 2011; Kazbekov and Qureshi, 2011; Roeing, 2016; Van Assche *et al.*, 2016a). Farmers benefit from these donor projects and state programs.

In addition, farmers receive knowledge from television, the internet, commercial service centers (which provide seeds, fertilizer and small-scale machinery) and advertisements. Having grown up in villages, farmers also continue to rely on their own experience and on personal networks, including neighbors and former *kolkhoz* agronomists for agricultural information and extension (interviews, 2013, 2015; Fey, 2012; Fey, 2016; Van Assche *et al.*, 2016b).

The Role of Agricultural Institutions

The Soviet system of knowledge production and sharing functioned well, organized through a complex network of agricultural ministries, agricultural universities, academies of agricultural sciences, research centers, *kolkhozes* and *sovkhozes* (Kazbekov and Qureshi, 2011; Shtaltovna, 2015; Hornidge *et al.*, 2016; Van Assche *et al.*, 2016a). Knowledge was generated in these institutions, based on state production targets, as well as in the specialized research institutes of the Academy of Agricultural Sciences and the MoA, with some variations across different

Soviet republics and years (Morgounov and Zuidema, 2001). Later, knowledge generation was channeled to the *kolkhozes/sovkhozes* through local and regional departments of the MoA. Every *kolkhoz/sovkhoz* had a set of trained experts—engineers, agronomists and accountants—who regularly attended courses to update their knowledge.

This system of knowledge production and sharing was changed by the post-Soviet transformation process. The Soviet network of education and research in Georgia has been partly sustained and partly reorganized or dissolved due to a lack of state subsidies (Beniwal *et al.*, 2010). The transformation period therefore had a negative effect on the quality of research and education in the country. The main concerns of Georgian scientists include a decline in research livelihoods (e.g. poor salaries, shrinking staff numbers); barriers to entering the English-dominated international scientific community; and the reorientation of research focus from a Soviet knowledge-production system to one that is market-oriented. Consequently, agricultural science has become an unattractive area for young people to either work or study in (interviews, 2013, 2015; Van Assche *et al.*, 2016b). One serious challenge that needs to be overcome is reorienting knowledge production from a small number of large, powerful and autonomous *kolkhozes* to a large number of small, vulnerable, under-financed and under-skilled farms (Shtaltovna *et al.*, 2012; Van Assche *et al.*, 2016a). Finally, there is no mechanism to link and coordinate knowledge and innovation provided by national agricultural research (research institutes, academies of science, universities and others) and agricultural producers (Morgounov and Zuidema, 2001; Beniwal *et al.*, 2010; Shtaltovna, 2013; Shtaltovna, 2015).

Since 2013, the Georgian government has worked to strengthen agricultural research and establish a nationwide extension system throughout the country. After 22 years of independence, Prime Minister Ivanishvili sought to establish a national network of extension services, although the main trigger in this process was the EU rather than the Georgian government. The government established 54 ICCs around the country as a component of the rural and agricultural development program, to complement the Association Agreement signed between the EU and Georgia (EU, 2014). There are now 1100 staff and an office and a car in every district center (interview with S. Kereselidze, Head of National ICC Network, Tbilisi, 2015). The functions of ICCs are:

- Forming and supporting cooperatives.
- Providing consultations and training farmers on a diverse range of topics (e.g. state agricultural priorities, agricultural techniques and plant production methods, livestock management).
- Helping find solutions to farmers' issues and connecting them to the necessary resources.
- Spreading the word about different MoA programs currently available to farmers.
- Connecting farmers and central government.
- Following up on farmers who receive state subsidies.

The main sources of knowledge for ICCs are information and literature provided by the newly established Scientific-Research Center at the MoA (discussed below); donors, projects and NGOs that provide extension training; the private sector; and new research from other post-Soviet republics (e.g. Ukraine). Often, staff at ICCs rely on their own practical experience, and they constantly learn from farmers (interviews, 2015).

Established only 2 years ago, ICCs are still in the process of formation and face challenges in becoming multiprofile experts that can meet the many problems of farmers. The ICCs currently lack the personnel to reach huge numbers of farmers and are burdened by extra administrative tasks issued by the government. ICC personnel also require capacity building to develop the professional expertise demanded by their clients. Such skills include learning to be client oriented; connecting farmers with other actors in the value chain, such as input suppliers and processors; searching for sales opportunities; and assisting in receiving loans.

They also receive little to no feedback from farmers on whether the consultation provided was of any use (interviews, 2015). Finally, future funding for the ICCs has not been secured and their sustainability is still questionable, especially given the political context.

However, establishing the ICCs is a great breakthrough in Georgian agriculture. As a positive effect of this initiative, more attention is now being paid to the agricultural sector than ever before, and much effort is being expended by donors, the government and other actors to make the ICCs a success. Furthermore, the EU Association Agreement has played an important role in advancing the reorganization process in terms of national research and the education system.

At present, the following agricultural scientific institutions can be regarded as the main knowledge and innovation sources for the functioning of agriculture in Georgia: the Academy of Agricultural Sciences, the National Agricultural University, the Technical University of Georgia and the Scientific-Research Center of the MoA.

The Academy of Agricultural Sciences has been one of the main sources of knowledge for the newly established extension centers and the umbrella institute for most agricultural research bodies in Georgia for many decades. This institution has many scientists who formerly worked with the Soviet Union and have high levels of expertise, but on average they are 70 years old. In addition, the academy is mainly involved in research but not in teaching, although it does have a few research institutes that provide training. Thus, the knowledge held within the academy might be regarded as outdated (Shtaltovna and Van Assche, 2013). Nonetheless, the Academy of Agricultural Sciences remains active. For example, it has contributed to the national strategy of agricultural development for 2015–2020 by developing a training program on various themes for the newly established extension service. In addition, the academy is making efforts to open up new research programs through international collaboration (interviews, 2013, 2015).

The National Agricultural University, together with the majority of its former research institutes, was privatized in late 2012. Since then, the university has been modernized by focusing mainly on providing a better quality of education, keeping the best staff, significantly improving infrastructure and facilities, and making its educational offerings fee-based to afford to provide better teaching (interviews, 2013, 2015). Such dramatic developments, and the political circumstances under which it was privatized, have brought about changes to the university. On one hand it has gained popularity with students, while on the other hand there is now barely any dialogue between the MoA and the National Agricultural University. The government refuses to cooperate with or utilize either the educational programs or research efforts of the National Agricultural University, due to perceptions of the university's quality. While in some areas of agricultural education the overall level of professionalism has increased significantly, in other areas of agricultural education and research it has declined considerably. For a university in Georgia, it is extremely hard to maintain even a minor part of what used to be offered by the Soviet agricultural research institutes (Shtaltovna and Van Assche, 2013). This has implications for the future of agricultural extension, at least if it is to be based partly on new Georgian research.

Furthermore, the National Agricultural University does not intend to link with farmers because 'it is a task of government' (interview with L. Baramidze, Agricultural University, Tbilisi, 2015). The view is that while knowledge and innovation produced at the university can and should be transmitted to agricultural producers, this is the responsibility of the government, not the universities. Conversely, the main focus of the university is on providing strong education and research, innovating, etc. As for the link between the university and production, the National Agricultural University provides different services that could be of interest to producers, such as laboratory testing, breeding and selling different varieties of fish and innovation patents.

Given the situation regarding agricultural education and research in Georgia, the government wants to create a new system.

Two attempts have been made in this respect. The first is at the Technical University of Georgia, which has opened a number of agricultural faculties and is attempting to bring students and lecturers together. However, it will take time and resources before the university establishes itself as a bona fide agricultural institute.

Second, in 2014–2015 the Scientific-Research Center of the Ministry of Agriculture was established at the MoA. It has 12 departments (MoA, n.d.) covering all areas of research that used to be carried out as part of the Soviet scientific system, in addition to newer elements such as extension, international relations, risk assessment, standardization, quality control and agricultural economics. The staff come mainly from the previous Soviet system. In addition, the Center attempts to hire young people from the Agricultural University, Technical University, Telavi Agricultural Institute and the Caucasus Institute. According to the Center's Director: 'Now, it is at a stage of rehabilitating what was lost in previous years, since the end of the Soviet Union, and adaptation to current needs' (interview with L. Ujmajuridze, Director of the Scientific-Research Center of the MoA, Tbilisi, 2015).

To summarize, Georgian research institutes faced many barriers as the country moved into a market economy in the post-Soviet, post-conflict period. Nevertheless, many new institutions are currently being established, which highlights the amount of progress made in the country's agricultural research and education system.

The Role of Donors and Projects

Reconstructing and strengthening agricultural extension systems is essential to post-conflict agricultural development (UNEP, 2002; Zaur, 2006; Longley *et al.*, 2007; USIP, 2013). As noted above, extension is a new phenomenon in Georgia. The concept was introduced and financed by donors as a part of agricultural and rural development projects in the post-civil war period, at which time numerous international donor organizations offered development relief aid (Goldberg *et al.*, 2011; Kazbekov and Qureshi, 2011). Domestic NGOs were funded by donors through various development projects and also provide agricultural knowledge to farmers. Even in this context, extension is rarely the main goal of many development projects but rather is often one component of larger agricultural and rural development objectives. Thus, in addition to the national research and education institutions discussed above, there are a number of donor-related extension providers in the country: international donor organizations, NGOs and, more recently, the ICCs organized by the government.

A number of extension-related projects are funded by USAID, one of which is the New Economic Opportunities (NEO) program. Among other project responsibilities, NEO implements a value chain approach to agriculture, of which extension is a part. The value chain strategy targets beekeeping, fruit growing (strawberries, raspberries, apples, vegetables, grapes and hazelnuts), and livestock husbandry, and is disseminated among 85 communities in the country, accounting for about 500 villages. To promote knowledge generation, NEO runs training sessions on modern technologies and provides grants for a select number of farmers to start up demonstration plots. Farmers who are not covered by the project can also visit and learn from the plots. NEO also provides farmers with nursery seedlings, saving them money on importing or purchasing seedlings on the open market. Farmers who own the greenhouses later sell their products to other farmers. As part of the value chain approach, NEO connects farmers with potential consumers, mainly Georgian hotels and food outlets. NEO also helps farmers to access microcredit, with loans ranging from 500 to 15,000 lari (1 lari = US$0.42 in August 2015). The program also provides collateral for farmers, demonstrating to the bank that farmers are involved in the food supply chain and will be able to sell their products.

In addition, NEO cooperates with farm service centers, supplying them with specialized equipment for soil and leaf testing and helping them to improve their services to farmers. Soil testing is provided free of charge, but when a farmer purchases fertilizer or seeds from the center's shop, the price includes a consultation fee and a charge for the testing equipment. In Gori, a meteorological station for the farm service center was installed with the help of NEO. If there is a risk of adverse weather, pests or plant diseases, information is sent out via SMS to the 1000 farmers subscribed to the service. NEO has also focused significantly on supporting the development of the private sector, believing this to be a more sustainable approach in a country like Georgia (interviews with Kirk Ramler, Chief of Party of NEO, Tbilisi, 2013, 2015).

Another project financed through USAID is the Strengthening Extension and Advisory Services in Georgia (SEAS) project, implemented by the Modernizing Extension and Advisory Services (MEAS) project at the University of Illinois. The goal of SEAS is to support the efforts of the Georgian MoA to further develop its recently implemented public agricultural extension and advisory service system (MEAS, n.d.). A major component of the SEAS project is the development of staff training, mentoring and continuing education programs for MoA employees who work directly with farmers to improve their agricultural knowledge and to better promote technology transfer. The overall project objective is accomplished through the use of activity-funded technical advisors, who work closely with the MoA in developing and implementing a multi-year strategy to enhance the capacity of ministry leadership and relevant field employees to provide effective, public-funded agricultural extension and advisory services to farmers in Georgia.

Through its activities, SEAS also cooperates closely with MoA, as well as with 54 extension centers, the National Agricultural University and the Scientific-Research Center. By doing so, it plays an important role in reinvigorating the connections between these actors (interview with Ben Mueller, SEAS project, Tbilisi, 2015), which is crucial in building a viable extension system. This activity brings long-standing US experience in extension to the country, the main aim being to identify farmers' problems and help them find solutions to these issues.

Although the SEAS project is still relatively new, it has promising goals and is one of a number of donor initiatives making a positive contribution to building a public extension system in Georgia. Furthermore, with the support of the SEAS project, the national government has initiated donor coordination meetings to consolidate efforts around projects dealing with agriculture and extension. Participants in the group include the Dutch embassy in Georgia, the Food and Agriculture Organization of the United Nations (FAO), the Swiss Development Agency, USAID and the World Bank. Whereas the MoA should be a leading party in this process and tell donors what should be done and where, its leadership is currently lacking.

A number of foreign development agencies and international organizations have played a positive role in extension during the first post-conflict stage of transition in Georgia and other post-Soviet countries (Salukvadze, 1999; Roeing, 2016), as they have incentivized and mobilized many actors, such as NGOs, and prompted action at the local level (Shtaltovna, 2015). Given the importance of agriculture and bearing in mind the problems in rural areas, the provision of extension services by development projects is a significant contribution to rural development in the country. Nonetheless, despite making a great contribution to constructing extension services, there are a number of shortcomings in the work of these agencies. For instance, many of the projects are short term, the efforts of donors are not coordinated and the projects do not always meet local needs. Donors, when implementing development projects, play their geopolitical roles in the region (Roeing, 2016), and to this end they are new and important actors in the knowledge and development field.

The Role of NGOs

Many NGOs emerged in the country as a response to societal problems caused by the civil war and the dissolution of the Soviet Union, as state institutions were unable to solve problems of agricultural and rural development or to implement donors' projects (Shtaltovna, 2015). International projects and donors mobilized many actors through established NGOs, some of which have become a platform for active and knowledgeable people, including former socialist specialists in diverse fields. Additionally, NGOs fill a big gap in the rural economy of Georgia, similarly to other ex-Soviet states (Shtaltovna 2015; Roeing, 2016). By implementing development projects, NGOs and their staff often provide services and agricultural advice that were previously delivered by *kolkhozes* but also try to solve many other problems in the socio-economic transformation process.

One example of an NGO providing extension in Georgia is the biological farming association Elkana, a Georgian NGO founded in 1994 (Elkana, 2011). At that time, villages in the country were completely devoid of means of production and also suffered from an acute shortage of information and knowledge. No government or public institution was capable of assisting farmers with qualified farming advice, and modern technologies were very few and far between in the country, so Elkana was established to assist farmers. Activities were initially limited to providing advisory services for farmers. Today extension is carried out through its advisory and economic division, and the association provides consultation services to over 900 members on issues relating to farming and business planning. Elkana's key extension services include:

- Planning to convert farms to biological methods of farm management.
- Conducting economic planning and optimization of farms.
- Providing consultations for organic farming.
- Consulting on business planning and marketing issues.
- Supplying seed to farmers and supporting the planting of traditional/local varieties.
- Conducting training and workshops on organic agri-production, business planning, community mobilization, rural tourism and biological safety.

NGOs in Georgia face several challenges. The essential factor in the survival of any NGO in Georgia is funding, as farmers are not yet ready to pay for consultations, mainly due to the small amount of readily available money. Furthermore, there is a high level of competition for donors' money, as there is little or no support for the NGO sector from government sources. This creates financial issues that affect the extension services NGOs can provide. To sustain their efforts, NGOs adopt different strategies such as selling agricultural tools and seeds/seedlings, providing text message-based weather information and other kinds of paid services.

NGOs also struggle to impact many farmers. Donor organizations provide strong backup for NGO activities, thus helping them to contribute to the rural economy. However, the coverage by projects is sparse. Despite the significant work done by NGOs, they cannot replace the state, nor can they create the environment of stable institutions and predictable law enforcement needed to create anything resembling a capitalist democracy (Van Assche *et al.*, 2013; Van Assche *et al.*, 2016a, 2016b).

The Role of the Private Sector

There are minimal private sector extension or advisory services in Georgia. However, two additional sources of knowledge and innovation for farmers are agricultural shops and farm service centers (*agro kartli*). As well as offering a range of inputs and tools for agriculture, they also provide agricultural consultations. These are mainly targeted at promoting some of their products. There are hardly any independent agronomists left now whom farmers could contract for agricultural advice.

Conclusion

Georgia is characterized not only as a post-conflict country, but also as a country in transition from socialism to capitalism, which is important when discussing the role of extension in rebuilding after the conflict. This factor adds an extra layer of difficulty when analyzing post-conflict extension and considering potential options for future development.

Nonetheless, extension can play a role in rebuilding agriculture in post-conflict and post-transition countries such as Georgia. Before its relatively recent introduction, the Soviet system of knowledge production and sharing differed from Western concepts of extension. Although the process was not specifically referred to as 'extension', the knowledge-transfer goal was similar in that it was a functioning agricultural knowledge and information provision system, organized through a complex network of agricultural ministries, agricultural universities, academies of agricultural sciences, research centers, *kolkhozes* and *sovkhozes*. Until recently, many research institutes, as well as the Academy of Agricultural Sciences and the National Agricultural University, functioned in a similar manner, providing some elements of extension in the process. However, these institutions were forced to change, mainly because they previously served the needs of collective agriculture and therefore had to shift direction towards a market-driven economy. In addition, the financing of these institutes has declined dramatically since the end of the Soviet era. Despite these hurdles, former Soviet research institutes still contribute to the overall picture of knowledge and innovation production in Georgia.

After the cessation of a number of civil wars in post-Soviet Georgia, extension was introduced and provided by international donor agencies and NGOs, as a component of larger agricultural projects. Through development projects, international donors provide knowledge, inputs and services for farmers. While implementing a wide range of agricultural projects, donors and NGOs fill an important gap in services that were previously provided by the Soviet agricultural system. Despite their successes, donors and NGOs should be expected to complement rather than substitute for government agricultural services (Van Assche *et al.*, 2013; Van Assche *et al.*, 2016b). The experience of other post-Soviet countries (e.g. Kazakhstan, Tajikistan, Ukraine and Uzbekistan) demonstrates that, without governmental will, there can be no development in any sector.

Similarly, in Georgia extension (and the agricultural sector as a whole) was not a priority following independence. However, since 2013 and with the signing of the collaboration agreement with the EU, changes have been made in Georgian agriculture for the first time since 1991. The regional ICCs were established to provide support to farmers and are already producing impressive results. Credit is also due to donor organizations which assisted the MoA with staff training and by providing different kinds of support to the ICCs. Successes by the ICCs and other extension-related efforts have pushed the MoA to support the ICCs, as well as to meet the requirements of the EU agreement, with dynamism and enthusiasm. The same triggers have given impetus to further reorganization of the national research and educational system, including the establishment of the Scientific-Research Center at the MoA as well an agricultural faculty at the Technical University. There is also space for private extension, which has emerged in recent years, and for these two bodies to cooperate (interview with Ben Mueller, Director of SEAS project, Tbilisi, 2015). These are good signs.

The Georgian extension system provides a good example for other countries seeking to pursue a similar course. The key learning point here is that the process was externally driven, in this case by the Association Agreement with the EU. However, to maintain these good intentions, the background processes should not be overlooked. For example, several unresolved problems remain concerning the lengthy process of reorganizing and funding the national agricultural research system. No funding is currently provided to libraries and research in the Academy of Sciences or other national research institutions. The situation has been exacerbated by the ongoing political conflict between the National Agricultural University and the MoA, which affects the generation of agricultural research and knowledge.

Regarding ICCs, it will take some time before they become real extension providers, and people who became extensionists still have to be retrained as agricultural experts to proactively help farmers find solutions to their problems. The task of these ICC personnel is both extremely important and very difficult, since they have to collect widely scattered knowledge from various places and actors (as previously described) and deliver it to a new customer—the 'farmer'—who may be located across the length and breadth of Georgia, and who is currently neither ready nor able to pay for the services provided. Time and resources are required to address these concerns.

Given that extension is a new notion in Georgia, it needs to be very context-specific, fit with the political and historical context of the country and meet the needs of different kinds of farmers, especially those who are smallholders and vulnerable if Georgian agriculture is to continue operating and growing in the future. At this point the question of sustainability arises, given that the EU is the main trigger in the process of public extension system development. After its goals have been achieved, Georgia will need to find justifications for maintaining the system, supporting the ICCs and providing services to farmers. Sustaining the dynamism and enthusiasm generated in the post-conflict and post-transition period is crucial, not just for Georgia but for other countries facing similar situations.

References

Beniwal, S., Maru, A., Khalikulov, Z. and Ahmadov, H. (2010) *Central Asia and the South Caucasus: Challenges, Opportunities, Priority Needs and Actions Required for Improving Agricultural Research for Development.* Central Asia and the Caucasus Association of Agricultural Research Institutions (CACAARI), Tashkent.

Bliss, F. (2012) *Agrarian Reform of the Republic of Tajikistan – A Roadmap for Local Governance and Agriculture Management Institutions Reform.* Food and Agriculture Organization, Dushanbe.

Djanibekov, U., Van Assche, K., Boezeman, D. and Djanibekov, N. (2013) Understanding contracts in evolving agro-economies: farmers, dekhqan and networks in Khorezm, Uzbekistan. *Journal of Rural Studies* 32, 137–147.

Ebanoidze, J. (2002) *Current Land Policy Issues in Georgia.* Economic and Social Development Department, FAO, Rome.

Elkana (2011) Farmers' service. Available at: www.elkana.org.ge/farmer/index.php?action=0&lang=eng (accessed 23 January 2017).

EU (2014) Association Agreement between the European Union and the European Atomic Energy Community and their Member States, of the one part, and Georgia, of the other part. *Official Journal of the European Union* 30.8.2014. Available at: eeas.europa.eu/georgia/pdf/eu-ge_aa-dcfta_en.pdf (accessed 23 January 2017).

Fey, L. (2012) Agricultural knowledge among Georgian small-scale farmers. A qualitative study in the district of Gori, Shida Kartli. MA thesis. University of Bonn, Germany.

Fey, L. (2016) Which knowledge, which agriculture? Local farmers and agricultural development in Georgia. In: Hornidge, A.-K., Shtaltovna, A. and Schetter, C. (eds) *Agricultural Knowledge and Knowledge Systems in Post-Soviet Societies.* Peter Lang, Bern.

Gogodze, J., Kan, I. and Kimhi, A. (2007) Land reform and rural well being in the Republic of Georgia: 1996–2003. Discussion paper 1.07. Hebrew University of Jerusalem, Jerusalem, Israel.

Goldberg, I., Goddard, J.G., Kurioakose, S. and Racine, J.L. (2011) *Igniting Innovation—Rethinking the Role of Government in Emerging Europe and Central Asia.* World Bank, Washington, DC.

Hornidge, A.-K., Shtaltovna, A. and Schetter, C. (eds) (2016) *Agricultural Knowledge and Knowledge Systems in Post-Soviet Societies.* Peter Lang, Bern.

Kazbekov, J. and Qureshi, A.S. (2011) Agricultural extension in Central Asia: existing strategies and future needs. IWMI Working Paper 145. International Water Management Institute, Colombo.

Lerman, Z. (2008) Agricultural development in Uzbekistan: the effect of ongoing reforms. Discussion Paper. Hebrew University, Jerusalem, Israel.

Lerman, Z. and Sedik, D.J. (2009) Sources of agricultural productivity growth in Central Asia. Presentation. Agricultural and Applied Economics Association 2009 AAEA and ACCI Joint Annual Meeting, Milwaukee, Wisconsin.

Longley, C., Christoplos, I., Slaymaker, T. and Meseka, S. (2007) Rural recovery in fragile states: agricultural support in countries emerging from conflict. Natural Resource Perspectives 105. Overseas Development Institute, London. Available at: www.odi.org.uk/sites/odi.org.uk/files/odi-assets/publications-opinion-files/10.pdf (accessed 23 January 2017).

MEAS (n.d.) Strengthening extension and advisory services in Georgia (SEAS). Modernizing Extension and Advisory Services Program (MEAS), University of Illinois. Available at: www.meas-extension.org/home/associate-awards/strengthening-extension-and-advisory-services-seas-activity (accessed 23 January 2017).

MoA (n.d.) Structure: Ministry of Agriculture of Georgia. Available at: www.moa.gov.ge/En/Structure (accessed 23 January 2017).

MoA (2015) *Strategy for Agricultural Development in Georgia 2015–2020*. Ministry of Agriculture, Tbilisi.

Morgounov, A. and Zuidema, L. (2001) *The Legacy of the Soviet Agricultural Research System for the Republics of Central Asia and the Caucasus*. International Service for National Agricultural Research, The Hague.

Mosashvili, G. (2007) ICT technologies in the agrarian sector of Georgia. Department of Information, Georgian Academy of Agricultural Sciences, Tbilisi.

Pavliashvili, J. (2009) *Preferences for Rural Credit Systems and their Impact on the Implementation of Credit Unions in Georgia*. Niedersächsische Staats- und Universitätsbibliothek Göttingen, Göttingen, Germany.

Roeing, T. (2016) Creating business mentalities: knowledge and rural development in Georgia from a discursive perspective. In: Hornidge, A.-K., Shtaltovna, A. and Schetter, C. (eds) *Agricultural Knowledge and Knowledge Systems in Post-Soviet Societies*. Peter Lang, Bern.

Salukvadze, J. (1999) Land reform and land management in post-Soviet Georgia. *Anthropology of East Europe Review* 17(2), 4–12.

Shtaltovna, A. (2013) Servicing transformation. Agricultural service organisations and agrarian change in post-Soviet Uzbekistan. PhD thesis, LIT Verlag, Münster, Germany.

Shtaltovna, A. (2015) Knowledge gaps and rural development in Tajikistan. Agricultural advisory services as a panacea? *Journal of Agricultural Education and Extension* 19 January, 1–17. DOI: 10.1080/1389224X.2014.997257

Shtaltovna, A. and Van Assche, K. (2013) Knowledge gaps and knowledge flows in agriculture in Georgia. Workshop report, 18 April, Academy of Agricultural Sciences of Georgia, Tbilisi. Available at: www.zef.de/uploads/tx_zefportal/Publications/ashtaltovna_download_Scientific%20WSTbilisi_130417.pdf (accessed 23 January 2017).

Shtaltovna, A., Van Assche, K. and Hornidge, A.-K. (2012) Where did this debt come from? Organizational change, role ambiguity and development in rural Khorezm, Uzbekistan. *Internationales Asienforum* 43(3/4), 179–197.

Trevisani, T. (2008) Land and power in Khorezm—farmers. Communities and the state in Uzbekistan's decollectivisation process. PhD thesis. ZEF/Free University of Berlin, Berlin.

UNEP (2002) *Agricultural Development and the Cost of Conflict*. United Nations Environment Programme, Nairobi.

USAID (2011) Economic prosperity initiative: sector assessment report. Prepared by Deloitte Consulting LLP. USAID Caucasus.

USIP (2013) A new opportunity: agricultural extension as a peacebuilding tool. United States Institute of Peace, Washington, DC. Available at: www.usip.org/publications/2013/01/new-opportunity-agricultural-extension-peacebuilding-tool (accessed 22 March 2017).

Van Assche, K., Shtaltovna, A. and Hornidge, A.K. (2013) Visible and invisible informalities and transformation. Lessons from transition countries: Georgia, Romania, Uzbekistan. In: Giordano, C. and Hayoz, N. (eds) *Informality in Eastern Europe: Structures, Political Cultures and Social Practices*. Peter Lang, Frankfurt, pp. 89–118.

Van Assche, K., Shtaltovna, A. and Hornidge, A.-K. (2016a) Local knowledge and expert knowledge in rural transition: Georgian wine production. In: Hornidge, A.-K., Shtaltovna, A. and Schetter, C. (eds) *Agricultural Knowledge and Knowledge Systems in Post-Soviet Societies*. Peter Lang, Bern.

Van Assche, K., Hornidge, A.-K. Shtaltovna, A. and Boboyorov, H. (2016b) Epistemic cultures in transition: agricultural expertise development in Tajikistan, Uzbekistan and Georgia. In: Hornidge, A.-K., Shtaltovna, A. and Schetter, C. (eds) *Agricultural Knowledge and Knowledge Systems in Post-Soviet Societies*. Peter Lang, Bern.
World Bank (2013) World development indicators: Georgia. Available at: http://data.worldbank.org/country/georgia (accessed 23 January 2017).
Zaur, I. (2006) Agriculture and conflict: a conceptual framework for development. MA thesis. Tufts University, Medford, Massachusetts. Available at: http://dl.tufts.edu/catalog/tufts:UA015.012.DO.00148.

12 Agricultural Extension in Post-Conflict Guatemala: Achievements and Challenges in Building a New National System

Vickie A. Sigman*
Freelance Consultant, Middleton, Idaho, USA

Introduction

Guatemala and the Western Highlands

Guatemala is an exceptionally beautiful Central American country with a population of around 14 million people, approximately 40% of whom are indigenous Mayans (USAID, 2011). Mayans are by far the largest group of indigenous peoples in Guatemala and are the majority population residing in the Western Highlands. There are extremely small groups of indigenous Xinca and Garifuna who reside in southeast Guatemala and the Caribbean Coast, respectively.

Unfortunately, Guatemala's beauty is not reflected in its current indicators of well-being. About half of the population lives in poverty and an equal percentage of children under 5 years are chronically malnourished (USAID, 2011). Years of strife and violence have left their mark on Guatemala, particularly in the Western Highlands, a mountainous area of high altitudes, volcanoes, forests and lakes. Feed the Future (FTF), the US government's global hunger and food security initiative, operates in five departments in the Western Highlands: Totonicapán, San Marcos, Huehuetenango, Quetzaltenango and Quiche. The sixth department, Solola, is not a priority FTF department. Collectively, these departments have the highest density of poor and food-insecure people in the country (USAID, 2011). Their poverty and health indicator rates are among the very worst in the world. About three-quarters of the Maya population live in poverty and over two-thirds of children under 5 years are chronically malnourished (Dworkin, 2015). The reasons for such dismal statistics are complex. According to Dworkin (2015) from the USAID/Guatemala Mission, they include not only poverty but also insufficient knowledge of nutritional needs and how to meet them, historic social exclusion and corruption, as well as low levels of investment in the region.

Agriculture is a key to Highland livelihoods even though most of the population is landless or land poor. Guatemala has the most inequitable and concentrated distribution of land ownership in Central America (USAID, 2010). Large farms (2.6% of the total number of farms) control 65% of agricultural land, while small farms (88% of all farms)—the majority of which are located in the Western Highlands—control only 16% of agricultural land (Lastarria-Cornhiel, 2003). In spite of this, and due to severely limited

* Corresponding author e-mail: vsigman@sigran.com

alternatives, the main occupation of at least one family member in 70% of Western Highland households is farming (48%) or working as a farm laborer (22%) (Angeles *et al.*, 2014).

The dominant Western Highlands farming system is maize-based, with most cultivation occurring on deforested and eroded hillsides. The large majority of Highland farmers, predominantly Mayans, are subsistence maize–bean farmers (Buena Milpa, n.d.a). Maize has been cultivated in Guatemala by indigenous peoples since ancient times (Wellhausen *et al.*, 1957). The production methods they use—including the use of hoes, machetes and planting sticks—have changed little over the centuries. Farmers tend to store their seed from year to year and yields are low.

According to hearsay, Highland farmers may be planting less maize or paying less attention to the maize they plant these days due to competing demands on their time, and other factors. Planting less maize could be due to smaller landholdings because of land distribution within the family or use for non-agricultural purposes. At the same time, research by the Strengthening Partnerships, Results, and Innovations in Nutrition Globally (SPRING) project (2015) in the Western Highlands indicates that since maize is the basis of the diet, most families feel it is essential to grow maize to help ensure that the family will have something to eat. In short, the more maize they grow and are able to store successfully, the less money they will need for maize purchases. Even for those who grow most of their own maize, purchase of maize is their single largest food expense (SPRING, 2015). The SPRING study suggested that decisions about production are usually taken jointly by men and women in the household, confirming that both play important roles in maize production. Women are largely responsible for backyard farming, producing a variety of agricultural and livestock products (Ruano and Zambrano, 2006).

Background to the Conflict

Land and ethnicity are central themes in Guatemala's frequently turbulent history. Under Spanish rule, indigenous peoples—mainly Mayan—were forcibly removed from their homes and lands and required to provide labor to the Spanish conquerors and colonists (Lovell, 1988; Kramer, 1994). The establishment of large plantations and the emergence of debt peonage began during this time (Lovell, 1988), with the indigenous population coerced onto ever-smaller plots at higher elevations (USAID, 2010). This period also saw the beginning of the mixed-race Spanish/indigenous mestizo population, referred to as Ladinos and commonly meaning non-indigenous Guatemalans. The 1800s, while an important period for some liberal economic reforms, were beset by constant conflicts between liberals and the conservative landholding class who came to dominate politically, economically and socially. Over time, most indigenous land was expropriated, resulting in the concentration of land in the hands of the very few at the expense of the great majority (Brown *et al.*, 2005).

In 1952, the newly elected President Jacobo Arbenz initiated a large-scale land reform designed to spur Guatemala's economic development. He looked towards ending the latifundio system (large landholdings by the elite few), which included lands held by the United Fruit Company (one of the largest landowners in the country), and distributing land to 'thousands of peasants' (United Fruit Historical Society, n.d.: 2). He ran abruptly foul of the wealthy landholding minority, who controlled extensive sugar, coffee, banana and livestock plantations, and of the US government, which became involved on the side of United Fruit. Against the backdrop of the Cold War, the USA viewed Arbenz as a communist, and the toxic mix of large-scale landholder and US interests along with Cold War politics led to a 1954 Central Intelligence Agency (CIA)-backed attack on the Arbenz government and its consequent collapse (United Fruit Historical Society, n.d.). Subsequently, the Arbenz reforms were reversed. Within 6 years, Guatemala's internal armed conflict had begun. A root cause of the conflict was lack of access and unequal distribution of land (Brown *et al.*, 2005; USAID, 2010), which was also a principal cause of poverty in Guatemala (Lastarria-Cornhiel, 2003).

Thirty-six years of conflict: 1960–1996

The Guatemalan conflict is complex, underpinned as it is by the systematic historical discrimination and social exclusion of the Mayan population. This became the driving force behind the efforts of anti-government forces. The beginning of the 1960s saw the growth of a leftist, socialist political movement that led to deadly confrontations between the government and indigenous people, students and the Church (Pillay, 2007). The horrors of the ensuing civil war are detailed in 'Guatemala: Memory of Silence' by the Commission for Historical Clarification (CEH), established in 1994 by the United Nations (CEH, 1999). Over the years of the conflict, an estimated 200,000 men, women and children—including babies—were killed and an additional 50,000 disappeared (CEH, 1999; Levinger, 2015).

There are enormously varying estimates suggesting that up to 1 million people were displaced in the 1980s as a result of the government's scorched-earth policy against the Mayan people (CIA, 2008; Rothenberg, 2012). Over 600 massacres occurred. There was torture, rape and execution. Whole villages were burned, including houses, crops and animals; and, in their own words, Mayans were hunted down like animals (CEH, 1999; Levinger, 2015). The CEH in its extensive review of testimony and documentation found that 93% of all atrocities committed were by the government—primarily the army—and 3% were by the guerrillas (CEH, 1999). The Commission stated with clarity that genocide against the Mayan people occurred, particularly during 1981 and 1983 at the zenith of atrocities (CEH, 1999). The great majority of the victims—83%—were Mayan (Rothenberg, 2012).

The Peace Accords and Contemporary Guatemala

Armed conflict ended with the signing of the Peace Accords in December 1996. One of the tenets of the accords was access to land and rural development (Klick, 2015). Unfortunately, the Peace Accords have yet to meet expectations. While some progress has been made (e.g. the reintegration of war-affected populations, less repression and more open expression of views in public and private), Guatemala has been unable to implement many of the agreed-upon components of the accords (Pillay, 2007; Klick, 2015).

Guatemala continues to suffer from 'pervasive racism towards indigenous peoples on the part of both the economic elite and the non-elite Ladino population' (USAID, 2012: 4). Political discord continues with rampant corruption at the highest levels of government (Malkin, 2015). There has been far less structural change than is needed to break the hold on the country by the powerful few (Pillay, 2007).

Security remains a major issue, with concerns about military involvement in internal security and crime and violence high. In a 2010 survey, close to one in four Guatemalans reported having been the victim of a crime in the previous year (Taft-Morales, 2014). A feature of post-conflict Guatemala is illegally armed groups which commit or threaten to commit violent criminal acts. They are reportedly an arm of an interconnected, yet hidden, powerful body of Guatemalans who have influence both within and outside the government and use their political connections and violence to protect themselves from prosecution in cases of corruption, drug trafficking and other forms of organized crime (Peacock and Beltrán, 2003; Pillay, 2007). The International Commission against Impunity in Guatemala (CICIG), created in 2006 in collaboration with the UN, is charged with assisting Guatemala to investigate and prosecute serious crime and corruption in the country (McKibben, 2015). Ensuring security and strengthening judicial and legal systems is critical to securing peace in post-conflict countries (Dobbins *et al.*, 2007; UN Department of Peacekeeping Operations, 2010).

Pre-Conflict Agricultural Extension

Preceded by the establishment of the Ministry of Agriculture in the early

1900s, agricultural extension was formalized in the 1950s. Extension focused on non-formal education and the provision of agricultural advice in support of smallholder farmers. The original model was influenced by the US Cooperative Extension Service model, in that extension teams were comprised of an agricultural extensionist, an educator for the home and a youth club promoter (Ortiz *et al.*, 2011; FAO, 2014). However, unlike in the USA, research and extension were not under the umbrella of a university but rather each in their own respective institution.

In the 1960s, the extension focus turned to larger commercial farmers (Hernández, 2014). This was reversed in the 1970s by a return to serving smaller farmers through a new multi-institutional approach that included refocusing extension on smallholder farmers and the establishment of agricultural financing and marketing institutions. Within the Ministry of Agriculture, extension was assigned to the General Directorate of Agricultural Services (DIGESA), with livestock extension under a different directorate. In the 1970s and 1980s, links between DIGESA and the National Agricultural Research Institution (ICTA) were strengthened through working together to test, among others, improved seed and fertilizer recommendations (Ortiz *et al.*, 2011).

Extension activities were carried out in some areas during the years of the conflict. Budget cuts in the 1980s, precipitated by structural adjustment along with the violence of the armed conflict, led to the total closure of Guatemala's public sector agricultural extension service in 1990. It was believed that the private and civil society sectors would fill the gap left by public sector extension. This did not happen (Ortiz *et al.*, 2011), although some donor-funded projects and non-governmental organizations' (NGOs) extension activities continued after 1990. The subject of extension was dropped from the curricula of universities and agricultural schools at this time.

Re-Establishing Agricultural Extension: The System

Development

Nearly 20 years after public sector extension was disbanded and 12 years after the signing of the 1996 Peace Accords, the Ministry of Agriculture, Livestock, and Food (MAGA) moved towards re-establishing a public sector National Agricultural Extension System (SNEA). This began around 2008 and was given further impetus by the government's 2009 *National Framework for Comprehensive Rural Development*. Ultimately, most components of SNEA were incorporated in a newer, larger system of extension, the National Rural Extension System (SNER), established in 2013 (see Hernández, 2014). The basic difference between the two systems is that SNER provides for a pluralistic extension system (including public, private and civil society sector service providers), whereas SNEA only dealt with public sector extension. The following describes the SNER system, specifically the public sector Rural Extension Service (RES). It is worth noting that SNER was established 26 years after public sector extension was disbanded and 17 years after the formal end of armed conflict.

Framework

Figure 12.1 shows the SNER framework. It is anchored by the RES, which falls under MAGA. Inputs to the system include technical information and strategy, incentives and financial resources. The primary outputs, representing the overall purpose of the system, are self-managed farmer organizations, technically competent producers and families with a higher quality of life. The framework links together research and educational institutions, such as the national research institution, agricultural universities and schools, farmers' organizations, and the specialized directorates of MAGA including

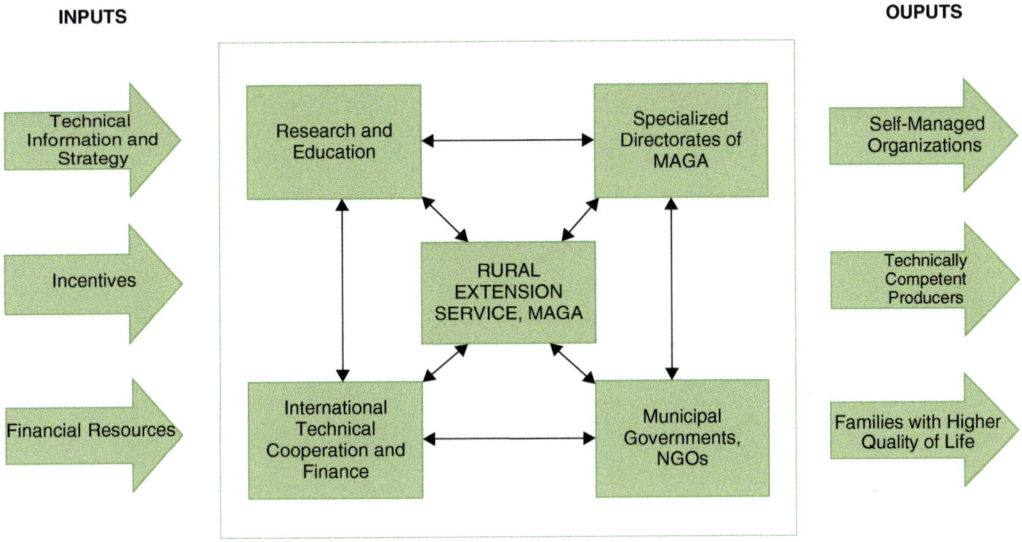

Fig. 12.1. Framework of the National Rural Extension System (SNER). From MAGA (2013).

horticulture, animal health and agricultural infrastructure.

The specialized directorates, functioning as subject matter specialists, provide technical backup to extension activities. International assistance is incorporated in the system along with municipal government services and support by NGOs.

Public sector extension

Currently, public sector extension drives the SNER system, although it is designed as a pluralistic system. Within SNER, the RES is the administrative home for all MAGA extensionists, who are typically referred to as MAGA rather than RES extensionists. Other actors in the pluralistic system are discussed later in this chapter.

Staffing and coverage

Within the context of the public sector, Guatemala has 22 departments (roughly equivalent to a US state) and 334 municipalities (equivalent to a US county). At the department level, the Department of Regional Coordination and Rural Extension channels and coordinates all MAGA services, including extension. At the municipal level, a Municipal Extension Agency (AME) is responsible for field-level extension and is staffed by a team of three MAGA extensionists in each municipality. There is a Rural Development Extensionist (usually male) who has administrative/supervisory responsibilities, supports farmers' group development and interacts with non-public actors in the extension system. A Family Farming Extensionist (typically male) focuses on participatory assessments of the agriculture-related needs of farmers and training volunteer extension promoters (local farmers called *promotores*) to support farmers' group activities. The third team member is a Healthy Household Extensionist (almost always female) who addresses issues related to the home, including backyard gardening and nutrition, and works with farmers' groups that consist primarily of women.

All AMEs are reportedly staffed; thus, MAGA currently has around 1000 field extensionists who have been brought on board since 2013. Qualifications required for field extensionists range from 12 years of education to a Bachelor's degree. Although Guatemala is in the process of decentralizing various government functions, MAGA has not been decentralized, and extensionists at the municipal level report to regional- or national-level MAGA officers rather than to municipal officials. There are exceptions to this however. Municipal officials, particularly mayors, play a large role in extension in some municipalities.

Even though there are three extensionists in every municipality, the number of farm families who are eligible to receive extension services, estimated at 1.5 million, is far greater than MAGA extensionists alone can serve. SNER thus embraces an approach to extension that is designed to broaden its coverage to the extent possible.

Approach

SNER promotes a farmer-to-farmer approach which, at its most basic level, involves training extensionists, who subsequently support and develop the capacity of volunteer extension *promotores*, who then work directly with groups of farmers to improve their livelihoods. The approach is less linear than it may seem, with feedback and dialogue among the different elements of the system and other capacity development activities occurring alongside training. As further detailed below, SNER promotes: (i) organizing farmers' groups through establishing Centers of Learning for Rural Development (CADERs); (ii) developing the capacity of volunteer extension *promotores* selected to provide guidance and capacity development to the CADERs; and (iii) using a cascading training system to develop capacity and technically backstop the system as a whole. The farmer-to-farmer approach improves coverage because integrating volunteer *promotores* in the system adds considerably to the number of farmers that extensionists can reach. It also localizes extension processes by substantively engaging local volunteers in important positions within the system. Figure 12.2 depicts the approach.

CADERs are neither physical locations nor facilities. They are groups of people organized around community challenges and opportunities. More specifically, members of CADERs share an interest in improving their agricultural or livestock production, processing and/or marketing and livelihood activities. CADER members learn by

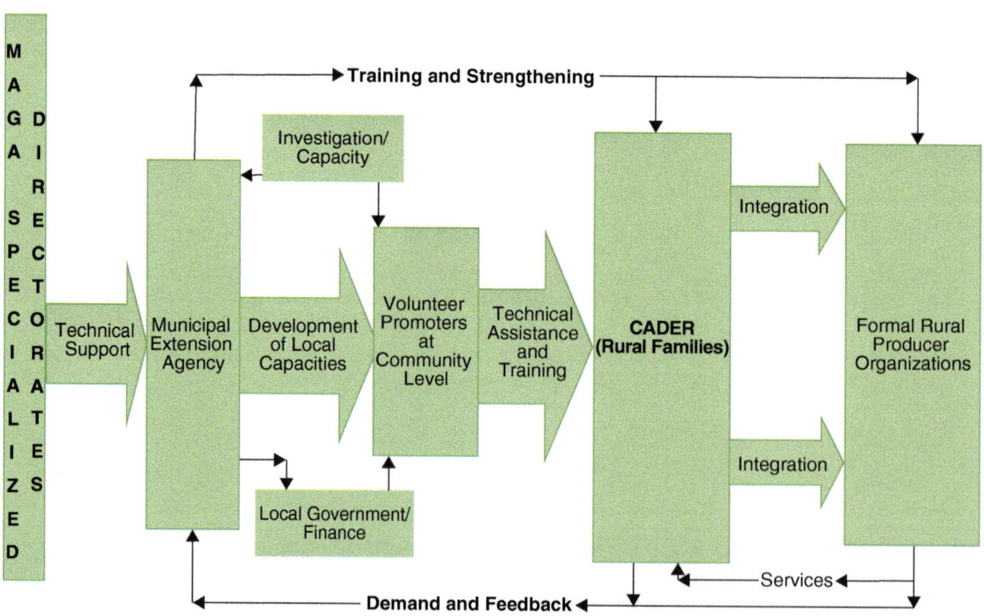

Fig. 12.2. SNER approach: municipal level. From MAGA (2013).

doing—building on their existing knowledge and acquiring and applying new learning and skills. The SNER framework suggests a minimum membership of ten members per CADER. Many CADERs are much larger.

Municipal-level extensionists are charged with organizing CADERs and with training CADER *promotores*. The group *promotores* and an extensionist lead a set of participatory diagnostic exercises to identify the needs of farmers in the CADER. These needs are prioritized by the CADER and used to formulate a group plan. Group plans are consolidated into a Municipal Agricultural Plan, which is part of the Municipality Development Plan, an official document of the municipality. Thus, SNER is a demand-driven system. The expectation is that ultimately CADERs that share interests will join together to form larger, more formal, rural producer associations which are able to provide extension and other services to their members.

Volunteer extension *promotores* are selected by their group. They are involved in the capacity development activities led by extensionists and, in turn, relay the knowledge, skills and attitudes they learn back to their group. While *promotores* working with public sector CADERs are not paid, they may receive incentives such as inputs for demonstration plots and training as well as recognition in their communities.

Capacity development

The capacity development of public sector extensionists takes place through a cascading training system that begins at the national level. National-level staff provide training to extensionists at the department and/or the municipal level. Department-level specialists from the specialized MAGA directorates also train municipal-level extensionists. Municipal-level extensionists then train *promotores* who train their CADER groups. Annual capacity development plans are developed at the various levels.

Capacity development for extensionists includes both technical agriculture and extension methodology and covers subjects ranging from crop production, home gardens, conservation agriculture and CADER plan development to demonstration plot establishment and facilitation. MAGA has led the development of several comprehensive guidebooks for use in training extensionists and for use by extensionists in their everyday work (MAGA, 2014; MAGA and FAO, n.d.).

Other actors in the pluralistic extension system

SNER is a *system* for extension and although it is largely driven by the public sector at the present time, it is designed to provide entry points for public, private and civil society sector engagement. Ideally and by design, initial entry points for the various actors are through interactions with the staff of municipal-level extension agencies (the AMEs). Municipal-level development plans and CADER group plans then help to identify where there is a match between what the community wants (as expressed in these plans) and what public, private or civil society sector extension actors have to offer.

There are various actors in the pluralistic extension system. Notably, not all extension service providers in Guatemala use the Spanish word for extension (*extensión*) to describe their activities, nor does everyone who works in extension in Guatemala identify themselves as extensionists. The reasons for this likely relate, at least in part, to the history of conflict and continued tensions between civil society and the government, leading some providers to distance themselves from association with the government. The Spanish words used by some extension providers to identify their work, such as '*acompañante*', do not translate well into English and thus 'extension' and 'extensionists' are used in this chapter.

An overview of extension research, specific extension-related projects and the key NGO extension service providers contributing to the pluralistic extension system in the Western Highlands is provided below. There have been reports of some private sector engagement in extension in the

Western Highlands but this type of engagement primarily takes place in other parts of the country and involves commercial crops.

Research

Institute of Science and Technical Agriculture (ICTA)

ICTA, Guatemala's national research institution, did not close in the 1990s along with extension and other government departments but continued to operate, albeit on a more limited scale, due to its status as a semi-autonomous institution. ICTA's work during the armed conflict was constrained by security concerns related to the ongoing violence.

ICTA's purpose is to contribute to Guatemala's agricultural development through science and the development of technologies for sustainable agricultural production systems. Its clients fall under the three categories of farmers Guatemala characterizes as: (i) less than subsistence; (ii) subsistence; and (iii) surplus.

ICTA and MAGA extensionists have experience working together in carrying out field activities. ICTA also works with some NGOs and seeks linkages with bilateral donors, private sector extension and other sectors of the government. For example, prior to SNER, ICTA engaged in projects with the Japan International Cooperation Agency (JICA) and the Food and Agriculture Organization of the United Nations (FAO) to train extensionists in production technologies. ICTA has also trained Rural Health Workers from the Ministry of Health and agricultural staff of various NGOs. ICTA currently has a program underway to train a new generation of researchers. The program gives some consideration to the role of extension in agricultural development.

ICTA scientists have had considerable success in substantially improving commercialized maize yields in the Pacific Coast area of Guatemala using hybrid seed and other inputs. However, they have had much less success in improving maize yields in the Highlands. This prompted the development of a small program focused on local knowledge systems and traditional and alternative systems of food production. The program is championed by a senior ICTA scientist interested in agricultural innovation systems.

For many years, ICTA has used a phased approach to conducting research and the institute expects to continue to do so, although the approach is currently under review to identify any needed modifications. The research process has feedback and evaluation loops and is informed by agricultural and socio-economic studies and information. The process involves five phases:

- Phase One: Experimental work is carried out on experimental stations to generate new technologies under ICTA-controlled conditions.
- Phase Two: On-farm experiments continue to generate and adapt technologies following their evaluation by ICTA.
- Phase Three: On-farm testing occurs under farming conditions with farmers evaluating the technologies. ICTA assesses the acceptance level by farmers of the technologies.
- Phase Four: Technologies are promoted by farmer collaborators who use the technologies on their own farms. Typically, ICTA begins involving extension services during this phase.
- Phase Five: Production is expanded as greater numbers of farmers adopt the technologies. This phase is facilitated by extension services.

Value could be added to the process by involving extension services during Phase Three. This would provide hands-on learning opportunities for extensionists, which would improve their capacity to understand and discuss the technologies with farmers. Extension could also facilitate and provide further feedback to ICTA regarding farmer perceptions of the strengths and potential weaknesses of the technologies under review.

The Buena Milpa project

'Buena' in Spanish means 'good'. Milpa is a Mesoamerica cropping system that traditionally produces maize, beans and squash.

Milpa also refers to a field. The Buena Milpa project, a USAID FTF initiative, began in 2015 and is led by the International Maize and Wheat Improvement Center (CIMMYT). The objective is to contribute to improving food security and decreasing malnutrition, while increasing the sustainability and resilience of maize-based farming systems in the Western Highlands of Guatemala (Buena Milpa, n.d.b).

The conceptual framework for the project is the agricultural innovation system (AIS). This is defined by the project as 'a concept that describes the organization of interacting and evolving groups of stakeholders (farmers, extension agents, researchers, institutions, companies, etc.) that work together around a common agricultural problem to stimulate innovation and change towards solving the problem' (Buena Milpa, n.d.a: 6).

The project focuses on improving aspects of maize-based farming systems, including conserving genetic diversity, natural resource management and farm system diversification. It emphasizes social inclusion to ensure that Mayan women and men and other marginalized groups are included in Buena Milpa activities. It is meant to be a scaling out activity, which implies a major emphasis on extension activities. Previously studied technologies, practices, processes and models are to be scaled out (Buena Milpa, n.d.b). The underlying rationale is to move away from a linear technology transfer and delivery model whereby researchers develop technologies that extensionists then disseminate to farmers, and towards an acknowledgment that innovation emerges from interactions among multiple actors and has organizational and institutional as well as technical dimensions (Nederlof *et al.*, 2011).

Buena Milpa staff do not work directly with farmers. The project has developed an innovation platform (IP) whose members implement, along with farmers, the research and extension work of the project. Buena Milpa refers to this body as an 'innovation network' to distinguish it from the term 'innovation platform' used by a larger CIMMYT-led project in Mexico (MasAgro). The latter functions somewhat differently from the Buena Milpa network but informed the development of the Buena Milpa project. The Buena Milpa network performs functions largely as an IP and is referred to as such in this chapter.

Referred to as collaborators, IP members are not subcontracted entities per se, although the project manages a small grants program to finance collaborative activities. Collaborators include representatives from the national research institution ICTA, MAGA extension services, universities and various NGOs providing extension services. The overarching idea is to use the strengths of different members of the IP, including farmers and their groups, to jointly learn, refine, test and promote the uptake of innovations (technologies, practices and processes). For example, IP members are involved in carrying out bean seed variety validation trials, establishing community-based maize seed banks and diversifying maize-based farming systems by integrating them with amaranth and poultry.

ICTA leads several on-farm trials related to Buena Milpa focus areas to fine-tune existing technologies and practices. Over time, as challenges arise in the implementation of field activities by IP members, ICTA will be a major contributor to developing research-informed solutions.

The primary role of Buena Milpa is that of innovation systems broker (also referred to as innovation facilitator). Innovation brokers are persons or organizations that purposefully catalyze innovations by bringing together actors and facilitating their interaction (Klerkx and Gildemacher, 2012). Collaborators in the IP have their own extension approaches, which Buena Milpa strengthens through training and other capacity development activities. IP collaborator approaches to extension largely mirror the SNER approach, whereby extensionists are trained to work with *promotores* who then work with organized groups of farmers.

FAO

In line with its global mission to eradicate hunger in the world, the FAO works closely with Guatemala's Secretariat of Food and Nutrition and is a major ally of MAGA.

The FAO assisted in the development of the SNER system and promotes its implementation. It has supported institutional development to strengthen MAGA's capacity to function at the center of the SNER system and has supported the Ministry in various other extension activities. These include, for example, designing a capacity development plan for MAGA extensionists, training them and developing learning materials for extensionists to use. The FAO and MAGA have previously implemented several extension projects in the Western Highlands, including a small farmer livelihoods project.

Tropical Agriculture Research and Learning Center (CATIE)

CATIE is a regional institution, headquartered in Costa Rica and, similar to a US land grant university, offers graduate education in agriculture. CATIE is also involved in research and outreach programs. Organizationally it has three divisions: an Educational Division, which operates its graduate school and training programs; a Division of Research and Development, which manages various projects around the themes of agriculture, forestry, livestock, environmental management and climate change; and a Division of Outreach and Development, which administers CATIE's Country Offices, Communication and Policy Office, and Management and Service Offering Unit. CATIE has a country office in Guatemala.

CATIE, in partnership with MAGA and with funding from Norway, implements an innovative extension-oriented project with the lengthy title 'Knowledge management for sustainable rural development innovation in Guatemala: strengthening family agriculture and farmer economy' (CATIE, 2013). The project operates in the Central Highland's departments of Alta Verapaz, Baja Verapaz and Chimaltenango. Among its objectives are to strengthen SNER and to design and implement a new structure within the SNER system called Local Systems of Extension (SLEs).

An SLE is essentially a coordinating unit for pluralistic extension actors who come together at the municipal level to coordinate all extension activity in the municipality. Each SLE is led by the municipal-level extension agency and is closely aligned with the work of the Municipal Commission of Food and Nutrition Security. SLEs were introduced to increase synergies, improve information sharing and reduce duplication of services among extension actors.

CATIE's Knowledge Management project has its own staff of extensionists who work in concert with MAGA extensionists. To date, the project has developed the capacity of around 70 MAGA extensionists in extension methodology, agrobiodiversity, gender issues in extension and seed bank development. Close to 400 *promotores* have been trained, the majority of whom are women (60%). Using Field School methodology, the project works with around 400 CADERs. The project has helped the CADERs to develop their group and municipal development plans with subsequent training based on their identified priorities. The majority of CADER members are also women. CATIE is considering a study to determine the factors influencing women's participation as *promotores* and as CADER members. A working hypothesis for this chapter is that, due to emigration and migration of men from communities, the role and participation of women in agriculture is becoming increasingly important and visible.

Más Fríjol

'Más fríjol' means 'more beans' in Spanish. Tortillas and beans have been the major staples of Guatemalan diets for many decades. Served together in the right quantities they are a good form of protein in diets even though they do not provide certain essential micronutrients. Currently, Guatemalans are eating fewer beans and traditional vegetables than previously, which negatively impacts their nutritional status. The price of beans has increased in recent years, and few improved varieties are available which are adapted to the microclimates of the Highlands. In addition, there have been changes in dietary preferences with less nutritious snack food growing in popularity.

The Más Fríjol project, a USAID/Guatemala FTF activity implemented by Michigan

State University, is designed to address these production and consumption-related challenges. Project objectives are to increase the production of beans in the Western Highlands and to improve the nutritional quality of diets through increased consumption of beans in combination with other nutritious vegetable and animal-sourced foods. The project uses a training-of-trainers approach. Más Fríjol provides training for project partners on nutrition-sensitive agriculture, focusing on the production and value of beans in the diet and ways to increase dietary diversification. The partners, who include MAGA extensionists, public sector health workers and staff from NGOs, then work with farmers' groups and their families to increase the production and consumption of beans.

Food for Progress

Food for Progress, a US Department of Agriculture-funded initiative, which began in 2012, is another activity in USAID/Guatemala's portfolio of FTF programs. Counterpart International implements the project along with its partners MAGA, Guatemala's University of San Carlos School of Agriculture, the Zamorano Pan-American Agricultural School in Honduras (established in the early 1940s by the president of the United Fruit Company) and the University of California at Davis. The project's goal is to increase sustainable agricultural knowledge and to improve livelihoods among rural communities in the Western Highlands.

Food for Progress supports local farming cooperatives and CADERs, farmer financing via cooperation with a local credit union and the strengthening of SNER. The project carries out diagnostics of cooperatives, assists them with organizational development and with developing business plans and provides targeted training. A cost-sharing arrangement with a Guatemalan credit union has expanded financial services to cooperative farmers, with more than US$3 million in loans made over the life of the project. Food for Progress has a small grant program to support CADERs and MAGA extensionists in carrying out their field activities.

Extension training opportunities, limited in the first instance, were much diminished during the armed conflict. In 1990, the extension curriculum was totally removed from university offerings. To address this gap and as part of its support for SNER, the Food for Progress partners have developed a certification training program for SNER extensionists. This 4-month training program emphasizes development policies and plans relevant to agriculture and extension, extension methodology (including adult learning and monitoring and evaluation), low-cost agricultural technologies, and food security and nutrition. Four cycles of training have been delivered since the beginning of the project with 283 certified extensionists graduating. The project held talks with MAGA in a pre-emptive move to address the issue of whether certified extensionists would continue to hold their MAGA positions over time, rather than potentially being removed after their annual contracts expired. This resulted in MAGA signing an agreement to keep certified extensionists on staff, barring disciplinary problems. Project estimates are that 80% of certified graduates continue to work in MAGA extension posts, although some believe the rate to be much lower. Discussions are underway with Guatemala's San Carlos University to determine the feasibility of including the certification training in its diploma-level offerings.

The Food for Progress project has a small cadre of extensionists who work with volunteer *promotores* and CADERs using SNER's farmer-to-farmer approach. The incentives for *promotores* to participate include training, possible inputs for demonstration plots and increased status in the community. Food for Progress emphasizes recognizing the accomplishments and contributions of men and women *promotores*, finding that such acknowledgment from municipal and MAGA authorities, as well as from the project, plays an important role in encouraging and sustaining *promotores*' participation. Over three-quarters of farmers benefiting from Food for Progress activities are women.

NGOs

ASSOCIATION OF ORGANIZATIONS OF THE CUCHAMATANES (ASOCUCH). Established in 1994, the Association of Organizations of the Cuchamatanes has grown to be an umbrella organization which works with numerous associations and cooperatives with the aim of empowering rural men and women, farmers and youth—a majority of whom are indigenous Maya. ASOCUCH has five focus areas:

- Building a strong network of local organizations and developing social capital.
- Improving local competitiveness and promoting local economies.
- Promoting natural resource management, community tourism and local environmental management.
- Stimulating citizen participation and promoting gender equity.
- Strengthening financial sustainability and efficient management.

ASOCUCH provides various services to its member groups. These include administering project finances and subcontracts, representing the organizations in fora such as the National Indigenous Committee on Climate Change and annual planning of activities. The organization has an online monitoring and evaluation system to track its various activities including extension services.

ASOCUCH supports agricultural extension activities primarily through training extensionists. Specialists train extensionists who work in ASOCUCH member associations or cooperatives. Training spans 1 year and covers technical subjects and extension methodologies, including diagnostics to identify farmer problems. Each association or cooperative has two extensionists, paid either by the entity or through a project managed by ASOCUCH. These extensionists—about 10% of whom are women—work directly with farmers' groups. They speak both Spanish and local Mayan languages.

The farmers' groups associated with ASOCUCH are involved in vegetable, potato, onion, coffee and sheep production, as well as reforestation and other natural resource management activities. Some groups have participated in the Collaborative Program on Participatory Plant Breeding in Mesoamerica. This collaborative program is a regional initiative, is funded by various donors and is designed to strengthen the participation of farmers in the conservation and utilization of genetic resources, breeding (selection, validation and seed production) and the sustainable management of crops.

COOPERATION FOR RURAL DEVELOPMENT OF THE WEST (CDRO). Cooperation for Rural Development of the West emerged in 1984 in response to the most devastating years of the armed conflict. Its history begins with indigenous leaders of two Mayan communities in Totonicapán. Looking to maintain their indigenous identity and to help their communities, these leaders created an approach to development based on Mayan precepts of the interconnectedness and interdependency of their communities and the related principles of total community participation and locally driven development. This approach continues to inform CDRO activities and is a unique feature of its identity.

CDRO's mission is to promote the comprehensive development of rural communities in western Guatemala using its unique approach and to empower communities to establish their own organizations in harmony with nature. Changes in the cycles of the organization's growth and retrenchment are largely a function of its donor funding base. It currently supports development activities in over 40 communities, working with around 20 organizations.

CDRO's activities target a wide range of subject matter areas, including organizational capacity development, education, health and natural medicine, natural resource management and food security. Gender is a cross-cutting issue in all of these areas.

CDRO's extension activities are implemented using a farmer-to-farmer approach similar to that employed by SNER. Agricultural technicians from CDRO work directly with volunteer *promotores* and community groups. CDRO has its own training system whereby *promotores* are trained in an agroecology school for 2 days every 2 months.

Participants are provided with transportation and food. Follow-up training for *promotores* is delivered on a periodic basis.

CDRO technicians and/or *promotores* train farmers' groups in various technical subjects. Demonstration plots are frequently used teaching tools with *promotores* and groups establishing plots using inputs usually provided by CDRO and the farming community.

A majority of farmer group members are women, particularly in groups led by women *promotores*. Because they are unable to make a living from the land available to them, and the extremely limited employment opportunities in their communities, many men—youth in particular—have emigrated from the country. Others migrate to undertake seasonal wage labor on Guatemalan plantations. Over 1 million Guatemalans live and work—legally and illegally—in the USA (World Bank, 2014). The remittances they send home to their families are very important sources of income for those remaining behind. It has been estimated that remittances represented 10% of Guatemala's GDP in 2013 (World Bank, 2014). The majority of remittance recipients are rural women, and around 50% of remittance monies are used for consumption with the remainder used for human and physical investments (World Bank, 2014). Such investments might include investing in children's schooling—by paying school fees—or financing improvements to agricultural production. Emigration and migration seem to be resulting in agriculture at the household level becoming more feminized. Nonetheless, it has been noted that as group activities prove to be successful, the men who live in the community become increasingly interested in participating.

LEGAL AND SOCIAL SERVICES (SERJUS). Legal and Social Services was founded in 1987 in response to the armed conflict and to support indigenous communities to overcome the years of violence and repression. It is now an umbrella social activist support and coordination organization for community-based organizations located, for the most part, in the Western Highlands. SerJus also works with organizations in other regions of Guatemala and Central America. It has an exceptionally strong orientation towards supporting the under-represented Mayan people.

SerJus provides a wide array of support and coordination services to its constituent organizations. This ranges from economic development activities and legal advice to dissemination and capacity development on theoretical and methodological approaches to community organization, social empowerment and popular education as influenced by Paulo Freire (1972), which encourages individuals to take control of their own learning towards advancing social change (Barroso, 2002). SerJus works towards a just, equitable and democratic society through organizing and developing the capacity of communities to develop themselves and empowering them to influence government at local, municipal, regional and national levels.

SerJus is involved in a number of agricultural development activities and has its own agricultural extension staff. Many SerJus member organizations also have agricultural extension activities. SerJus is overtly opposed to the use of genetically modified seeds and participates in efforts to improve local maize varieties. Its agricultural extension approach is similar to SNER's. SerJus extension staff work directly with volunteer *promotores*, who engage directly with farmers' groups. *Promotores* install on-farm demonstration plots to assess technologies and practices and use the plots as teaching tools for their groups.

Although SerJus utilizes a farmer-to-farmer approach that has much in common with the approach used by NGOs and SNER, there are substantive underlying differences that inform SerJus extension services. These include SerJus's history, its political orientation, its vision of development and its emphasis on popular education. These elements influence how SerJus extensionists work and with whom. These differences tend to constrain possibilities of positive working relationships between SerJus and the government, which have social justice and empowerment perspectives that have historically (and presently) been at odds.

SAVE THE CHILDREN (SAVE). Save the Children opened its Guatemala office in 1999 and ever since has been helping Guatemalans to overcome the impact of the armed conflict through various education, health, nutrition, livelihoods, natural resource management, and democracy and governance programs. Its farmer-to-farmer extension approach is similar to that of SNER. Two key people constitute the extension services at the community level: an agricultural leader (akin to an extensionist) and a *promotore*. The agricultural leader works in two or three communities, each of which has two or three *promotores*. Under the supervision and with support from the agricultural leader, each *promotore* works directly with around 30 farmers. The leaders and *promotores* are supported by Save's technical specialists, who provide capacity development in both technical subjects and extension methodologies. Capacity development is based on adult non-formal education methodologies and emphasizes learning by doing. Manuals and brochures are available for use in capacity development activities.

Key Achievements

Strengthening the provision of agricultural services by public institutions and other organizations involved in the agricultural sector is an essential part of the post-conflict reconstruction and normalization process (Dobbins *et al.*, 2007; USAID, 2009). Guatemala has made, and is expected to continue to make, important progress in this regard.

The creation of a national extension system took only about 3 years, no small endeavor. In a very short time, MAGA managed to recruit and place over 1000 people in extension posts across the country. There is now a cadre of public sector extensionists who have experience and have received some capacity development. The extension system explicitly includes farming and farm household extension services, which address production for markets and consumption as well as family nutrition. The latter is not typically included in government extension services and represents a laudable component of the system.

While quantitative data are lacking regarding the quantity, quality and benefits of SNER's achievements in extension, the achievements are real: CADERs have been organized; farmer needs assessments have been carried out; group and municipal development plans have been formulated and have informed extension activity; and extension activities have been implemented with farmers and farm families. There is a network of *promotores* who have been trained and who now work directly with farmers and their households. Extension materials to backstop extension fieldwork are available in Spanish and in several Mayan languages. SNER continues to evolve, and its evolution is a dynamic process that will be informative and interesting to follow as an example of a post-conflict national pluralistic extension system.

Within the framework of the national system, new ways for research and extension to work together are being collaboratively pursued by the national research institution ICTA and an innovative scaling project focused on improving maize-based farming systems in the Western Highlands. The project facilitates an IP that brings together researchers, extensionists and other development workers from government institutions, NGOs and the academic community. IP members apply themselves to agricultural problems in which they all have a stake and that are unlikely to be resolved by any one group acting alone. This type of collaboration represents a major achievement. International and regional actors are actively supporting extension through building the capacity of various groups, including government extensionists, local *promotores* and organized farmers' groups. Linkages are being made with Guatemalan institutions of higher education to reintroduce extension as an area of study in the university curriculum. Attention is being paid to the importance and improvement of Mayan traditional foods and to improving nutrition as well as to increasing production.

Civil society has become an increasingly dynamic force in Guatemala. NGOs, which have a long history of working for the Guatemalan people, continue to build social capital and empowerment in communities. These NGOs, several of which were founded and are maintained by indigenous Mayans, provide key extension-related services to farm men and women. For example, to empower members and strengthen the market position of farmers, they promote the development of cooperatives and associations and support the training of local *promotores* who directly work with farmers' groups to improve farm family livelihoods. Several NGOs are especially adept at involving women in their agricultural extension programs. In light of the difficult context of pre- and post-conflict Guatemala, all of these achievements are noteworthy.

Key Challenges

Although it has been almost two decades since the signing of the Peace Accords, many extension challenges can still be linked back to the armed conflict. Some are similar to challenges faced by extension systems in other post-conflict countries; others are more specific to Guatemala. These challenges—described below—are interconnected. For the purposes of this chapter they are categorized by the relative strength of their effect on the government extension service (SNER).

Challenges to SNER

Politicization

Politicization is viewed as endemic in Guatemalan institutions and organizations and is certainly not unique to MAGA. Nonetheless, the most frequently heard critique of public sector extension is that it is highly politicized. This is a significant challenge shared by many post-conflict countries (Hoove and Scholtbach, 2008). Political leaders and their associates are often in a position to control access to resources due to limited accountability. Political patronage and cronyism in government institutions frequently continue in post-conflict situations (Hoove and Scholtbach, 2008). In Guatemala, there are claims that staff positions in public sector extension are distributed based on party affiliations, with the result that extension staff who do not have the requisite qualifications are hired and subsequently perform poorly. Reports indicate that staff are compelled to campaign for, or otherwise promote, a given party or candidate or official. The distribution to farmers and farm families of agricultural inputs, or other goods such as basic foodstuffs, in which extension is involved is also politicized, with those in the 'right' party receiving them and others less so. Politicization tends to generate resentment and intensify distrust among the different actors in the extension system.

System components

The Municipal Extension Agency (AME), which is central to the effective operation of SNER, is weak. This reflects a characteristic of post-conflict contexts, that of fragile government institutions (Hoove and Scholtbach, 2008). Concerns range from AME's lack of interaction with other official bodies at the municipal level to its limited interaction with NGOs. Public sector extensionists report to MAGA at the regional or national level, which could limit their interactions with authorities at the municipal level. NGOs may not see the value in cooperating with the government extension agency. On the contrary, based on historical reasons related to the armed conflict, they may see cooperation with government as inhibiting the potential success of their activities. Nonetheless, within the SNER framework, AMEs are the initial entry points for NGOs to work with communities. The purpose of SNER as a pluralistic national system is undermined when there is limited interaction and limited collaboration among extension providers operating in the same location.

Capacity development opportunities

On balance, extensionists receive less capacity development than they need to efficiently and effectively carry out their jobs.

Their ability to apply extension methodologies and plan their work programs is reportedly quite limited. While teaching and learning materials are available, some extensionists receive materials without explanation or training in how best to use them. Again, this crucial limitation can be related to the armed conflict. There were singularly few opportunities for extension capacity development, particularly from 1990 until very recently. There was no public extension service through which to channel capacity development, and agricultural extension had been removed from the higher education curriculum. This points to one of the indirect costs of conflict, the lack of human resource development.

Staff stability

It is not clear that all public sector extension posts are affected by the challenge of staff stability. However, municipal-level posts definitely are. The large majority of extension staff are currently not civil servants and they do not enjoy job security. They are typically hired on an annual basis with annual renewal unassured. In 2016, a new government will be elected in Guatemala and it is highly likely that there will be many changes in public sector staff. This means a break in continuity in both staff and the work of extension from year to year and as administrations change. New staff will require training, and the benefits of previous investments in staff training may not accrue to public sector extension and those they serve if current trained staff are not rehired or retained. While there is discussion of using performance metrics to inform hiring decisions, the performance evaluation system is very weak and ineffective at this time. Although the government document describing SNER calls for permanent positions for all public sector extension staff (MAGA, 2013), this is currently not the case.

Experience shows staff stability to be a multifaceted issue. While the popular response is to make all extensionists civil servants, this will not have the desired result unless there is unqualified assurance that there are sufficient program funds to support their work. The post-conflict context exacerbates these complications due to constrained budgets, the weakness of the civil service and the need to rapidly build a new SNER system.

System-wide challenges

Politicized maize

In Guatemala, there are sensitivities related to genetic property rights and seed patents that impact maize seed development (Bilaterals, 2014) and there have been cases of communities not welcoming institutions that work on maize breeding (CIMMYT, 2015). This has led to the politicization of maize in parts of the country, particularly in the Highlands. Due to the centrality of maize in Mayan culture, indigenous populations in the Western Highlands may view maize breeding activities as a threat to their native maize and thus to their culture.

Distrust and lack of confidence

Often, the actors in the SNER system do not trust or have much confidence in each other. Underlying this distrust and lack of confidence is the reality that Guatemala is a post-conflict country with all the complications associated with that status. Furthermore, NGOs are generally better resourced than MAGA extensionists, which tends to create tensions between the two. The question of whether international donor resources should flow to government institutions or NGOs is a sensitive one. There are NGOs that refuse to work with public sector extension based on their direct or indirect experiences during the armed conflict or because they have differing philosophies of development. Others do not engage with public sector extension due to the challenges associated with doing so. Some extensionists advise that it is difficult to develop trust with the farm families they work with and that trust is critical to their work. This appears to be less a case of distrusting specific individuals working in public sector extension than of the government they represent. Public mistrust of government institutions is characteristically high in post-conflict contexts (Hoove and Scholtbach, 2008; Klick, 2015, Panić, 2008), and this is reportedly the case in Guatemala.

Coverage

Estimates suggest that MAGA extension can reach 20–30% of the 1.5 million farm families who should benefit from extension services. As a pluralistic system, SNER is designed to increase coverage and increase over time the number of farmers and other stakeholders that it serves. However, the coordination of extension actors and activities in the extension system is quite limited as are collaborative extension activities. This is the case even though MAGA endorses the establishment of strategic alliances with others in the system. At the same time, MAGA has yet to develop the capacity and establish its legitimacy to coordinate actors in the system, nor is a national extension policy in place to guide extension in general and support such coordination. Distrust, stemming in part from years of conflict, challenges aspirations to collaboration and the establishment of alliances among extension actors.

Inclusion

Increasing the participation of indigenous and Ladino women and of indigenous men in extension remains a challenge. The need to strengthen women's participation in economic and social development is enshrined in the 1996 Peace Accords, as are the rights of indigenous peoples. Some extension programs have made important progress towards including women as beneficiaries, and it is likely that many clients of current programs operating in the Western Highlands are indigenous Mayan women. Even so, there are few indigenous women extension agents working specifically in agriculture. Notably, women serve as Healthy Household Extensionists, who focus on essential home-based activities, including backyard gardening and nutrition. Currently, limited numbers of indigenous women hold mid- to high-level positions in MAGA, donor projects or extension-oriented NGOs, with this situation for Ladino women being somewhat better than for indigenous women. The pool of indigenous women qualified for these posts may be comparatively small given that females have less chance than males to pursue and complete their education. Thus, gender and ethnicity converge—such that being female reduces the likelihood of being in school—as does being indigenous (Edwards, 2002). The total enrollment of non-indigenous students is minimally higher than that of indigenous students at the primary level, much higher at the secondary level and overwhelmingly higher at the university level (Edwards, 2002).

Resources

The resources available for public sector extension are inadequate to provide the competitive salaries, level of capacity development, facilities (offices and equipment) and communications support needed to develop and secure a vibrant, effective and sustainable system. According to a number of NGOs, donor funding has decreased substantially from the levels obtained around the time of the Peace Accords. This, while unfortunate for the NGOs, is a feature of post-conflict funding: large initial investments that decrease over time.

Recommendations

The recommendations that follow are linked to the challenges confronting extension. As with the challenges, the recommendations are decidedly interconnected. While the history, causes, nature and outcomes of conflict differ among countries, some post-conflict countries are likely to share experiences similar to those of Guatemala. In such countries, these recommendations will resonate among those involved in the complicated work of rebuilding extension systems.

Politicization

Corruption is generally understood to mean the misuse of public services for private gain (World Bank, 1997); in other words, the abuse of entrusted power for private gain (Spector *et al.*, 2015: 1). Nationally, corruption is fought through CICIG (the International Commission Against Impunity in Guatemala), which has investigated and tried several of Guatemala's high-level officials, including former presidents (WOLA, 2015). While patronage and cronyism at MAGA are not of the same magnitude as the criminality of the CICIG investigations, their actions send a message that corruption can be addressed, and its cost to those involved can be very high.

USAID has developed a guide for anti-corruption programming that suggests a wide range of actions to target corruption (Spector *et al.*, 2015). The anti-corruption actions that relate most directly to extension include standardizing government agency operations to reduce opportunities for corruption, and increasing transparency and public awareness of government operations, enabling civil society to identify and report corruption (Spector *et al.*, 2015). The first action applies to MAGA's contractual arrangements. Standardizing these arrangements, along with developing mechanisms and capacities to monitor standardization, offers a reasonable approach to addressing the politicization of MAGA.

The second action—increasing transparency and involving civil society—is more problematic as it could be risky for civil society actors. Developing a reform-oriented strategy offers a plausible way to address transparency and civil society involvement (de Asis, 2000) within the context of extension. Strategy development should involve a participatory process that brings together public sector reformers and civil society to highlight corruption-related problems in extension and ultimately develop and implement an action plan to address problems identified. Diagnostic work by an independent firm or group to identify specific key problems arising from corruption minimizes the risk to local actors.

Problem identification should be followed by a participatory workshop where the results are presented and action plans are developed collaboratively between government and the community (de Asis, 2000). In Guatemala, this could be carried out at department and municipal levels with the government extension entities and civil society. Strategy development and implementation requires political will, technical assistance and financial support. Given the impact of politicization on effective extension, a justifiable case can be made for using extension and NGO resources and technical assistance to support these processes. This leaves the question of political will in limbo. Where sufficient political will exists, there are strategic tools that can contribute to diminishing the effect of politicization on development and delivery of extension services.

Distrust and lack of confidence

Corruption is closely related to distrust, since it is the abuse of entrusted power for private gain (Spector *et al.*, 2015). Thus, a path towards trust requires decreasing corruption, as discussed above. Lack of confidence in public sector extensionists is also related to corrupt hiring practices, to some degree, also discussed above. Still, creating trust and instilling confidence are thorny challenges that actors in the SNER system face. Recommendations include investigating studies of trust and collaboration theories, and approaches to identify experimental ways to build trust among SNER stakeholders (see Wood and Gray, 1991; Kramer and Tyler, 1996; Mattessich *et al.*, 2001; Sloan and Oliver, 2013).

At the local level, a key recommendation is to create safe haven space for increased interaction among various extension stakeholders: NGOs, projects, farmers and government extensionists. In practice, this could include creating a citizens' committee to advise on extension; holding community-based fora with communities to review and discuss extension activities; convening small group discussions between extension and its stakeholders to discuss issues of trust and confidence and their impacts on mutually shared goals; experimenting with team-building exercises; and actively modeling trust in relationships. Broader recommendations, based on global experience, point to addressing a deficit of trust following conflict by ensuring inclusivity in the peacebuilding process, building institutions and sustaining international support (Ban, 2012).

System components

To address weaknesses in municipal-level AMEs, it is recommended that investments in this component of the SNER system be prioritized. This includes clarifying—through job descriptions—the expected interactions of AME staff with other government entities at the municipal level and earmarking funds to support such interactions. To encourage NGOs to support and engage with AMEs, it is recommended that SLEs be introduced at the municipal level. SLEs, prototyped by CATIE and discussed earlier, provide a tested mechanism for engaging all extension actors in a municipality to carry out their work in a coordinated manner.

Capacity development opportunities

MAGA's extension services currently have capacity development plans which are constrained by funding gaps. Thus, the recommendations focus on leveraging existing training and reallocating some capacity development funds. For example, almost all projects and NGOs provide capacity development for the extensionists involved in their specific activities. It is recommended that MAGA request that NGOs and projects routinely include government extensionists in such initiatives as co-trainers or participants, providing financial support if necessary.

High-level MAGA advocacy to reintroduce the study of extension in university curricula would add considerable weight to ongoing efforts and is recommended. Cross-training of agricultural Family Farming Extensionists and Healthy Home Extensionists is also recommended to expand their capacities and add complementary content to the focus of their respective activities. Finally, it is recommended that the use of ICT for capacity building be explored and that a pilot program to test ICT learning and teaching innovations be put into place. Extension learning and teaching materials—developed by other actors in the global extension system including the MEAS project, regional extension networks and the Global Forum for Rural Advisory Services (GFRAS)—are now readily available on the internet and may be used to support ICT-based capacity development.

Staff stability

The major recommendation with regard to staff stability is to commission a study of the cost/benefit of the two current scenarios—annual contracts or the incorporation of extension agents into the civil service— and use the results to inform decisions on this issue. It is also recommended to strengthen the performance management system so that hiring decisions can be based on evidence of performance.

Politicized maize

While politicized maize will not be an issue in all post-conflict countries, similar culture-related issues may exist with other crops, livestock or agricultural practices. Recommendations to facilitate extension work with Guatemalan farmers and communities in maize breeding and breeding-related activities include using best practices to approach targeted communities and emphasizing transparency in community relationships. For example, prior to initiating activities, involve respected community members or leaders who have participated in maize breeding activities in introducing and explaining any proposed activities to the community; obtain the permission of local authorities to carry out participatory activities and involve them in the activities; and convene community meetings to answer questions the community may have, explain benefits of activities and discuss what will be done by whom and when.

Coverage

Effectively and efficiently responding to the extension-related needs of vast numbers of farmers and others along the value chain is challenging for most countries, especially in post-conflict contexts. Building on the SNER framework and significantly expanding the use of ICTs are the recommendations put forward to address the coverage challenge in Guatemala.

As a framework for the extension system, SNER is anchored by the government extension service. To improve system coverage, it will be necessary to develop the capacity of MAGA to coordinate the system. This is in line with the earlier recommendation to introduce SLEs which would function as coordination mechanisms at the municipal level. The formulation of a specific national extension policy, collaboratively developed by extension stakeholders, is also recommended. Among other issues, the policy should clarify the roles of the public, private and civil society sectors in extension activity with an emphasis on articulating coordination and collaboration guidelines.

The current use of ICTs to address issues of coverage and to support extension activities is minimal. This is due in part to the limited ICT capacity of actors in the extension system. Investing some capacity-developing funds to increase the use of ICTs in extension is recommended, as is pilot testing of approaches designed to expand coverage and improve impact. The private sector offers a potential source of co-funding for ICT-related efforts through public–private partnerships. In concert with face-to-face extension, possible ICT applications include providing helpline services and technical, marketing and weather-related information to farmers; stimulating interactions between government and civil society; informing programming through rapid collection of data; and supporting monitoring and evaluation processes.

Inclusion

The overarching recommendation is to take measures to include women—both indigenous and Ladino, but prioritizing the inclusion of Mayan women—and indigenous men in all extension organizations and activities. To implement this recommendation: address inclusion in the proposed national extension policy; empower women to make decisions about extension programming that will affect them and their families; use and build on existing channels to reach women and indigenous populations by engaging with local indigenous NGOs; provide secondary- and university-level scholarships to increase the pool of qualified women to undertake field- and higher-level extension posts; design and support leadership development opportunities for women; review hiring practices to remove possible constraints to hiring qualified women; and develop cross-cultural and gender-sensitive training to be delivered to a broad audience of extension stakeholders. Finally, develop or use existing tools to monitor the diversity of extension staff and of participants in extension activity. A recommended tool is USAID's Women's Empowerment in Agriculture Index, which measures the inclusion of women in project activities as well as women's empowerment and agency.

Resources

> Building institutions and other peacebuilding tasks can take a generation. This highlights the need for sustained international political and financial support. It also underscores the importance of mutual accountability over the long term, which creates a more balanced partnership between donors and recipient governments.
>
> Ban Ki-moon, UN Secretary-General, 20 December 2012

Ultimately, governments are responsible for financing public services such as extension. In post-conflict countries, their capacity to do so contributes to their legitimacy and demonstrates that the peace process continues to move forward. Guatemala is progressing along these fronts. Still, many rural farm families are receiving less extension advice and fewer services than they need to measurably improve their livelihoods, and this is particularly so for the indigenous Mayan population. Additional funding is needed to strengthen Guatemala's extension system, through increasing the government's budget allocation to extension, reallocating existing funding or—in light of the UN Secretary-General's statement—through seeking international financial support.

References

Angeles, G., Hidalgo, E., Molina-Cruz, R., Taylor, T., Urquieta-Salomón, J., Calderón, C., Fernández, J.C., Hidalgo, M., Brugh, K. and Romero, M. (2014) Monitoring and evaluation survey for the Western Highlands Integrated Program: baseline 2013. MEASURE Evaluation. Available at: https://www.measureevaluation.org/resources/publications/tr-14-100 (accessed 23 January 2017).

Ban, K.-M. (2012) Overcoming post-conflict 'trust deficit' requires strategies based on inclusivity, institution-building, sustained support, Secretary-General tells Security Council. Press release. United Nations, New York. Available at: www.un.org/press/en/2012/sgsm14739.doc.htm (accessed 23 January 2017).

Barroso, M. (2002) Reading Freire's words: are Freire's ideas applicable to southern NGOs? International Working Paper Series, 11. Centre for Civil Society, London School of Economics and Political Science, London.

Bilaterals (2014) Guatemala: people reject Monsanto Law for threatening food. Available at: http://www.bilaterals.org/?guatemala-monsanto-law-rejected (accessed 23 January 2017).

Brown, M., Daly, J. and Hamlin, K. (2005) *Guatemala Land Conflict Assessment*. Management Systems International, Washington, DC.

Buena Milpa (n.d.a) *Work Plan 2015: Feed the Future—Fighting Poverty and Malnutrition through Fostering Sustainable, Resilient, and Innovative Maize-Based Farming Systems in Guatemala*. Buena Milpa, Quetzaltenango, Guatemala.

Buena Milpa (n.d.b) *Work Plan Buena Milpa 2015–2016: Complement to Originally Submitted Work Plan Approved in March 2015*. Buena Milpa, Quetzaltenango, Guatemala.

CATIE (2013) *Resumen proyecto: Gestión del Conocimiento para la Innovación del Desarrollo Rural Sostenible en Guatemala—Fortaleciendo la Agricultura Familiar y la Economía Campesina*. Centro Agronómico Tropical de Investigación y Enseñanza, Ministerio de Agricultura, Ganadería y Alimentación, Real Embajada de Nuruega, Guatemala City, Guatemala.

CEH (1999) Guatemala: memory of silence. Commission for Historical Clarification. Available at: www.aaas.org/sites/default/files/migrate/uploads/mos_en.pdf (accessed 23 January 2017).

CIA (2008) The 2008 World Factbook: Guatemala, history. Available at: www.allcountries.org/wfb2008/guatemala/guatemala_issues.html (accessed 23 January 2017).

CIMMYT (2015) *Performance Management Plan and Monitoring, Evaluation, and Learning Handbook*. CIMMYT, Texcoco, State of Mexico, Mexico.

de Asis, M. (2000) *Reducing Corruption at the Local Level*. World Bank, Washington, DC.

Dobbins, J., Jones, S., Crane, K. and DeGrasse, B. (2007) *The Beginner's Guide to Nation-Building*. RAND Corporation. Santa Monica, California.

Dworkin, M. (2015) Poverty and chronic malnutrition in Guatemala—an integrated solution to a complex problem. Available at: https://usaidlearninglab.org/lab-notes/poverty-and-chronic-malnutrition-guatemala-integrated-solution-complex-problem (accessed 23 January 2017).

Edwards, J. (2002) *Education and Poverty in Guatemala*. Guatemala Poverty Assessment (GUAPA) Program, Technical paper no. 3. World Bank, Washington, DC.

FAO (Food and Agriculture Organization) (2014) *Aprendizajes de las Transformaciones de los Sistemas de Extensión y Transferencia Tecnológica de Guatemala: una Propuesta de Lineamientos de Política para el Futuro*. FAO, Rome.

Freire, P. (1972) *Pedagogy of the Oppressed*. Penguin, London.

Hernández, J. (2014) *Caracterización del Sistema Nacional de Extension Rural en Guatemala, departamentos de Chimaltenango, Baja Verapaz, y Alta Verapaz*. Master's thesis. CATIE, Turrialba, Costa Rica.

Hoove, L. and Scholtbach, A. (2008) Democracy and political party assistance in post-conflict societies. Netherlands Institute for Multiparty Democracy, The Hague. Available at: http://nimd.org/wp-content/uploads/2015/02/Democracy-and-Political-Party-Assistance-in-Post-Conflict-Societies.pdf (accessed 23 January 2017).

Klerkx, L. and Gildemacher, P. (2012) The role of innovation brokers in agricultural innovation systems. In: World Bank (ed.) *Innovation Systems: An Investment Sourcebook*. World Bank, Washington, DC, pp. 221–230.

Klick, M. (2015) Toward context-sensitive state building for development: state-local complementarity in rural Guatemala. PhD thesis. University of Denver, Denver, Colorado. Available at: http://digitalcommons.du.edu/etd/342 (accessed 23 January 2017).

Kramer, W. (1994) *Encomienda Politics in Early Colonial Guatemala, 1524–1544—Dividing the Spoils*. Dellplain Latin America Studies No. 31. Westview Press, Boulder, Colorado.

Kramer, R. and Tyler, T. (eds) (1996) Trust in Organizations: Frontiers of Theory and Research. Sage, Los Angeles, California.

Lastarria-Cornhiel, S. (2003) Guatemala country brief: property rights and land markets. University of Wisconsin, Land Tenure Center. Madison, Wisconsin. Available at: https://www.nelson.wisc.edu/ltc/docs/guatemalabrief.pdf (accessed 23 January 2017).

Levinger, L. (2015) Guerra inconclusa: la voz de los sobrevivientes. Editorial Maya Na'oj, Guatemala City, Guatemala.

Lovell, W. (1988) Surviving conquest: the Maya of Guatemala in historical perspective. Latin American Research Review 23(2), 25–57.

MAGA (2013) El nuevo sistema nacional de extensión rural (SNER): un aporte para el desarrollo socioeconómico de las familias campesinas. Ministerio de Agricultura Ganadería y Alimentación. Guatemala City, Guatemala.

MAGA (2014) Cuadernos de extensionistas [Extension notebooks]. Ministerio de Agricultura Ganadería y Alimentación, Guatemala City, Guatemala.

MAGA and FAO (n.d.) Guía técnica del extensionista rural [Rural extension technical guide]. Ministerio de Agricultura Ganadería y Alimentación Guatemala and FAO, Rome.

Malkin, E. (2015) President Otto Pérez Molina of Guatemala resigns amid scandal. *The New York Times*. Available at: www.nytimes.com/2015/09/04/world/americas/otto-perez-molina-guatemalan-president-resigns-amid-scandal.html?_r=0 (accessed 23 January 2017).

Mattessich, P., Murray-Close, M. and Monsey, B. (2001) *Collaboration: What Makes it Work—A Review of Research Literature on Factors Influencing Successful Collaboration*, 2nd edn. Amherst H. Wilder Foundation, St. Paul, Minnesota.

McKibben, C. (2015) Corruption, impunity, and the International Commission against Impunity in Guatemala. Council on Hemispheric Affairs, Washington, DC. Available at: www.coha.org/corruption-impunity-and-the-international-commission-against-impunity-in-guatemala/ (accessed 23 January 2017).

Nederlof, S., Wongtschowski, M. and van der Lee, F. (eds) (2011) *Putting Heads Together: Agricultural Innovation Platforms in Practice*. Bulletin 396. Royal Tropical Institute (KIT), Amsterdam, the Netherlands.

Ortiz, R., Rivera, O., Cifuentes, I. and Morrás, E. (2011) *Estudio de Sistematización de Buenas Prácticas de Extensión en Guatemala*. FAO, Rome.

Panić, M. (2008) Post-conflict countries: aid effectiveness and permanent peace. United Nations, New York. Available at: www.un.org/en/development/desa/policy/wess/wess_bg_papers/bp_wess2008_panic.pdf (accessed 23 January 2017).

Peacock, S. and Beltrán, A. (2003) *Hidden Powers in Post-Conflict Guatemala: Illegal Armed Groups and the Forces Behind Them*. Washington Office on Latin America, Washington, DC.

Pillay, R. (2007) Evaluation of UNDP assistance to conflict-affected countries: case study Guatemala. United Nations Development Programme. Available at: http://web.undp.org/evaluation/documents/thematic/conflict/Guatemala.pdf (accessed 23 January 2017).

Rothenberg, D. (ed.) (2012) *Memory of Silence: The Guatemalan Truth Commission Report*. Palgrave Macmillan, New York.

Ruano, S. and Zambrano, A. (2006) Guatemala: country gender profile. Available at: www.jica.go.jp/english/our_work/thematic_issues/gender/background/pdf/e06gua.pdf (accessed 23 January 2017).

Sloan, P. and Oliver, D. (2013) Building trust in multi-stakeholder partnerships: critical emotional incidents and practices of engagement. *Organization Studies* 34(12), 1835–1868.

Spector, B., Winbourne, S. and Dininio, P. (2015) *Practitioner's Guide for Anticorruption Programming.* Management Systems International. Arlington, Virginia.
SPRING (2015) *Formative Research: Key Influencers of Household Food Access in the Western Highlands of Guatemala.* USAID/SPRING, Arlington, Virginia.
Taft-Morales, M. (2014) *Guatemala: Political, Security, and Socio-Economic Conditions and US Relations.* Congressional Research Service, Washington, DC.
UN Department of Peacekeeping Operations (2010) *Justice Update 1.* United Nations, New York.
United Fruit Historical Society (n.d.) Jacobo Arbenz (1913–1971). Available at: www.unitedfruit.org/arbenz.htm (accessed 23 January 2017).
USAID (2009) *A Guide to Economic Growth in Post-Conflict Countries.* USAID, Washington, DC.
USAID (2010) USAID country profile: property rights and resource governance—Guatemala. USAID, Washington, DC. Available at: http://usaidlandtenure.net/wp-content/uploads/2016/09/USAID_Land_Tenure_Guatemala_Profile_0.pdf
USAID (2011) *Guatemala: FY 2011–2015 Multi-Year Strategy* (Feed the Future Strategy). USAID, Washington, DC.
USAID (2012) *Guatemala Country Development Cooperation Strategy: 2012–2016.* USAID, Washington, DC.
Wellhausen, E., Fuentes O.A., Corzo, A. and Mangelsdorf, P. (1957) *Races of Maize in Central America.* Publication 511. National Academy of Sciences, National Research Council, Washington, DC. Available at: www.ars.usda.gov/sp2UserFiles/Place/50301000/Races_of_Maize/Races_Central-America_0_Book.pdf (accessed 23 January 2017).
WOLA (2015) *The International Commission Against Impunity in Guatemala: A WOLA Report on the CICIG Experience.* Washington Office on Latin America, Washington, DC.
Wood, D. and Gray, B. (1991) Toward a comprehensive theory of collaboration. *Journal of Applied Behavioral Science* 27(2), 139–162.
World Bank (1997) *Helping Countries Combat Corruption: The Role of the World Bank.* World Bank, Washington, DC.
World Bank (2014) *Guatemala Economic DNA: Harnessing Growth, with a Special Focus on Jobs,* 1st edn. World Bank, Guatemala City, Guatemala. Available at: https://openknowledge.worldbank.org/handle/10986/20400 (accessed 21 February 2017).

13 Synthesis: The Political Economy of an Extension Service Provider on the Frontline of Conflict

Ian Christoplos*
Danish Institute for International Studies, Copenhagen, Denmark

Introduction

This chapter looks at the lessons from this volume and other literature regarding the conditions for developing extension services in post- and chronic-conflict settings and the results that have been achieved. The analysis seeks to situate the findings from the preceding chapters in the context of what we know about recovery processes. In essence, this chapter seeks to understand the processes of rebuilding extension from a political economy perspective, highlighting the various paradigms of resilience that underpin these efforts.

The first section of this chapter looks at the historical context of the processes under way, particularly how the role of the state is perceived and pursued. The tensions between the economic heartlands and the hinterlands that frame agricultural policies are described, together with the factors that determine how pluralism and markets are perceived.

The following section scans the different meanings applied to the currently popular catchword 'resilience'. Although the chapters in this volume do not explicitly frame their analyses in terms of resilience, the different ways in which states, farmers and aid agencies pursue rebuilding extension implicitly reflect these underlying paradigms.

The next section looks directly at the empirical experience of the largely aid-driven extension reform efforts. Emphasis is on what can be learned from the relatively fragmented efforts to pick up the pieces of weakened bureaucracies and perversely incentivized market actors amid a scramble for political influence and resources. A critical perspective is applied to the applicability of 'best practice' in extension in light of these prevailing conditions.

The chapter concludes with a set of what may appear to be out-of-the-box recommendations, which draw more on lessons learned from capacity-development efforts in fragile states than on conventional agricultural extension toolboxes. The importance of a clear ethical stance is stressed.

Historical Context and the Role of Extension

Institutional histories

The chapters in this volume amply illustrate the importance of the historical legacies

* Corresponding author e-mail: ian@glemdev.com

regarding the role (or absence) of the state in directing agricultural change in general and extension in particular. As will be analyzed below, this is easily forgotten by those planning post-conflict programs in their search for new paradigms for resilience, whether in terms of bouncing back to 'normalcy' or in 'building back better'.

The cases here illustrate how the strength and fragilities of the state long before the most recent conflicts steer perceptions about what kind of state should be rebuilt. Intentions to re-establish agricultural services are also thus related to dreams of the past, be they related to 'scientific socialism' (Myanmar, Georgia, Tajikistan) or surprisingly similar developmentalist paradigms of the 1970s, when grand infrastructure programs, tied with massive investments in technology transfer (of which extension was a part), were expected to transform 'backward' societies into 'modern' ones through irrigation, mechanization and industrialization (Afghanistan, Iraq, Liberia). Even if the conditions for such ambitious transformations are in no way present in landscapes of chronic conflict, the dirigiste visions of authorities may be rekindled, especially with the arrival of large financial investments and donors eager to spend money quickly.

At the opposite end of the spectrum, the pre-conflict decline in the accountability of the state and its social contract with the population was often a precipitating factor in the conflict. This is particularly important with regard to marginalizing and disenfranchising the youth (Sierra Leone). Failures in pre-conflict extension may be just a symptom of this broader breakdown of legitimacy, but as a frontline service it is a sentinel indicator of this collapse.

One notable, and perhaps rather surprising, aspect of the role of the state is that the involvement of the military has not always just emerged during the conflict (Myanmar). The role of the military in agriculture may be 'off the radar screen' and profoundly disturbing for most of those involved in state reconstruction and rural development, but it is a fact on the ground that must be confronted. This includes both the continuing role of the armed forces in farming and land acquisitions and the assumptions around agriculture's contribution to processes of disarmament, demobilization and reintegration (DDR) of former combatants (South Sudan, Myanmar). Aid to agriculture is part of the calculations of the generals, who often have no intentions of returning to their barracks.

A finding in several of the chapters is the need to consider the role of the bureaucracy in countries where genuine extension may be unknown. If agricultural bureaucracies have traditionally focused on delivering inputs, buying produce (Iraq, Georgia, Afghanistan, Liberia) and/or just gathering data (the Democratic Republic of the Congo [DRC], Myanmar), then the extent of commitments to this unfamiliar role is likely to be weak. The concept of the bureaucracy actually serving farmers, as opposed to giving directives or extracting rents, may also be unknown. When extension is a concept that has been entirely introduced by donors (Georgia, Iraq, Tajikistan), this creates profound obstacles to moving beyond a donor-driven agenda. At best, generating commitments to extension may require considerable time, a rare commodity in the short-term funding windows of most chronic- and post-conflict programming.

Hopes may be placed in civil society and farmer organizations as advocates for such a shift to serving farmers, and in some cases there is some evidence of a historical role in this regard, but the cases suggest that a foundation for this is usually not present. Distrust towards collective action and non-governmental organization (NGO) competition for aid resources to undertake service provision stand in the way of hopes for a new role for civil society. Overall weaknesses in societal trust resulting from the horrors of the conflict may mean that the preconditions for collective action are not present, particularly where the state participated in these abuses (true for most of the cases, Sierra Leone in particular).

Similarly, the role of the private sector that is envisaged in ideals of pluralistic extension is generally not something that had existed in the past. Ideas about sharing of responsibilities with the private sector may be looked upon with considerable suspicion. The very concept of pluralism in agricultural and extension services is thus not something that can be revived, but must start from scratch.

Given the often massive influx of aid in post- and chronic-conflict contexts, it is easy to forget that states have differing histories of donor relations. All have a major relationship with donors, though with some it has involved confronting sanctions (Iraq, Myanmar, Afghanistan), others as 'donor darlings' (Liberia, Georgia). In some instances, donors have kept a certain distance from the state due to distrust in its readiness to support the population (Sierra Leone) and the general desire to maintain humanitarian principles during the conflict.

Some countries have been party to efforts to enhance aid effectiveness (Mozambique). Most, however, have been out of the loop of discussions on how to harmonize donor efforts and align them with national priorities. This isolation from the aid effectiveness agenda is again partly due to distancing from the state as part of applying the principles of neutrality and impartiality during the humanitarian phase.

These factors delineate how state sovereignty is perceived in relation to leading the development agenda and influence how governments attempt to manage aid flows. The cases presented here suggest a tendency to play aid agencies off each other, rather than evoking principles of harmonized and aligned aid provision.

Tensions between Heartlands and Hinterlands

Beyond the institutional histories, there are historical geographies that profoundly influence the space for agricultural transformation and, with this, extension. Grossly simplified, the visions and plans for post-conflict extension reflect how highly productive areas and hinterlands are perceived to contribute to and/or threaten national development. Differentiated extension strategies for the heartlands and hinterlands thus directly contribute to realizing national development, and in many respects even symbolize the distinctive social contracts that exist for different categories of citizens. In some cases, this dichotomy reflects concerns about feeding the population and/or contributing to export income through investments in high-potential areas, versus maintaining control over regions that have traditionally held little allegiance for the state-building project (Liberia and Myanmar stand out, but there are elements of this in all the cases). Furthermore, this dichotomy is also reflected in ethnic differences (Afghanistan, Iraq, Sri Lanka) and even who is recognized as a citizen (Myanmar, Sri Lanka).

The highly productive areas may require hydraulic bureaucracies (Afghanistan, Iraq, Myanmar, parts of Sri Lanka) capable of managing complex irrigation regimes and ensuring that farmers adapt their production calendars and choose crops accordingly. Control in the hinterlands often has more to do with preventing the production of illicit crops and smuggling (Afghanistan, Myanmar), or mere benign neglect of the agricultural sector due to more pressing priorities such as mineral extraction (Sierra Leone). In some countries the dichotomies are about plantations (e.g. perennials such as rubber) versus smallholder subsistence farming (Liberia).

Land laws and the extent to which the state has tolerated or even encouraged dispossession of land from poor smallholders (Myanmar, and a growing concern elsewhere) also reflect the types of support that are expected to be provided to large commercial and small subsistence farmers in different areas. Land consolidation may also be seen as a precondition for effective extension in countries where fragmentation of landholdings is extreme (Georgia, Tajikistan, also Bosnia—see Christoplos, 2007). However, promoting land consolidation is inevitably a very sensitive topic, perhaps too sensitive to address in delicate chronic- and post-conflict contexts.

These various factors have meant that rural insurgencies are often (but not universally) based in the hinterlands (Liberia, Myanmar, to some extent in the other cases), and approaches to agriculture in these areas reflect different approaches to counter-insurgency. Goals may be about either 'hearts and minds' or repression and land dispossession—approaches that are not always mutually exclusive.

An inevitable result is that virtually all the countries reviewed in this volume could not be said to have had a single extension service even before the conflict. They have rather had a patchwork of very different types and structures of agricultural services that reflect both the varying agro-economic systems and the intentions of the state in different localities.

Continuity and Change

The chapters in this volume present different perspectives on the ways in which conflicts have affected agricultural livelihoods. In most of the cases displacement has been severe. In some, a considerable proportion of the population remains in conflict zones (Afghanistan, Iraq, Myanmar), frequently retreating back to subsistence production. Subsistence may be the only option where populations have lost access to urban markets, which may have shifted to reliance on cheaper imports. Large-scale commercial farms generally collapse (Afghanistan, Iraq, Liberia, Sierra Leone), not only due to direct conflict but also as a result of disappearing subsidies, irrigation maintenance and labor. From other literature it is obvious that gender roles change, often with women taking on greater responsibilities, especially for subsistence production, while men are fighting or migrating (Coulter, 2009). In some countries the focus of agricultural development during the conflict shifts from former extensive systems to more intensive methods in limited safe zones (Iraq). The comparative advantages of illicit production increase in the hinterlands (Afghanistan, Myanmar). 'Conflict entrepreneurs' are able to profit through the changing and often great differences in prices of commodities and services. It would be fallacious to assume that markets disappear, but they certainly change in ways that do not bode well for a post-conflict return to normalcy, as war economies become entrenched and sustain themselves.

Another potentially long-term change emanating from conflict concerns accelerated rural–urban migration, particularly among the youth (emphasized in Sierra Leone and Sri Lanka, but presumably relevant in all). The chapters present different pictures of the extent to which youth may be interested in returning after the conflict (with an eagerness to return in Iraq, but less evidence of this elsewhere). Even if displaced people do return, their farming skills may be minimal after decades or generations of displacement (Afghanistan), or periods in the past when they worked as laborers on large farms (Liberia, Georgia, Tajikistan). Land distributions have meant that even those with professions totally unrelated to farming have received land and are expected to become 'farmers' (Georgia, Iraq, also Bosnia—see Christoplos, 2007). This does not mean they are necessarily interested in becoming skilled farmers, though. In some countries farming is seen to be a last resort livelihood (Georgia), or a hobby to provide a modest addition to other forms of employment (Bosnia—see Christoplos, 2007). The Sierra Leone chapter stresses the factor of 'forgetting by not doing' as generating massive challenges (and needs) for extension after displacement.

All of these factors converge in the needs, demands and assumptions regarding the role of extension. Already weak knowledge of markets among extension staff (in all cases) is aggravated by the fact that the new chronic- and post-conflict markets are different and opaque. Demand for training of displaced farmers (and their descendants in long-running conflicts) is so far beyond the prevailing capacity of extension staff that it is hard to see where to start. A comprehensive approach to training (both basic and in-service support) is required, but this cannot be achieved through short-term and ad hoc NGO-led early recovery efforts. At the same time, private extension for illicit crops may have expanded greatly, driven by enormous profits (the vast geographical expansion of opium in Afghanistan being an obvious example). Several of the chapters point out that the pressures on the societal fabric have generated distrust regarding the intentions of rent-seeking authorities who may be more interested in controlling rather than developing markets (Afghanistan, Georgia, Iraq, Sierra Leone, Liberia).

Contrasting Paradigms of Resilience

Resilience applied to post-conflict transitions

To understand the ways in which authorities, planners, donors and farmers perceive the expected and desired recovery process, it is important to unpack the different meanings of an overly popular catchword—resilience (Bené *et al.*, 2014). Perhaps the most common default assumption about what resilience means (but also the most misleading) is that society should and could be supported to 'bounce back' to a pre-conflict state. Many conflict-weary farmers and bureaucrats are nostalgic for a past 'normal' life. The changes described above raise major questions about the extent to which this is at all realistic, but this does not prevent people from placing their hopes on a return to normalcy. This may also generate distrust towards outsiders advising them that they should instead embrace a 'brave new world', exemplified by a shift to pluralistic extension services with many actors. There may be a desire to return to a strong state with a social contract that existed (or is imagined to have existed) in the past.

Those promoting these brave new worlds often draw their inspiration from concepts of resilience deriving from the ecological sphere (Bené *et al.*, 2014), wherein crises are seen as opportunities and pressures to improve and evolve in ways that address the factors that generated the crisis in the first place. Among politicians and aid agencies this is often referred to as 'building back better'.

However, this 'better' is a disputed and problematic concept. In some of the countries a continued securitization of the development agenda is part of this, as the role of the military is not diminished and security concerns influence what forms of recovery are favored (Myanmar, South Sudan). These aspects may be intertwined with the DDR agendas mentioned above. These concepts of resilience have also been extensively criticized as favoring those with the capacities to build back better, ignoring and marginalizing the poor who are written off as not being resilient enough for the programs on offer (Bené *et al.*, 2014). The reason for this is the tendency of resilience models to distract attention from the factors of power and politics that determine 'whose resilience counts' (Cannon and Muller-Mahn, 2010). Furthermore, it has been noted that outsiders pursuing a resilience agenda may mistake small attempts to cope with a desperate situation as signs of building back better, even if such forms of resilience do not contribute to 'better' livelihoods in the long run (Pain and Levine, 2012).

This myopic perspective on building back better may be aggravated by pressures to show quantitative and fast results (Christoplos, 2014), sometimes by focusing on 'picking winners' (Bosnia—see Christoplos, 2007). The tendency for such demands to lead to an elite bias in extension, as it is easier to achieve results with wealthier farmers, has long been recognized (Christoplos, 2010).

Institutional resilience

Extension services are generally withdrawn during conflicts, with some notable exceptions. Areas under government control, often near urban areas, may continue to receive support and may even be considered vital for food supplies for these areas (Liberia, Myanmar). Some NGOs keep extension going by providing their own services, either for dispersed internally displaced persons (IDPs) or those in camps, even if these services remain secondary to input provision (emphasized in Myanmar and Sri Lanka, but presumably in most cases). The post-conflict period is characterized by an ambiguous process wherein these NGOs either seek ways to merge their efforts with those of the government, or merely make token and hurried efforts to 'hand over' programming to woefully underequipped extension structures (Uganda—see Wairimu *et al.*, 2015).

For donors, building back better has sometimes been seen as a matter of using the recovery process as a vehicle for public administration reform; that is, getting rid of bloated bureaucracies and inappropriate subsidization schemes, a paradigm often not shared by local farmers, bureaucrats and politicians (Afghanistan, Iraq, Liberia). The efforts of post-conflict states to rebuild a social contract with their citizens may be stymied by donor demands for creating a 'lean' state. Governments looking to solidify public support are generally suspicious of vague references to a greater role for a private sector that they distrust and which may not be interested in licit investments in agriculture (Afghanistan, Iraq).

In general, it seems that the international community often speaks with a forked tongue regarding the trade-offs between different resilience models, sometimes calling for lean structures to create more efficient bureaucracies that reflect the principles of New Public Management, whereas at other times investments in public service provision are stressed as the foundation for state legitimacy and peace building. However, the empirical bases for the latter claims have been increasingly recognized as shaky (Denney *et al.*, 2015). New research suggests that the link between state legitimacy and service provision relates to the extent to which expectations exist about a role for the state in a given service (McLoughlin, 2014). This would suggest that health and education, where the role of the state is seen as self-evident, might be more important for legitimacy than agriculture in countries where farmers are largely accustomed to fending for themselves. The fact that extension was in some cases already failing to provide useful and effective services to farmers in the decades before the conflict (Guatemala, Sierra Leone) may reinforce these low expectations of the relative importance of extension.

In countries where agriculture has received considerable public investment in the past, this may have been in the form of infrastructure, input supply and marketing rather than extension per se. The cases in this volume display a wide variety of farmer expectations regarding extension and generally show that farmers have largely given up on (if they ever had hope for) accessing extension. This would suggest that the link with legitimacy is likely to be tenuous unless extension is linked to 'hard' services in terms of subsidized inputs or infrastructural rehabilitation.

Plans for a leaner bureaucracy may not just be about promoting efficiency and pluralism. These plans may merely reflect exasperation with the extreme weakness of the public service, paired with a recognition that, due to a lack of tax revenues and decimated human resource capacities, there is no longer enough left of the public service to even begin to be built back better. Even if there is distrust regarding the sustainability and accountability of the NGOs providing services, and recognition that the private sector is too weak and uninterested to invest in advisory services, NGO-led service provision may be seen as the 'least bad' option. Some resilience approaches reflect an unspoken approach that could be referred to as 'pluralism by default', giving less attention to public agencies while waiting (perhaps indefinitely, judging from the cases in this volume) for better alternatives to emerge for supporting public sector development in conjunction with broader economic and societal recovery.

Market drivers of resilience

Market-driven recovery is part and parcel of the pluralistic 'building back better' narrative. A major motivating factor behind this narrative is the assumption that there are huge, unexploited potentials for profitable agriculture (Liberia, Myanmar, to some extent in all cases). These 'low-hanging fruit' assumptions are often based on availability of underutilized land (Sierra Leone), proximity to major markets in more stable countries and the potential of recapturing export markets of the past (e.g. dried fruit in Afghanistan). Furthermore, there is often an assumption that China (and other emerging economies) will be interested in both purchasing agricultural products and providing much-needed investments.

Another important aspect of assumptions regarding market drivers is that there are potential entrepreneurs who, with support from extension (particularly market-oriented extension and business development services), will be able to rebuild the economy. These entrepreneurs include both commercial farmers and other value chain actors. The narrow focus on entrepreneurs in recovery programming has recently been criticized as ignoring how power and politics, rather than inherent entrepreneurialism, steer impacts on the livelihoods of the poor (Sri Lanka—see Kapadia, 2014). Efforts to pick winners in the agricultural sector often lack a clear theory of change regarding how the benefits accruing to those individuals will diffuse or trickle down to the rest of the population (Longley *et al.*, 2006; Bosnia—see Christoplos, 2007). Donor priorities often contribute to this blind spot.

The studies in this volume all note the weakness of the public sector as a driver of market-oriented recovery. There is a lack of faith that the public extension service can be reformed or trusted to pursue a market-oriented agenda (Iraq, Liberia). Sometimes this leads to hopes that value chain actors, including input dealers and traders, will act as opinion leaders and fill the breach (Georgia, Myanmar).

Resilience and aid

Resilience in aid programming has long been associated with linking relief, rehabilitation and development (LRRD) (Christoplos, 2014). According to its most rudimentary (and perhaps most common) interpretation, the LRRD challenge is about making the aid architecture work by ensuring that humanitarian programming and 'early recovery' contributes (or at least does not create obstacles) to development efforts. It is often about finding opportunities to work with government agencies, such as extension, to begin building their capacities even if programming remains dominated by aid actors.

A common criticism in this volume and elsewhere of LRRD in relation to agriculture is that food aid, distributions of free inputs and related efforts undermine the market drivers of resilience. These forms of aid may undercut the efforts of entrepreneurs or reinforce their dependencies and linkages with unsustainable and soon-to-be discontinued aid programs (emphasized in Liberia but inherent in all cases to some extent).

These fears, paired with a realization that humanitarian needs for such subsidies still prevail, have led to a search for compromise solutions. A major solution is to look for ways to link procurement, extension (sometimes) and quasi-markets for products and inputs—for example, the World Food Programme's Purchase for Progress initiative and associated warehouse receipt systems (Liberia). Another approach is to transform handouts of inputs into seed fairs, often using vouchers linked to labor input in public works (Uganda—see Remington *et al.*, 2002; Wairimu, 2014). Some of the analyses in this volume acknowledge that modest extension reform efforts could be effectively piggybacked onto NGO-led agricultural rehabilitation programs focused on input distribution (Sri Lanka). These are seen to add hard content to softer information provision, thereby enhancing acceptance (Mozambique).

As mentioned above, another aspect of such hybrid LRRD efforts is searching out the potential entrepreneurs who are expected to drive market recovery and providing them with considerable amounts of capital and investment (Bosnia—see Christoplos, 2007; Sri Lanka—see Kapadia, 2014). This is expected to both kick-start their production and turn them into entrepreneurial models for other actors in the value chain. A concern here relates to the tendency to focus resilience efforts on those who are already resilient, ignoring and further marginalizing others.

The backdrop for much of the LRRD processes in the cases in this volume is the declining role of NGOs as the focus of efforts shifts from rehabilitation to development. It is generally recognized that rebuilding public sector agricultural institutions will involve a downsizing of competing NGO structures. This is driven by concerns about the lack of accountability of NGOs, the assumed greater efficiencies of (lower-paid) public sector agencies and the need to rebuild the social contract between the state and its citizens. The cases here show that this is not a linear process. Some NGOs are agile in shifting from humanitarian to development modalities and thereby retain a justification for their presence (Sri Lanka, Uganda—see Wairimu *et al.*, 2015). Where government services are not being revived, but where agriculture is recognized as central to recovery processes, NGOs may retain a major role long after the conflict has ended (Sierra Leone). Some international NGOs hand over their service-provision roles to national NGOs and concentrate on raising funds for these services.

Public agencies, and particularly extension, frequently fail to fill the gap created by downsizing NGOs, and this LRRD process may therefore be less about the proclaimed 'handovers' and more about a collapse of services due to donor fatigue. Some NGOs may see the LRRD process as a shift out of service provision and into advocacy. Such shifts may not be welcomed by states that are increasingly distrustful of advice from foreigners. In many chronic- and post-conflict transitions, the space for international NGOs and local civil society is shrinking fast.

Finally, a chief (but often overlooked) aspect of LRRD efforts in general, and extension support in particular, is the hope that pilot projects mounted during the rehabilitation phase will be scaled up, scaled out and replicated. Explicit and implicit theories of change are built on assumptions that 'somebody' will learn from the many successful pilot efforts being mounted in the post-conflict transition and apply them at a scale that will make a considerable dent in the existing needs gap, become part of government plans and policies, or otherwise achieve sustainability and measurable impact. As discussed below, there is little evidence that these theories of change are accurate. Pilots are often followed by more pilots in a serial process, with uncertain ultimate ends. Expensive methods (such as Farmer Field Schools in Afghanistan, the DRC, Myanmar and Sierra Leone) are commonly promoted, despite recognized challenges in moving these methods beyond pilot phases even in non-conflict countries.

Patterns in Picking up the Pieces through Aid Interventions

An aid-driven agenda

An overarching finding of the cases in this volume is that, despite the extremely varied contexts and conflicts, rebuilding extension is an aid-driven agenda. Extension is apparently not a top priority for states, even in those cases where agriculture more generally is given primacy (Iraq, Liberia). Given the absence of political commitments and capacities to cover the recurrent costs of extension activities, aid-financed service provision is therefore the norm. Even if programs are justified based on claims of the importance of building sustainable institutions, the tendency is to invest the majority of resources into this direct service-provision role. This reflects the need in chronic- and post-conflict countries to find an appropriate balance 'between the exercise of capacity and building it' (Brinkerhoff, 2010: 69).

This is not to claim that national actors are passive. The cases provide ample examples of various forms of 'donor shopping', often by local authorities or individual Ministries looking to strengthen their own sphere of interest (Iraq, Myanmar, South Sudan). This leads to different forms of aid-driven 'Balkanization' as individual donors or NGOs 'take responsibility' for a given geographical area (Mozambique, Myanmar), value chain (Liberia) or population (Guatemala). This may be a pragmatic response to the need to manage shifting aid responsibilities in an LRRD environment where aid resources dominate. However, it leads to fragmented reform and capacity-development efforts. As noted above, the cases show little evidence of commitments to aid effectiveness principles. In the one exception that proves the rule, a sector-wide approach in Mozambique led to a disinvestment in extension as the Ministry was not very interested in extension.

These tendencies are aggravated by the ways in which aid agencies actually weaken local institutional structures by poaching the most qualified staff and creating enormous and disillusioning resource differentials between government (and even national NGO) structures and international organizations (Afghanistan, Liberia, Myanmar, also Sri Lanka after the South Asian tsunami—see Christoplos, 2006). Extension seems to be particularly affected as the best staff flee to organizations with far higher salaries, more vehicles and higher operational budgets so that they can do their jobs. Related to this, some of the cases note that employment opportunities emerge for 'retired professors' who are able to return to work in extension activities, sometimes as private consultants who are perhaps expected to form the nucleus of a future private extension service. Furthermore, in some cases extra work for NGOs may provide much-needed 'top-ups' to encourage grossly underpaid staff to remain in their jobs (Sierra Leone, Uganda—see Wairimu *et al.*, 2015).

Other disincentives to making concerted efforts to rebuild extension involve the search for quick wins and fast, measurable 'results'. Services are judged on what and how quickly they can deliver, not on their contributions to institutional sustainability. 'Instead of strengthening institutional capacities, many agencies still try to simply "do livelihoods" by handing out predetermined assets' (Hilhorst *et al.*, 2010: 1113). In fragile states where state capacities are weak, so-called 'technical advisors' are often pulled into managerial and operational roles, sometimes even outnumbering those who they are expected to advise (Danida, 2011; UN, 2013). In some respects, this is valid given the need to provide visible results so as to rebuild confidence in the state and to address the acute needs of the population (Christoplos *et al.*, 2014b), but it does have a downside. When ministries are treated as mere implementing partners wherein their capacities are rented rather than built, the prospects for ownership are extremely poor (Hilhorst *et al.*, 2010).

Confronting a changed(?) landscape

The fallacies described above regarding a simple return to normalcy are underlined by how programming is designed to respond to a very different agrarian landscape than that which existed before the conflict. Interventions involve confrontations between different policy paradigms. This is most obvious in the statist versus market-driven ideologies, where state institutions, particularly those governments with a socialist or developmentalist history, tend to distrust the New Public Management models often promoted by donors (Minogue *et al.*, 1998; Afghanistan—see Christoplos, 2004). The aforementioned nostalgia for a strong state is common and may stand in the way of so-called bottom-up methods (and even the idea of extension altogether) if there is a preference for a more dirigiste approach to the agricultural sector.

These ideological disputes often play out in parallel with jockeying for influence among different ministries, regions and individuals (Iraq, Myanmar, South Sudan). Extension, as a frontline service visible to the population, is sometimes seen as a valuable pawn in these power plays with potentially significant political benefits (Georgia). In some countries these power plays include the military, who are not interested in returning to their barracks, but rather in reaping some of the benefits of the post-war economy through direct involvement in agriculture (Myanmar, South Sudan).

An important aspect of the changed environment for extension is the transformations that have occurred in the agro-economy. Factors such as the emergence of new retail (especially supermarkets) are seen in some cases as an opportunity for farmers (Sri Lanka), presumably with appropriate extension support. Others (perhaps more realistically) see these developments as raising serious questions about the ability of smallholders to reach changing markets, even with the help of extension (Bosnia—see Christoplos, 2007). An important factor here is that local agriculture may have shifted (back) to subsistence (Liberia, Sierra Leone) and petty trading, often led by women, whereas the demands of supermarkets may require links to male-dominated formal value chains.

The gap between the state of existing structures on one hand, and the needs described in the cases on the other, raises attention to questions of what can be accomplished with the short-term projects described. Extension-to-farmer coverage in almost all of the cases was abysmal at the outset of the interventions, and the programs described do not seem likely to make a significant or sustainable dent in these figures. It would appear that many interventions lack realistic theories of change that reflect well-known findings regarding the time required for capacity development (Boesen and Therkildsen, 2004). Program designs instead appear to reflect the limits of quasi-humanitarian funding windows and exit strategies. The LRRD concept suggests that this can be addressed by designing these inevitably somewhat piecemeal efforts to focus on opportunities for modest contributions to rebuilding capacities for longer-term development. This is certainly possible but requires a reorientation from measuring results according to services delivered to instead emphasizing these capacity-development outcomes.

The lack of a realistic theory of change is aggravated by the fact that there are some cases where the recovery process involves retrenchment rather than expansion of the public sector (Georgia, Myanmar), or a continued decline in salaries and logistical capacities for public extension agents (Sierra Leone). Staff may be concentrated to higher potential areas (Myanmar) leaving the hinterlands to 'somebody else' (perhaps the humanitarian agencies).

Extension efforts are usually linked to efforts to respond to the needs and realities of returnees. These may be IDPs or refugees from neighboring countries who have not been farming for decades or even generations (Afghanistan, South Sudan). They may have been urbanized (Liberia, Sri Lanka). Some donor countries have hopes that agriculture will generate employment opportunities which can help reverse refugee flows (Bosnia—see Christoplos, 2007). Hopes are sometimes placed in engaging the well-educated diaspora in agricultural efforts (Somalia—see IFAD, n.d.; Hradsky *et al.*, 2011).

A further factor in the changed landscape is that of making decisions about what forms of agriculture to support. Some efforts reflect a desire to return to past large-scale commercial production on plantations to generate revenues and attract investors (especially China). Other countries recognize that building on smallholder food crop production is essential to address national food security concerns and to satisfy the demands of the large number of smallholders, some of whom have taken over former plantations. Renovation of irrigation schemes is a common priority that is expected to contribute to both.

In all these efforts there may be a clear need for some form of collective action in the form of cooperatives, water-users' associations (Central Asia—see Christoplos, 2012), or farmers' organizations, all of which are seen as entry points for extension. Where land fragmentation is severe, donors have sometimes seen cooperative promotion to be a precondition for effective extension services (Georgia). However, most of the cases note a strong level of distrust for most forms of collective action, as described above. In a notable exception, the Mozambique case highlights how private extension linked to out-grower schemes in cotton (and later tobacco) has been an effective way to work with a large group of farmers without the baggage of dealing with mutual distrust.

A final related aspect of the dynamics around smallholder versus plantation agriculture is that of land concessions and other forms of large-scale land acquisitions (particularly stressed in Myanmar, also relevant in Cambodia—see Christoplos *et al.*, 2014c). Many politicians and even the military see lucrative opportunities in the chaotic post-war context, with widespread displacement and uncertain land titles. Even if they are not ready to become agro-entrepreneurs themselves, they may pursue rents from provision of land concessions to investors. In addition to the direct financial benefits, they can show 'development' as forests are replaced with rubber plantations and other commercial crops. This can generate moral quandaries for international agencies when decisions on extension investment must be made that can aid and abet investments which could lead to further conflict and displacement, as well as environmental destruction, but where these farms may be the only way of working with the rural population in the affected areas.

Experience with applying 'best practice' in extension

The findings described above raise critical questions about the applicability of conventional assumptions regarding 'best practice' extension in chronic- and post-conflict contexts. Questions need to be asked about what works where and when, and what needs to be modified. It is important to consider the plausibility of standard theories of change in these contexts, and also the risks and ethical dilemmas that may come to the fore. The gap between needs and capacities in fragile states has often evoked a dysfunctional tendency among international agencies to come up with grossly unrealistic 'long list of "things that must be done"' (Grindle, 2007: 559). The search for ways to apply best practices is symptomatic of this tendency.

In general, the cases in this volume describe efforts to promote a shift in policy paradigms together with grand capacity-development efforts in what are clearly unrealistic time frames. The imperatives of crisis response take precedence over realism and application of well-known lessons learned about the long-term commitments required to achieve policy engagement and capacity development (Boesen and Therkildsen, 2004). The aims of these programs might be relevant if they were implemented within a more comprehensive commitment (particularly to capacity development) and incremental time frame, but are out of sync with both the immediate realities and the plausible influence of small and time-bound projects. The needs and problems of extension in conflict zones are not amenable to quick fixes, but apparently that is usually all that is on offer.

Pluralism is one of the concepts that frames much best practice advice in extension (Christoplos, 2010). The experiences described in this volume do not question this as a long-term goal, but the findings describe situations where pluralism is being achieved more by default than by design. Indeed, the goal of a pluralistic and decentralized extension service may even be window dressing for inevitable failures to control and coordinate the fragmented settings of pilot projects and political fiefdoms. Fragmentation may be inevitable, but the best way to move towards pluralism by design is to design and implement programs based on a recognition of their niche within wider agricultural innovation systems. This can create awareness of where related investments are needed, for example, in other aspects of value chains, research and training of extension staff.

Most of the programming described involves value chain development. In these programs extension is part of a package of subsidized services designed to link potential entrepreneurial farmers with other actors in the market. In many respects this is certainly appropriate, given the fractured market chains and the need for a broader perspective on extension's role in the agro-economy. The difficulties described in persuading extension services and their ministries to think in broader perspectives indicate the need for a value chain approach. However, the tools at hand are not necessarily the right ones. The weak, amorphous and frequently nefarious nature of the private sector suggests that deeper and blunter analyses of the actors and interests are needed. In some cases, this may involve looking for ways to 'use long spoons when dining with the devil'; for example, by developing markets without subsidizing warlords. In others it may mean acknowledging that the private sector is simply not there and we do not know when it may emerge. There may be potential to develop value chains through contract farming and out-grower schemes.

A final overarching aspect of applying best practices is that, even in stable contexts, these models have been developed and applied in a rather naive manner. Insufficient attention has been paid to the political economy of policy reform processes, both at the national level and within the micro-level dealings, which determine how the relationships between farmers and extensionists emerge. Faith in the applicability of policy templates is out of place in any context. Given the massive risks that exist in conflict situations and the diversity in the contexts described in this volume, such faith is absurd and dangerous.

Recommendations: Ethics and Evidence in Approaching Agriculture amid Chronic Conflict

This author suggests that an overarching recommendation for developing extension in chronic- and post-conflict contexts is to anchor plans in a solid political economy analysis, taking close account of the drivers of conflict. This should not just stop with a study. Tried and proven tools exist for applying these analyses in practical decision making in conflict zones, most notably the so-called 'do no harm' approach (Anderson, 1999). Lessons can be drawn from the provision of other services where such methods have been applied, as well as from broader research into the role of services in chronic- and post-conflict settings (SLRC, n.d.; Jackson, 2015; Mallett and Denney, 2015). Existing research on service provision in conflict can help extension planners to think outside of their boxes.

Through such analyses, it is possible to draw certain ethical red lines where the risks of feeding into conflict tensions, ethnic and gender discrimination, and inappropriate large-scale land acquisitions can be avoided. Choices in post-conflict agrarian transitions also have huge implications for the hinterland–heartland dynamics that generally contributed to the conflict to start with. Sometimes programs contradict their stated objectives when interventions, justified based on peace building, effectively exclude sections of the population that are prone to be drawn into or most affected by renewed tensions. Consulting with actors (such as think tanks) specializing in conflict dynamics can help extension planners avoid pitfalls and perhaps even contribute to reconciliation by bringing parties to the conflict together in market transactions.

Plans for strengthening extension services need to confront the facts on the ground—including food aid, subsidized input provision through agricultural rehabilitation programs and the proliferation of NGOs—and not just decry their negative effects. Support to agricultural extension is part of the LRRD architecture and not above it. A useful starting point is to bring the humanitarian agencies into a dialogue on what they have learned about opportunities for supporting agriculture through existing extension (even if this has been mostly for helping with distributions) and respecting the NGOs' understanding of vulnerabilities in the local population. It is important to avoid assumptions that 'the humanitarian phase is over, now it is time for development'. The two modalities can work side by side—transparency and cooperation are the most important tools to synchronize modalities to avoid undue salary differentials and subsidy levels. Modalities such as Community-Driven Development efforts promoted by the World Bank and others have often struggled with similar challenges and can provide some support in finding appropriate programming norms.

Efforts should shift from visions of best practice to principled but 'good enough governance' (Grindle, 2007). Here, too, broader and constructive engagement with the wide range of aid actors can provide a way to position these efforts to reflect the modest footprint of extension interventions and respect the fact that much of the population may remain acutely food insecure and not be ready to have their livelihoods built back better (Christoplos and Hilhorst, 2009). Solutions can be found in advocating for a broader palette of services wherein governments and donors are held accountable for both social protection and economic growth objectives. This may involve supporting both subsistence or petty trade-oriented women's agriculture, and more commercial agriculture which can create jobs that may be primarily for men. Too often, Ministries of Agriculture have been encouraged to ignore the most vulnerable sectors of the population, simply leaving them to humanitarian agencies or social services. Extension agencies, in turn, have ignored those populations which have not been interested in the services on offer as being 'laggards'. Humanitarian agencies are sometimes complicit in the failures where handouts are perceived to be more in their comfort zones, and where they lack trust that donors will support programming that questions narratives about markets as the solution. Efforts to analyze and focus on common concerns for the vulnerability of the population to both food security and market risks can provide a basis for more inclusive programming.

Another area where greater cross-learning and constructive collaboration may be found is by linking agricultural information system efforts with other communication initiatives. Extension efforts have focused unduly on technical fixes (including information and communications technology (ICT) and particular training methods) without looking enough at broader information flows. Media (social and conventional) are becoming part of peace building as a way of promoting greater transparency and public accountability. There are indications that, in a post-conflict context, attention may shift from using these interventions to raise attention to human rights abuses, to focusing instead on social and economic rights (Zimbabwe—see Christoplos *et al.*, 2014a). Existing community (and even pirate) radio stations may be eager to respond to their listeners' demands for information about how to secure their land tenure, what aid can offer and whether 'the Chinese are coming'. Extension programs do not need to reinvent the wheel if they can see how their content can be integrated into these wider efforts to create a more open post-conflict society. Several of the chapters bring out the importance of using available media outlets, but greater attention is needed to more ambitious efforts that not only provide technical information, but also create a more transparent public debate on priorities in agricultural development and enable farmers and their organizations to demand accountability from extension.

The tunnel vision in many interventions is indicative of a broader problem with the illusory nature of project-driven extension. The massive gap between needs and the size and scattered nature of the projects described suggest that there is a necessity to reflect closely and pragmatically on the sphere of influence of these modest interventions. By looking at how these projects fit into the overall processes underway, it is possible to judge how potential contributions can be maximized. This involves moving from pluralism by default to finding small entry points for pluralism by design wherein the existing capacities of different actors are accepted as the point of departure, but where there is a realistic vision and strategy for developing the differentiated capacities that will be needed in the future. It is also about recognizing the factors that may generate or obstruct ownership (by ministries, extension agents, local government and farmers) of the reform processes and models on offer.

By breaking out of project tunnel vision, it will also be possible to differentiate the winners and losers in the agrarian transitions that are underway. It is a way of targeting those who must be included in the desired post-conflict society, rather than targeting those who can most easily benefit from new markets. All the actors involved in post-conflict extension efforts have a responsibility to investigate the winners and losers in different agricultural recovery scenarios. A gender and youth lens that reflects lessons about how exclusion was an initial driver of the conflict is essential. It is also important to assess if and how returning migrants may (re)engage in farming and the extent to which they have 'forgotten by not doing', and therefore require particular attention. Furthermore, by looking at broader recovery trajectories and obstacles it is possible to assess how new retail and other changes in markets may have profound implications for who may potentially win and lose from agriculture in the future. Climate change demands additional foresight, as the vulnerabilities that existed before the conflict may not reflect the future.

Going beyond studying these issues, it is important that these analyses are used to advise (and even confront) governments and donors about the consequences of their decisions. Transplantation of market models, expectations about private sector-led efforts and other norms from more stable contexts are rarely appropriate. Decision makers in capitals and donor countries need to be held accountable for responding to the on-the-ground realities of poverty and vulnerability.

A final, overarching aspect of a realistic approach to post-conflict extension is to acknowledge the often dire state of public finances and the unlikelihood that state revenues will cover the costs of a 'handover' of responsibilities for recurrent costs. There is no magic formula to transform exit strategies into more than statements of 'goodbye and good luck', but a realistic perspective on project contributions to long-term processes is the place to start.

Central to shifting the focus to financial, human resource and other capacity constraints is for donors to recognize the disincentives they create when their 'results agenda' focuses only on the impacts of services and not on the capacity-development outcomes that will determine sustainable, longer-term effectiveness. If agencies receiving aid are judged based on indicators related to contributions to these capacities, this will be the foundation for an incentive structure to break out of the aid-financed service provision cul-de-sac.

These recommendations for combining ethics and pragmatism can be summarized as a call for a human rights-based approach (Sida, 2015) to extension amid conflict. Such an approach, built on principles of participation, transparency, accountability and non-discrimination, would highlight both the opportunities and the risks involved in using extension to strengthen the voice and capacities of farmers, as well as the accountability of states, civil society and donors. Furthermore, a more explicit focus on the principle of non-discrimination would bring out how extension can help initiatives avoid a narrow focus on resilience for those (men) who are most resilient, and thereby reduce the risk of marginalization which can feed into future conflicts.

References

Anderson, M.B. (1999) *Do No Harm: How Aid Can Support Peace—or War.* Lynne Rienner, Boulder, Colorado.

Bené, C., Newsham, A., Davies, M., Ulrichs, M. and Goodfrey-Wood, R. (2014) Review article: resilience, poverty and development. *Journal of International Development* 26, 598–623.

Boesen, N. and Therkildsen, O. (2004) *Between Naivety and Cynicism: A Pragmatic Approach to Donor Support for Public Sector Capacity Development.* Danish Institute for International Studies, Copenhagen, Denmark.

Brinkerhoff, D. (2010) Developing capacity in fragile states. *Public Administration and Development* 30, 66–78.

Cannon, T. and Muller-Mahn, D. (2010) Vulnerability, resilience and development discourses in context of climate change. *Natural Hazards* 55(3), 621–635.

Christoplos, I. (2004) *Out of Step? Agricultural Policy and Afghan Livelihoods.* Issues Paper. Afghanistan Research and Evaluation Unit, Kabul.

Christoplos, I. (2006) *Links between Relief, Rehabilitation, and Development in the Tsunami Response.* Tsunami Evaluation Coalition, London.

Christoplos, I. (2007) *Between the CAPs: Agricultural Policies, Programming, and the Market in Bosnia and Herzegovina.* HPG Background Paper. Overseas Development Institute, London.

Christoplos, I. (2010) *Mobilizing the Potential of Agricultural and Rural Extension.* Food and Agriculture Organization, Rome.

Christoplos, I. (2012) Agricultural information amid conflict: for whom and for what? In: Özerdem, A. and Roberts, R. (eds) *Challenging Post-Conflict Environments: Sustainable Agriculture.* Ashgate, Farnham, UK.

Christoplos, I. (2014) Resilience, rights, and results in Swedish development cooperation. *Resilience: International Policies, Practices and Discourses* 2(2), 88–99.

Christoplos, I. and Hilhorst, D. (2009) *Human Security and Capacity in Fragile States.* Occasional Paper 01. Disaster Studies, Wageningen University, Wageningen, the Netherlands.

Christoplos, I., Nilsson, A., Maradze, T. and Persson, K. (2014a) *Review of the Strategy for Swedish Aid Initiatives in Zimbabwe: January 2011–December 2014.* Sida Decentralised Evaluation 2014:31. Swedish International Development Cooperation Agency, Stockholm, Sweden.

Christoplos, I., Engstrand, K. and Liljelund Hedqvist, A. (2014b) Capacity Development Literature Review. UTV Working Paper 2014:1. Swedish International Development Cooperation Agency, Stockholm, Sweden.

Christoplos, I., Funder, M., McGinn, C. and Wairimu, W. (2014c) Human rights perspectives on climate change adaptation: civil society experiences and opportunities in Cambodia and Kenya. DIIS Report 2014:30. Danish Institute for International Studies, Copenhagen, Denmark.

Coulter, C. (2009) *Bush Wives and Girl Soldiers: Women's Lives through War and Peace in Sierra Leone.* Cornell University Press, Ithaca, New York.

Danida (2011) *Addressing Capacity Development in Danish Development Cooperation—Guiding Principles and Operational Steps.* Danida Technical Advisory Services, Ministry of Foreign Affairs of Denmark, Copenhagen, Denmark.

Denney, L., Mallett, R. and Mazurana, D. (2015) *Peacebuilding and Service Delivery.* United Nations University Centre for Policy Research, Tokyo, Japan.

Grindle, M. (2007) Good enough governance revisited. *Development Policy Review* 25(5), 553–574.

Hilhorst, D., Christoplos, I. and van der Haar, G. (2010) Reconstruction from below: a new magic bullet or shooting from the hip? *Third World Quarterly* 31(7), 1107–1124.

Hradsky, J., Boesen, N., Land, A., Baser, H., Guizzardi, S. and Sorgenfrei, M. (2011) *Perspectives Note on Capacity Development in Fragile Situations.* Organisation for Economic Co-operation and Development, Development Assistance Committee, Paris.

IFAD (n.d.) *Diaspora Investment in Agriculture (DIA) Initiative.* Financing Facility for Remittances, International Fund for Agricultural Development, Rome.

Jackson, A. (2015) *Politics and Governance in Afghanistan: Nangarhar Province.* SLRC Briefing Paper 9. Secure Livelihoods Research Consortium, Overseas Development Institute, London.

Kapadia, K. (2014) Sri Lankan livelihoods after the tsunami: Searching for entrepreneurs, unveiling relations of power. *Disasters* 39(1), 23–50.

Longley, C., Christoplos, I. and Slaymaker, T. (2006) Agricultural rehabilitation: mapping the linkages between humanitarian relief, social protection, and development. HPG Research Report 22. Overseas Development Institute, London.

Mallett, R. and Denney, L. (2015) *After Ebola: Towards a Smarter Model of Capacity Building.* SLRC Briefing Paper 12. Secure Livelihoods Research Consortium, Overseas Development Institute, London.

McLoughlin, C. (2014) When does service delivery improve the legitimacy of a fragile or conflict-affected state? *Governance: An International Journal of Policy, Administration, and Institutions* 28(3), 341–356.

Minogue, M., Polidano, C. and Hulme, D. (eds) (1998) *Beyond the New Public Management: Changing Ideas and Practices in Governance.* Elgar, London.

Pain, A. and Levine, S. (2012) A conceptual analysis of livelihoods and resilience: addressing the insecurity of agency. HPG Working Paper. Humanitarian Policy Group, Overseas Development Institute, London.

Remington, T., Maroko, J., Walsh, S., Omanga, P. and Charles, E. (2002) Getting off the seeds and tools treadmill with CRS seed vouchers and fairs. *Disasters* 26(4), 316–328.

Sida (2015) *A Human Rights Based Approach to Sustainable Livelihood Systems.* Swedish International Development Cooperation Agency, Stockholm, Sweden.

SLRC (n.d.) Secure livelihoods research consortium: researching livelihoods and services affected by conflict. Available at: www.securelivelihoods.org (accessed 23 January 2017).

UN (2013) United Nations guidance note for effective use and development of national capacity in post-conflict contexts. United Nations Inter-Agency Team on National Capacity Development, New York.

Wairimu, W. (2014) Food insecurity and agricultural rehabilitation in post-conflict northern Uganda. In: Christoplos, I. and Pain, A. (eds) *New Challenges to Food Security: From Climate Change to Fragile States.* Routledge, London.

Wairimu, W., Christoplos, I. and Hilhorst, D. (2015) From crisis/relief to development: the policy and practice of agricultural service provision in northern Uganda. *Agriculture and Human Values* 1–14. DOI: 10.1007/s10460-015-9665-0

Index

Page numbers in **bold** type refer to **figures**.
Page numbers in *italic* type refer to *tables*.